ALLEZEIT WACHE 1842

# The Crack Tip Opening Displacement in Elastic-Plastic Fracture Mechanics

Proceedings of the Workshop on the CTOD Methodology
GKSS-Forschungszentrum Geesthacht, GmbH,
Geesthacht, Germany, April 23-25, 1985

Editor: K. H. Schwalbe

Sponsored by

Studiengesellschaft zur Förderung der Kernenergieverwertung in Schiffbau und Schiffahrt e.V. (KEST), Hamburg

Stiftung Volkswagenwerk, Hannover

Springer-Verlag
Berlin Heidelberg NewYork Tokyo

Prof. Dr.-Ing. K. H. Schwalbe
GKSS Forschungszentrum Geesthacht
Institut für Werkstofftechnologie
Max-Planck-Straße
2054 Geesthacht, FRG

ISBN-13:978-3-642-82820-1 e-ISBN-13:978-3-642-82818-8
DOI: 10.1007/978-3-642-82818-8

Library of Congress Cataloging in Publication Data

Workshop on the CTOD Methodology (1985 : GKSS-Forschungszentrum Geesthacht)
The crack tip opening displacement in elastic-plastic fracture mechanics.
1. Fracture mechanics--Congresses.
I. Schwalbe, Karlheinz. II. Title.
TA409.W67 1985 620.1'126 86-10066
ISBN-13:978-3-642-82820-1 (U.S.)



2161/3020-543210

# Preface

In April 1985 two workshop meetings were arranged in two consecutive weeks:
the

Second CSNI Informal Workshop on Ductile Fracture Test Methods

was organised and hosted by OECD, Paris, and chaired by F.J. Loss. It took place on 17th-19th April 1985. It dealt primarily with experimental techniques in elastic-plastic fracture mechanics and standardisation of $J_{Ic}$ and $J_R$-curve tests. In order to enable overseas participants to attend this meeting and the

Workshop on the CTOD Methodology

the latter was scheduled for 23th-25th April 1985 at the GKSS-Forschungszentrum Geesthacht.
Thus, a number of participants took part in the debates on the merits of the CTOD concept, having the state-of-the-art of the J-integral philosophy freshly in mind.

In this respect, the twin arrangement was very fruitful.
Three days were planned for the meeting, in order to have sufficient time for presenting and discussing 27 contributions.
Because of the workshop-type of the meeting several contributions on ongoing research were presented which were not intended to be published.

The Studiengesellschaft zur Förderung der Kernenergieverwertung in Schiffbau und Schiffahrt e.V. (KEST) and the Stiftung Volkswagenwerk generously sponsored the workshop, which is gratefully acknowledged.
Thanks are due to GKSS-Forschungszentrum Geesthacht for hosting the meeting.

Geesthacht, April 1986

K.-H. Schwalbe

# Introduction

Fracture mechanics offers two concepts for the treatment of a cracked body which behaves in an elastic-plastic manner: the J-integral and the CTOD concept. Both concepts are of course equivalent, but different methods for determining the material properties and for transferring them to structural parts have been developped.

Under certain circumstances, the J-integral is a path-independent contour integral characterising the crack tip field uniquely. The determination of the critical value, $J_{Ic}$, which is a measure of the material's resistance against the initiation of growth of a pre-existing crack, can be achieved by test methods which are already standardised in some countries (U.S.A., Japan, China) or which are in the stage of standardisation in other countries. For the J-integral as a driving force parameter, the EPRI handbook collects J solutions for a number of crack configurations. If the structural configuration being considered cannot be identified with one of the cases compiled in the handbook or if the stress-strain behaviour of the material differs significantly from a power law, individual numerical calculations have to be carried out.

The J-integral is also used to correlate a certain amount of crack growth although this is in principle not in accordance with the assumptions made for the derivation of J. Consequently, only relatively little crack growth is supposed to be J-controlled.

The CTOD concept assumes that a characteristic displacement at the crack tip is a unique measure of the crack tip field and hence of the fracture behaviour of a cracked body. The user is provided with a concept consisting of a test standard (BS 5762) and the so-called CTOD Design Curve. The document BSI PD 6493:1980 contains (among other details) guidelines on how to apply the CTOD Design Curve to practical cases. Application of the Design Curve leads normally to conservative assessments of the acceptance of defects.

Whereas the J-integral is mostly used to characterise the fracture behaviour of materials in the presence of a ductile tearing mode of crack growth, the CTOD concept is primarily applied to the ductile-to-brittle transition region of structural steels.

Thus, two different worlds have been developped as has also been demonstrated by the discussions at the two meetings (see Preface).
This is an unnecessary complication and it causes confusion to the less experienced user. Practical application of fracture mechanics will certainly have a greater chance to be accepted also by non-experts if parallelisms of this kind can be avoided (and they can!).

Unfortunately, there is no specific displacement at the crack tip which could be regarded as "the natural crack tip opening displacement". This was once again confirmed by the workshop discussions: at least seven different definitions of the CTOD were presented, which is certainly a weakness of the concept.

It may be beneficial to the reader of the present proceedings to bear in mind that the workshop discussion can be structured as outlined below. This structure can be used like a grid which can be laid on the various papers in order to get a reasonable overview on what was presented.

- As mentioned above, the CTOD is very much a matter of definition. Each of the definitions presented at the workshop makes sense, but it is important to establish correlations between the various CTODs.
  The symbols used in the presentations, i.e. $\delta_{45}$, $\delta_o$, $\delta_5$, $\delta_T$, $\delta_t$, $v_R$..., show that there is much to do in order to unify the various philosophies.

- The remote techniques for determining a critical CTOD do not seem to yield values which correspond in all cases with the displacement at the very crack tip.

- There is an increasing tendency for correlating stable crack growth with the CTOD, i.e. for determining δ-R-curves in contrast to J-R-curves. This seems to be attractive since some limited findings suggest that there are less stringent size requirements for a δ-R-curve than for a J-R-curve. On the other hand, a recently introduced modification of the J-integral seems to improve the ability of J to correlate crack growth.

- Although in principle both the J-integral and the CTOD philosophies and their techniques are compatible, this compatibility must be demonstrated by as many practical examples as possible in order to make visible that there is no real borderline between these two worlds. Empirical and theoretical correlations between J and $\delta$ show that there can indeed be a unique relation between these quantities.

- Like J, the CTOD can be predicted as a driving force parameter by full finite element calculations. It would, however, be very attractive if simple formulae could be developped in order to simplify the practical application. In this context, the Dugdale model - although strongly simplifying the reality - is still being used.

- Application to a structural part can be done as follows:
  - The driving force parameter, $\delta_{app}$, can be calculated numerically and compared with the material property, $\delta_{mat}$ .
  - The CTOD Design Curve yields conservative estimates for the acceptance of defects; it is not aimed at predicting failure. This drawback is balanced by the bulk of practical experience showing that the Design Curve works for the purpose it was developped for.
  - Procedures are under development which are easy to use (without numerics) but which are nevertheless capable of predicting failure situations with reasonable accuracy. By incorporating the R-curve methodology in such procedures one can take advantage of the increase of the material's resistance against crack growth with the crack growth process.

Thus, the merits of the CTOD concept can be characterised by slightly easier measurement of $\delta$ compared to J - in particular in the presence of crack growth - and by the availability of simple application tools.

# Participants

| | | |
|---|---|---|
| Aeberli, K.E. | FRG | GKSS-Forschungszentrum Geesthacht GmbH<br>2054 Geesthacht |
| Amstutz, H. | FRG | Technische Hochschule Darmstadt<br>6100 Darmstadt |
| Andrews, W.R. | USA | General Electric Company,<br>Schenectady, NY 12345 |
| Blauel, J.G. | FRG | Fraunhofer-Institut für Werkstoffmechanik<br>7800 Freiburg |
| Burget, W. | FRG | Fraunhofer-Institut für Werkstoffmechanik<br>7800 Freiburg |
| Cornec, A. | FRG | GKSS-Forschungszentrum Geesthacht GbmH<br>2054 Geesthacht |
| Dawes, M.G. | UK | The Welding Institute<br>Abington<br>Cambridge CB1 6AL |
| Dormagen, D. | FRG | Rheinisch-Westfälische Technische Hochschule<br>5100 Aachen |
| Ernst, H. | USA | Westinghouse Research & Development Center<br>Pittsburgh, PA 15235 |
| Fontaine, A. | F | IRSID, 78105 St Germain en Laye |
| Fossati, C. | I | CISE, 20090 Segrate Milano |
| Garwood, S.J. | UK | The Welding Institute<br>Abington<br>Cambridge CB1 6AL |
| Heerens, J. | FRG | GKSS-Forschungszentrum Geesthacht GmbH<br>2054 Geesthacht |

| | | |
|---|---|---|
| Hellmann, D. | FRG | GKSS-Forschungszentrum<br>Geesthacht GmbH<br>2054 Geesthacht |
| Heuser, A. | FRG | Rheinisch-Westfälische<br>Technische Hochschule<br>5100 Aachen |
| Hoffmann, M. | FRG | Technische Hochschule<br>Darmstadt<br>6100 Darmstadt |
| Kanninen, M.F. | USA | Southwest Research<br>Institute, San Antonio<br>Texas 78284 |
| Köttgen, K. | FRG | Technische Hochschule<br>Darmstadt<br>6100 Darmstadt |
| Loss, F.J. | USA | Materials Engineering<br>Associates<br>Lanham MD 20706 |
| McCabe, D.E. | USA | Westinghouse Research &<br>Development Center<br>Pittsburgh, PA 15235 |
| Mattheck, C. | FRG | Kernforschungszentrum<br>Karlsruhe<br>7500 Karlsruhe |
| Morawietz, P. | FRG | Rheinisch-Westfäl. TÜV<br>4300 Essen |
| Newman, J.C. | USA | NASA Langley Research<br>Center, Hampton VA 23665 |
| Schmitt, W. | FRG | Fraunhofer-Institut<br>für Werkstoffmechanik<br>7800 Freiburg |
| Schwalbe, K.-H. | FRG | GKSS-Forschungszentrum<br>Geesthacht GmbH<br>2054 Geesthacht |

# Contributions to the Workshop

---

Session: Crack Tip Examination

---

| | |
|---|---|
| W. Schmitt | Numerical Evaluation of CTOD: 2D and 3D Applications. |
| H. Amstutz<br>T. Seeger | Problems of Numerical CTOD Analysis. |
| J. Heerens<br>A. Cornec | Experimental and Theoretical Investigation of Crack Tip Blunting. |
| C. Mattheck<br>D. Munz | Theoretical Calculation of CTOD Using a Dugdale Model Including Strain Hardening. |
| M. Hoffmann<br>T. Seeger | Dugdale Solution for Strain Hardening Materials |
| M.F. Kanninen | ASTM Computational Round Robin. |

---

Session: Experimental Techniques

---

| | |
|---|---|
| D.E. McCabe | Geometry Dependence of Hinge Methods. |
| D.E. McCabe | The ASTM Round Robin. |
| K.-H. Schwalbe | Introduction to GKSS work on CTOD. |
| D. Hellmann | Compatibility of $\delta_5$ with BS 5762. |
| J. Heerens<br>K.-H. Schwalbe<br>D. Hellmann<br>A. Cornec | On the Relationships Between Various Definitions of the Crack Tip Opening Displacement. |

Session: CTOD and Crack Growth Including J-CTOD Relationship

J. Newman — The Crack Tip Opening Displacement ($v_R$) Resistance Curve Method.

W. Andrews — CTOD and J Relationships for a Growing Crack.

D.E. McCabe — Results of SEN and DEN CTOD Measurements.

H. Ernst — The Correlation Between the Schwalbe Type CTOD and Modified J.

D. Hellmann — Correlation of Stable Crack Growth with CTOD ($\delta_5$) under Plane Stress Conditions.

A. Cornec — Correlation Between $\delta_5$ and $\delta_{45}$.

Session: Application

M.F. Kanninen — Use of a CTOD Criterion for Unstable Crack Propagation and Crack Arrest in Ductile Materials.

M.G. Dawes — Progress in CTOD Test Methods and Application to Welded Joints.

W. Burget, J.G. Blauel — CTOD Testing of Welds (Procedure, Critical Values, Residual Stresses, Welding Qualification Test.

B. Marandet, A. Fontaine — Statistical Analysis of CTOD and $J_{Ic}$ Values in SAW C-Mn Steel Welds.

D. Dormagen, W. Dahl — On the Application of the CTOD Concept.

S.J. Garwood — The Use of CTOD Methods in Fitness for Purpose Analysis.

K.-H. Schwalbe — The Prediction of Failure Situation Using the CTOD ($\delta_5$)Concept.

# Contents

# Crack Tip Examination

# Numerical Evaluation of Crack Tip Opening Displacements: 2D and 3D Applications

W. SCHMITT and T. HOLLSTEIN

Fraunhofer-Institut für Werkstoffmechanik (IWM)

Wöhlerstr. 11, 7800 Freiburg, Federal Republic of Germany

Summary

The numerical tools to compute elastic-plastic crack tip loading parameters J and CTOD are well established for arbitrary geometry and loading situations also including temperature gradients and stable crack propagation. The evaluation of J is possible with only minimum requirements concerning mesh refinement and numerical effort. The determination of CTOD requires increased but not prohibitively large effort concerning mesh refinement and evaluation time. The correlation of J and CTOD postulated from continuum mechanics considerations could be verified numerically and experimentally.

## Introduction

Several concepts have been proposed to describe the process of crack initiation and propagation in ductile materials. Every concept assumes a suitable loading parameter L, which characterizes the intensity of the stress and strain field in a relevant vicinity of the crack front. The onset of crack extension takes place if L exceeds a critical material value $L_c$. If stable crack extension occurs prior to catastrophic failure, this extension is controlled by the respective (material) crack resistance curve L(Δa). For linear elastic and brittle material the stress intensity factor $K_I$ is the fracture controlling parameter. Also, the singular stress and strain field around the crack tip is controlled by $K_I$. In the case of ductile material behavior the J-Integral concept and the CTOD-concept have found a wide range of applications.

Path-independent integrals for two-dimensional crack problems were introduced independently by Cherepanov [1] and Rice [2]. Especially Rice's work gave significant impulses towards the application of the J-Integral for elastic-plastic problems. In [2] the equivalence of J with the energy release rate is demonstrated for elastic materials and also the proportionality of J with the crack tip opening displacement CTOD is shown.

The practical application of the proposed fracture concepts rely on both the experimental procedure to measure the material resistance in simple laboratory tests and the possibility to calculate the respective loading parameter for any geometry and loading situation. Test standards for the experimental evaluation of the J-Integral and of J-resistance curves have been established e.g. by the ASTM [3], for the CTOD-measurement those standards are given e.g. by the BSI [4]. Numerical methods, most commonly used in this context is the finite element method FEM, are usually indispencable to evaluate the applied values of J or CTOD for arbitrary geometry and loading situations. To ease the application of the J-concept for practical engineering problems, a series of J-solutions of 2D and 3D crack problems have been compiled in a Plastic Fracture Handbook [5].

## Application of the Finite Element Method (FEM)

The J-Integral and the CTOD can be considered as quantities belonging to the continuum mechanics description of a structure and are derived from deformations, strains or stresses in the cracked structure. Therefore standard numerical techniques to compute stress fields in arbitrary structures are applicable also for fracture problems. One of the most powerful methods which is widely applied in elastic plastic fracture mechanics is the finite element method ([6,7]). The efficiency of the FEM with respect to fracture problems may be significantly increased in two ways:

1. To model the appropriate singularities of the crack tip stress field, special crack tip elements [8] may be employed or the collapsed element technique ([9,10]) may be applied. Here, one face of a standard isoparametric quadratic element is degenerated to a point or a line. To model linear elastic problems, the midside nodes of the adjacent sides are shifted into a quarter-distance from the crack tip. Thus a $r^{-1/2}$-singularity in the strains is obtained. Only one node is used at the crack tip or the multiple nodes are tied together. For elastic-plastic problems, the crack tip is modelled by independent multiple nodes and the midside nodes of the adjacent sides remain at their original position. Thus, a $r^{-1}$-singularity in the strains is obtained. The independent degrees of freedom of the crack tip nodes result in non-zero crack tip opening displacement in the numerical model.

2. To calculate the J-Integral or energy release rates the FEM offers the very effective virtual crack extension method [11]. The application of this method generally requires program modifications. If the evaluation of J is not required in the course of a FEM-analysis, e.g. to control the amount of crack propagation, the virtual crack extension method can also be applied in a post-processing program.

In the following, numerical evaluations of CTOD performed within several research projects are compiled and compared with experimental results wherever possible.

## Evaluation of Single Edge Notched Tension (SENT) and Compact (C) Specimens in the Elastic-Plastic Range

The experimental evaluation of the fracture toughness $J_{Ic}$ or the critical CTOD requires that the correlation between measured quantities, like forces or displacements at very few points, and the crack tip loading parameters J or CTOD are known. Calibration experiments are performed and theoretical/ numerical concepts are developed and checked with these experiments not only to correlate the measured quantities with the crack tip parameters, but also to qualify the numerical methods for the application to arbitrary structures. A number of investigations along this line have been done for C- and SENT-specimens [12,13].

### C-Specimens of Structural Steel St E 460 (Ni-V)

A series of standard C-specimens with thickness 20 mm and 40 mm and relative crack lengths a/W between 0.5 and 0.7 were tested at room temperature. For one experiment 2D and 3D elastic-plastic FEM analyses were performed. Fig. 1 shows very good agreement of the force-displacement curves of the experiment and the 3D-analysis.

This analysis was further evaluated with respect to the crack opening displacement in the center plane and at the surface (Fig. 2). Through the use of collapsed element with multiple independent crack tip nodes the calculated linear variation of the crack opening displacement with distance from the crack tip is very smooth.

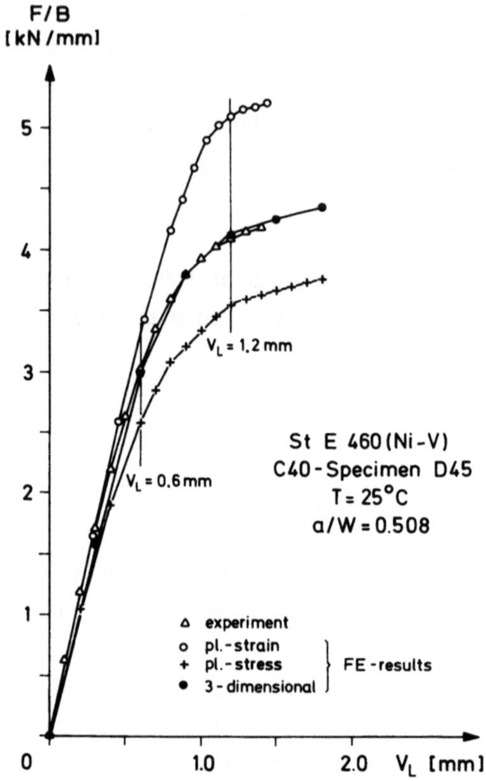


Fig. 1. Compact-specimen: force-displacement diagrams

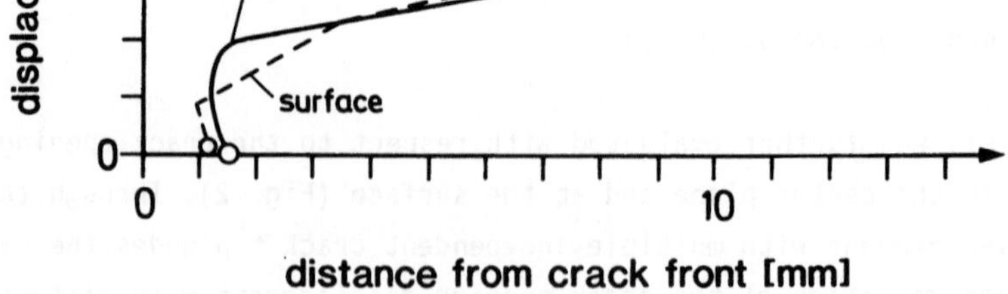


Fig. 2. Compact-specimen: crack opening displacement

The decrease of COD at the surface near the tip is due to the necking of the specimen near the crack tip. This is demonstrated by a 3D-plot of the displaced mesh near the crack front (Fig. 3) and by the plot of COD across the thickness at different distances from the crack front (Fig. 4), both taken from a similar analysis.

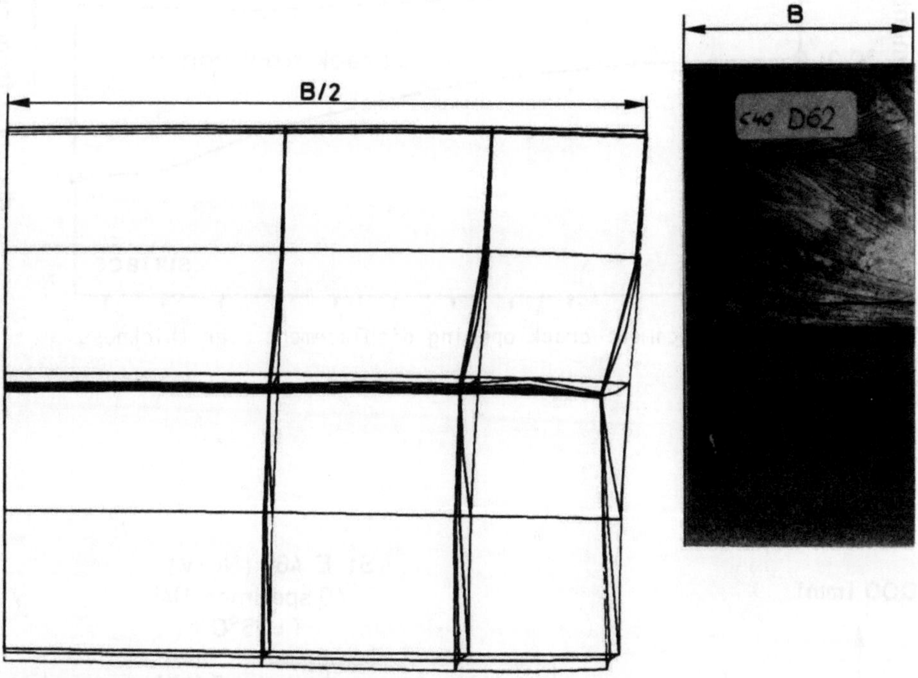


Fig. 3. Compact-specimen: deformed area around crack front

The comparison of the calculated COD at the surface (e.g. Fig. 2) with measurements at two load levels is given in Fig. 5. Beside the almost perfect agreement two problems are to be seen here: First, the experimental evaluation - the displacements of the crack flanks are determined photographically - near the crack tip breaks down. Second, from Fig. 4 it is clear that the COD and also the CTOD determined at the surface is smaller by a factor of about 2 compared with the COD and CTOD prevailing across the major part of the specimen thickness. So, a CTOD (= $\delta$) can be defined by the extrapolation of the COD to the position of the original crack tip. This definition allows to determine CTOD easily experimentally as well as numerically.

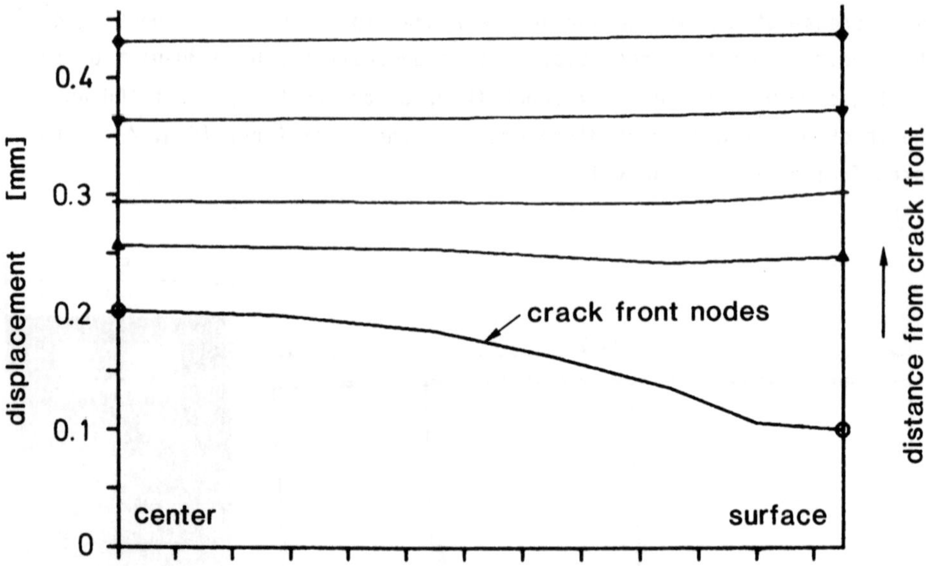


Fig. 4. Compact-specimen: crack opening displacement over thickness

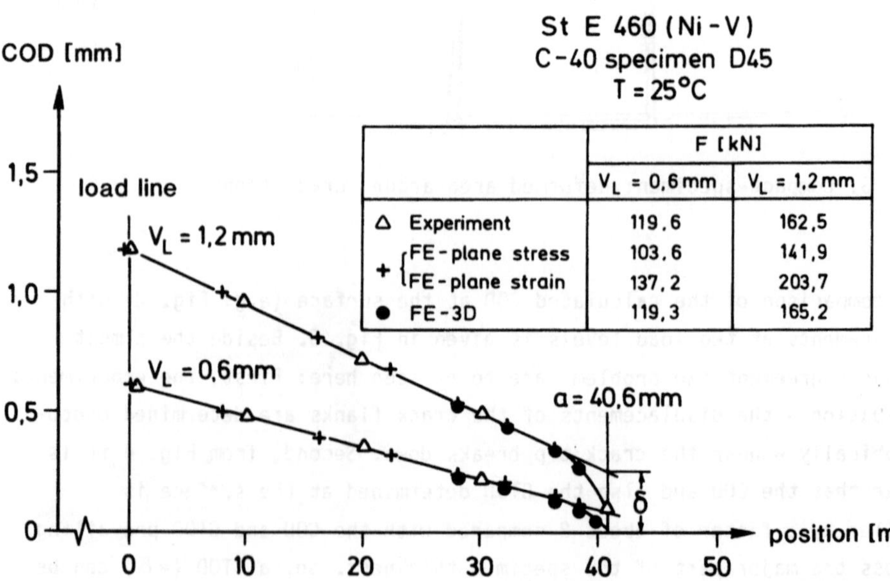


| | F [kN] | |
|---|---|---|
| | $V_L$ = 0,6 mm | $V_L$ = 1,2 mm |
| △ Experiment | 119.6 | 162,5 |
| + { FE-plane stress | 103.6 | 141,9 |
| + { FE-plane strain | 137.2 | 203,7 |
| ● FE-3D | 119,3 | 165,2 |

Fig. 5. Compact-specimen: crack opening displacement at the surface

If the experimental evaluation of the surface displacements is made, the near tip displacements should not be considered. All experimental and numerical CTOD-values discussed in this section have been obtained by linear extrapolation not taking into account near tip displacements.

To verify the correlation $J = \beta R_{po,2}$ CTOD the results of eight experiments and of the numerical evaluation of one experiment (2D and 3D) are plotted in Fig. 6. For the 2D-analyses the factor came out to be 1.18 and 1.88 for plane stress and plane strain, respectively. The experimental results show very little scatter and agree well with the result of the 3D-analysis. Here, the factor of proportionality was found to be about 1.55.

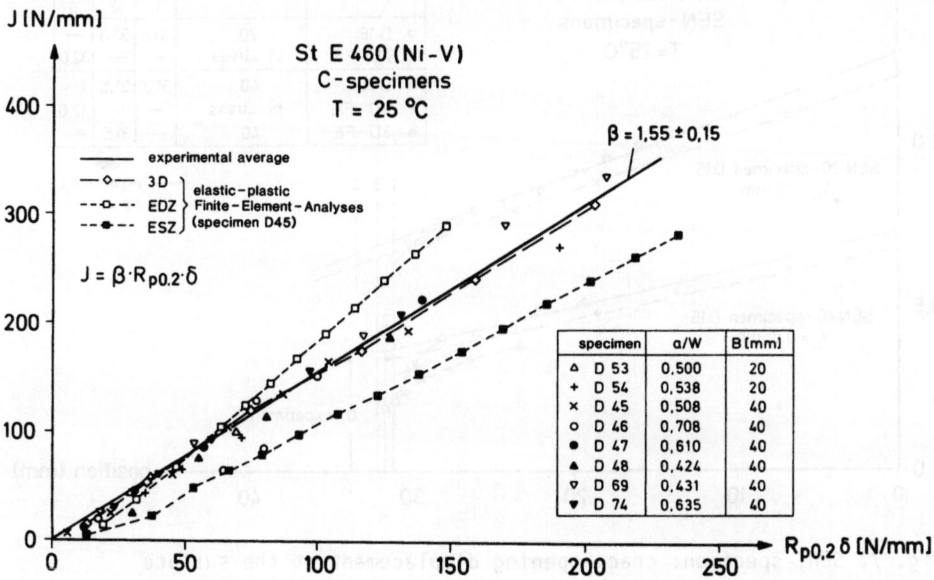


Fig. 6. Compact-specimen: correlation between J and CTOD

SENT-Specimens of the Structural Steel St E 460 (Ni-V)

For this geometry the elastic-plastic deformation and the failure behavior were expected to be different from that of the compact specimen loaded predominantly in bending. In order to avoid bending effects at all, the SENT-specimens were loaded not by bolts but by stiff hydraulic clamping grips with concentric flutes. It turned out, however, that this loading configuration introduced a strong influence of the details of the loading conditions that needed a more thorough analysis [14]. To obtain a comparably good agreement between analysis and experiment as in the previous case the complete loading device had to be included in the model. For calculations where the loading device was not included ("infinite angular stiffness") the overall force-displacement curves of the experiment and the analysis did not agree. Even when the COD was compared for the same amount of specimen elongation the COD agreed only qualitatively (Fig. 7).

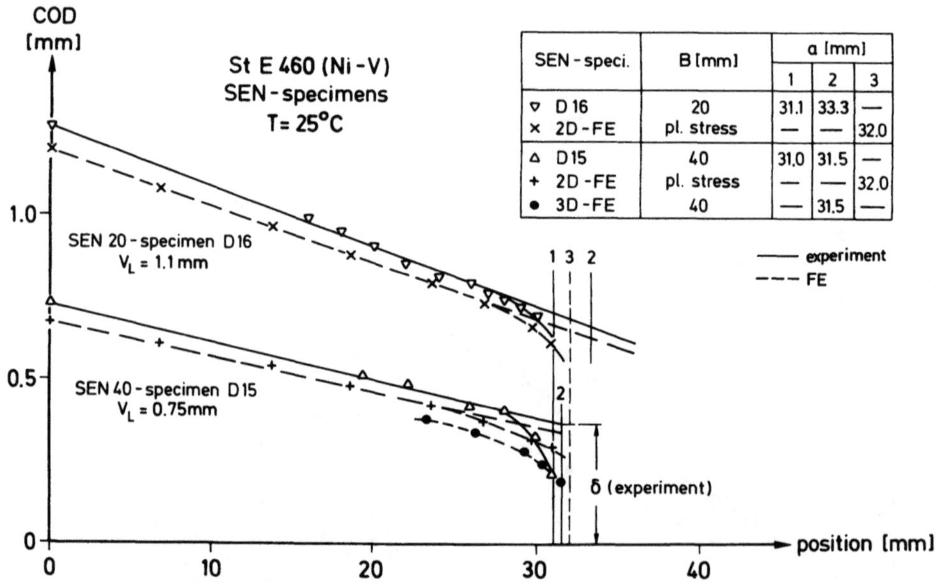


Fig. 7. SENT-specimen: crack opening displacement at the surface

However, the same trends as with the compact specimens can be observed, allowing the CTOD to be obtained by linear extrapolation of the linear part of the crack flanks to the position of the original crack tip. The J-CTOD values from these simulations lie just above the experimental scatterband (Fig. 8) yielding factors of proportionality of 1.12 (plane stress), 1.35 (plane strain) and 1.29 (3D) evaluated at $J = 300$ N/mm [12]. The differences between the β values for the C- and SENT-specimens may be due to geometry effects e.g. bending versus tension.

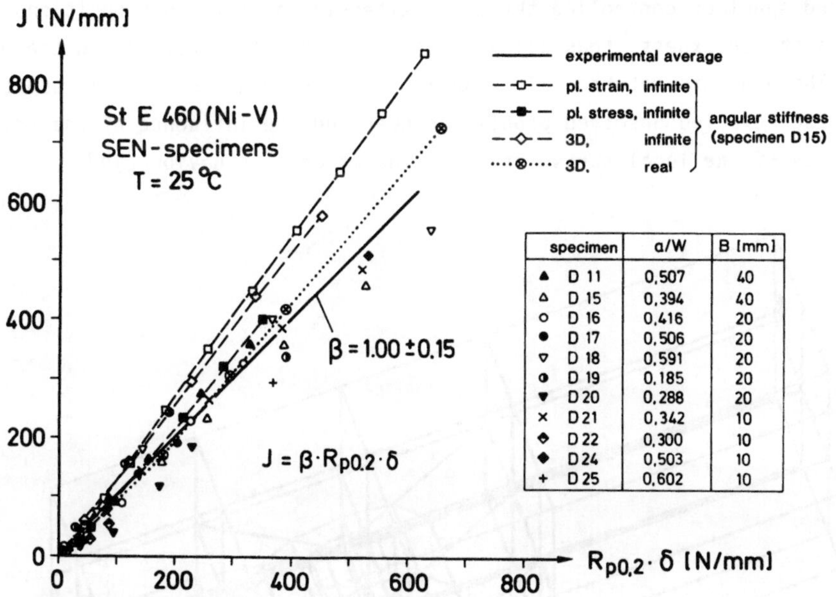


Fig. 8. SENT-specimen: correlation between J and CTOD

For a 3 D-simulation where the loading device was included in the model ("real angular stiffness") the numerical and the experimental curves (force-displacement and especially J-CTOD, Fig. 8) are in good agreement. For J = 300 N/mm the factor of proportionality determined from the numerical simulation is 1.03 which is in excellent agreement with the factor of 1.00   0.15 determined from the experiments.

Compact Specimens with Stable Crack Extension: 3D-Evaluation

At the IWM a series of J-R experiments were performed on different steels at different temperatures using compact specimens with and without side-grooves (e.g. [15,16]). The numerical tools were provided and tested to simulate such experiments in 2D and 3D (e.g. [17,18]). As a first step in the chain of transferability of results from small specimens to structures two specimens taken from the same plate were tested and analysed [19].
In order to examine whether the J-R curve obtained from the side-grooved specimen can be used to control the crack extension in the numerical simulation of a J-R experiment with a smooth specimen a three-dimensional FEM-analysis of the smooth specimen was performed. The J-R curve from the side-

grooved specimen controlled the crack extension at different positions across the thickness, thus allowing for crack front tunnelling in the model. The analysis matched fairly well with experiment in several respects but also revealed unsolved problems concerning the influence of the triaxiality of the local stress state on the material behavior [19].

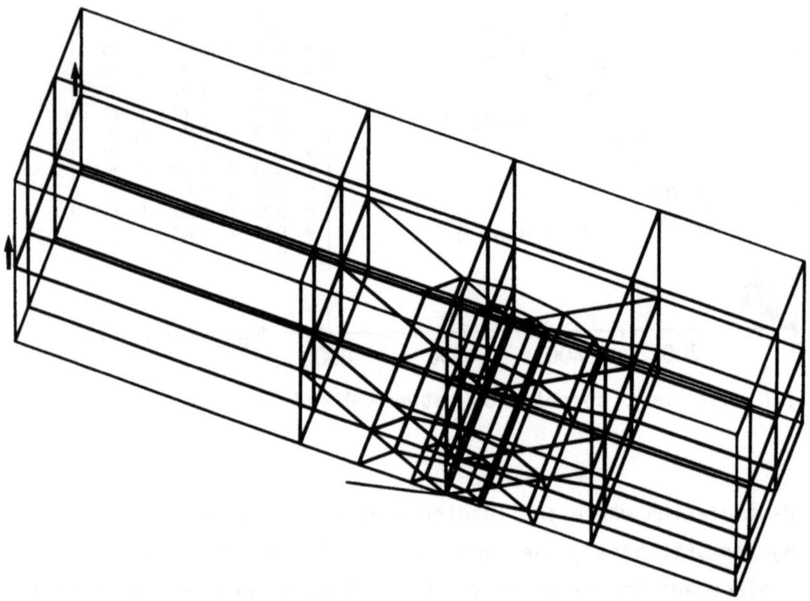

Fig. 9. Compact-specimen: FEM-mesh

Fig. 9 gives a view of the FEM-mesh, Fig. 10 shows the variation of the J-Integral over specimen thickness for different load levels, and Fig. 11 presents the calculated propagation of the crack. A multilinear fit of the stress-strain data of a steel A 542 (0.2 % proof stress $R_{po.2}$ = 533 MPa, ultimate tensile stress $R_m$ = 670 MPa) was taken for the calculations [19]. The numerical evaluation of COD and CTOD in the center and at the surface of the specimen is given in Fig. 12 for the last load step. For the sake of simplicity, the distance is measured from the original position of the crack front. Except at the surface in the near tip range, the linear variation is predominant as in Figure 2. Because the extrapolated displacement practically coincides with the respective displacement of the crack tip node in the center of the specimen, a direct evaluation of CTOD from the crack front nodes (Fig. 13) is possible.

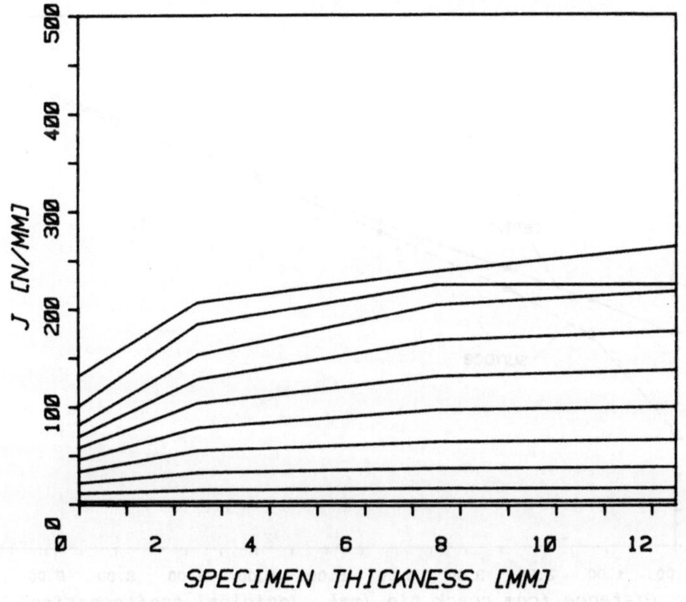


Fig. 10. Compact-specimen: variation of J over thickness

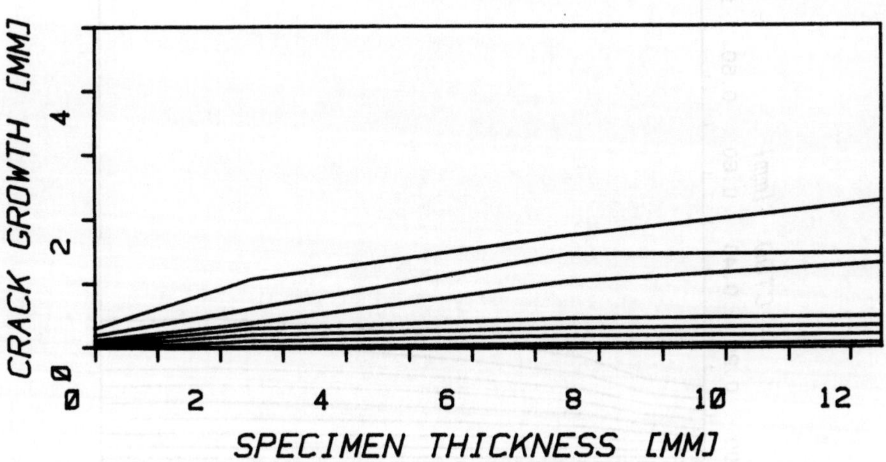


Fig. 11. Compact-specimen: calculated tunnelling of the crack front

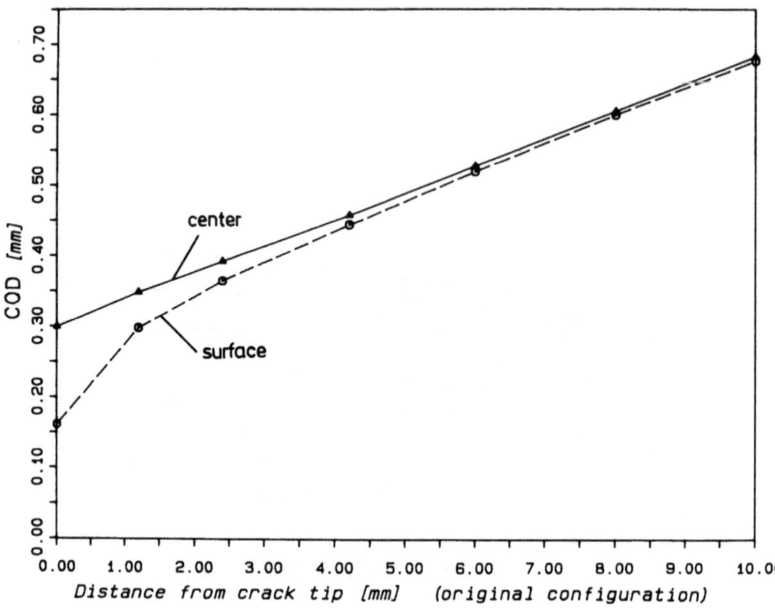


Fig. 12. Compact-specimen: crack opening displacement

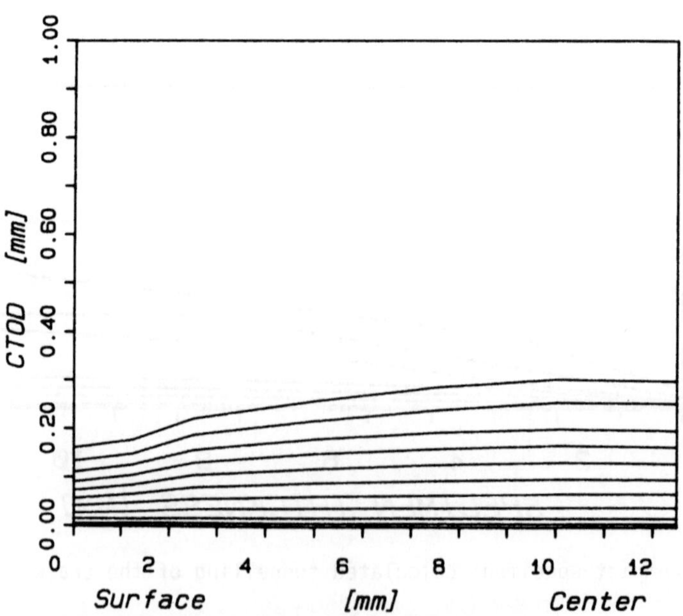


Fig. 13. Compact specimen: direct evaluation of CTOD

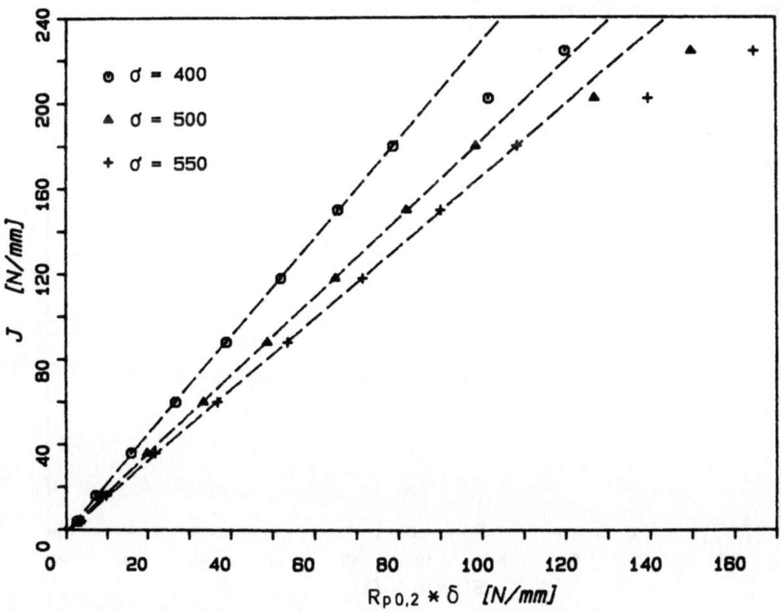


Fig. 14. Compact-specimen: correlation between J and CTOD

The linear correlation of J and CTOD is confirmed in Fig. 14 also for load levels far above $J_i$ (here = 50 N/mm). Only for a very large amount of crack extension the curves tend to bend over toward higher CTOD. The interpretation of the slope of the curves poses some problems. If the nominal yield stress of 400 MPa is used, the slope ($\beta \approx 2$) exceeds by far the experimental and numerical results described in the previous chapter. The explanation may be found in the true stress - true strain curve (Fig. 15) of the material. In the analyses in the previous chapter due to the coarse FEM-idealisation and due to the fact that these analyses had been stopped before onset of crack propagation the effective plastic strains contributing to J in the analyses were small. With a significant amount of crack propagation in this analysis it must be assumed that also larger strains do contribute significantly to the behavior of the numerical model and that therefore the "effective" yield stress in the correlation should be increased. This has been done in Fig. 14 with the result that for an assumed yield stress of 550 MPa a slope $\beta$ = 1.65 is obtained which is in excellent agreement with the theoretical prediction by Rice [2].

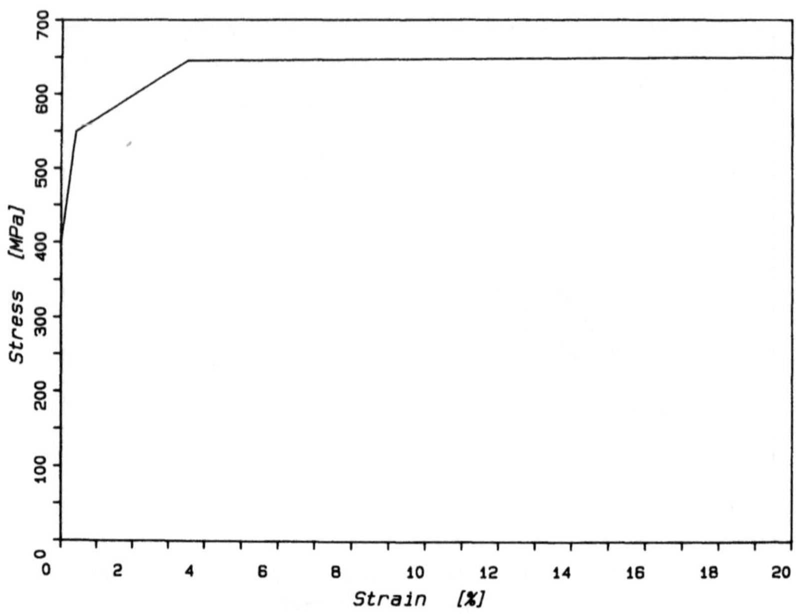


Fig. 15. Compact-specimen: stress-strain curve

## Thermal Shock Analysis of a Nozzle Corner Crack

In the framework of the german reactor safety program HDR a natural flaw in a nozzle corner subjected to repeated thermal shocks was analysed [20]. The flaw was modelled in 3D as detected and sized by nondestructive examination. Elastic and elastic-plastic FEM analyses were performed to evaluate the variation of crack tip loading parameters $K_I$ and J along the crack front. Fig. 16 shows the J-Integral for different times in the transient. Fig. 17 and Fig. 18 show the variation of the COD at the nozzle surface and at one position for the depth direction. Due to temperature loading the variation is no longer linear, so the evaluation of CTOD by extrapolation is not possible. But the shape of the curves indicate that the displacement of the crack tip node may be used as CTOD. The variation of CTOD along the crack front for different times in the transient is plotted in Fig. 19. It can be seen that CTOD shows the same qualitative behavior as J (Fig. 16). Due to the complex geometry and loading situation modelled in this analysis a direct comparison of J and CTOD has not been attempted.

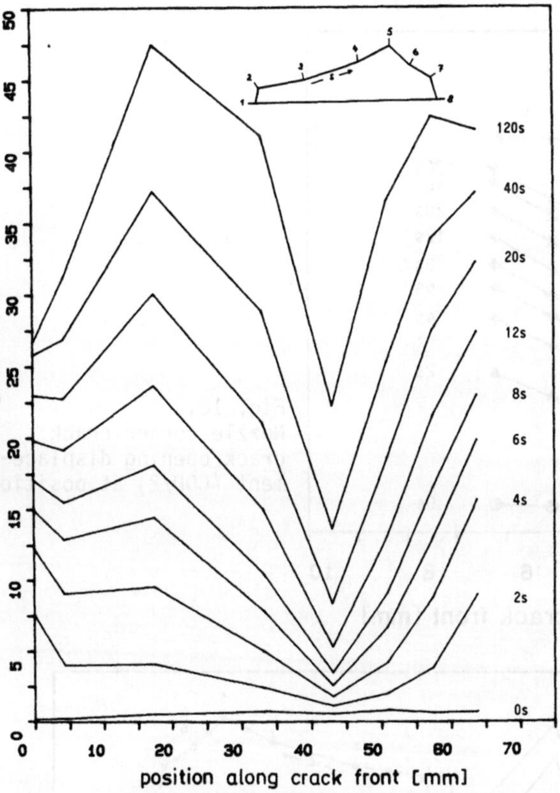


Fig. 16.
Nozzle corner crack:
J along crack front
at different times

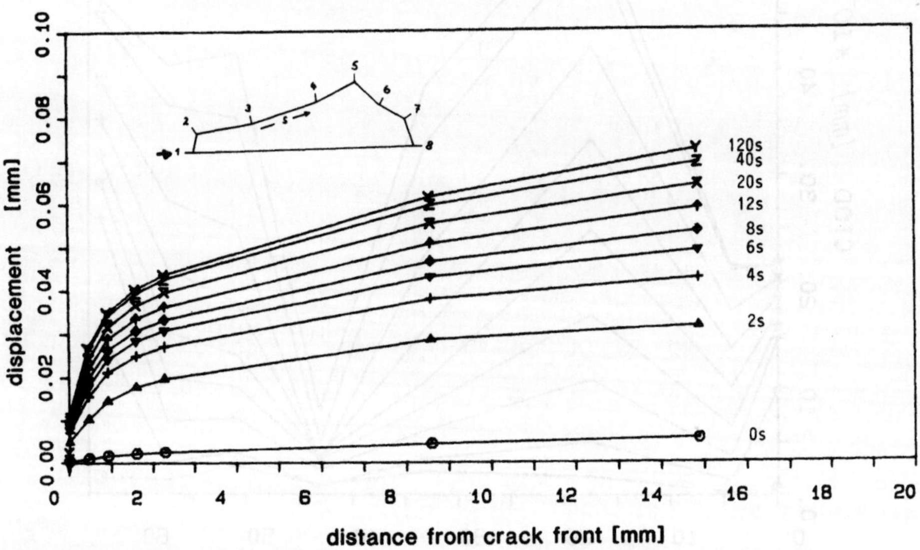


Fig. 17. Nozzle corner crack: crack opening displacement (COD/2) at the surface

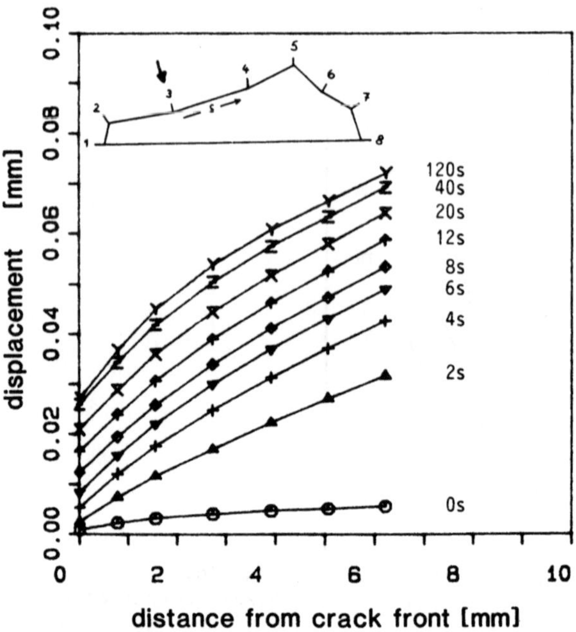


Fig. 18.
Nozzle corner crack: crack opening displacement (COD/2) at position 3

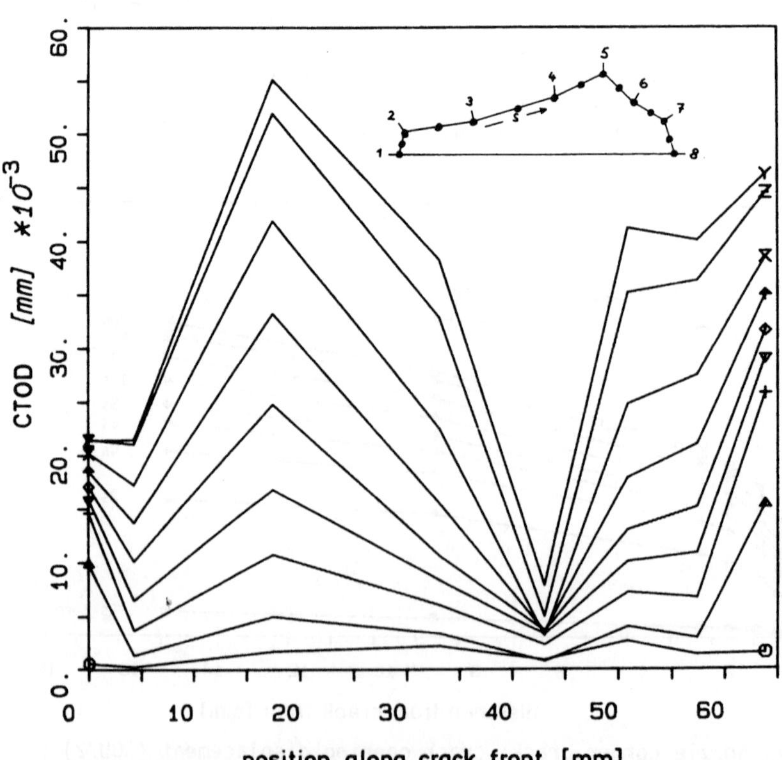


Fig. 19. Nozzle corner crack: direct evaluation of CTOD, parameter: time

## Conclusions

The numerical tools to compute elastic-plastic crack tip loading parameters J and CTOD are well established for arbitrary geometry and loading situations also including temperature gradients and stable crack propagation. The evaluation of J is possible with only minimum requirements concerning mesh refinement and numerical effort. The determination of CTOD requires increased but not prohibitively large effort concerning mesh refinement and evaluation time. The correlation of J and CTOD postulated from continuum mechanics considerations could be verified numerically and experimentally.

## References

1. Cherepanov, G.P.: Crack propagation in continuous media, J.Appl. Math. Mech. (PMM) 31 (1967) 476-485

2. Rice, J.R.: A path independent integral and the approximate analysis of strain concentrations by notches and cracks, J.Appl. Mech. 35 (1968)

3. ASTM E 813-81: Standard test for $J_{Ic}$, a measure of fracture toughness, Annual Book of ASTM Standards, Part 10, Philadelphia (1981)

4. BS 5762: 1979: Methods for crack opening displacement (COD) testing, British Standards Institution (1979)

5. Kumar, V., German, M.D. and Shih, C.F.: An engineering approach for elastic-plastic fracture analysis, report no. NP-1931, EPRI, Palo Alto (1981)

6. Zienkiewicz, O.C.: The finite element method, Mc Graw-Hill, London (1977)

7. Bathe, K.J. and Wilson, E.L.: Numerical methods in finite element analysis, Prentice-Hall, New Jersey (1976)

8. Tracey, D.M.: Finite elements for the three-dimensional elastic crack analysis, Nucl. Eng. Design 26 (1974) 282-290

9. Henshell, R.D. ans Shaw, K.G.: Crack tip finite elements are unnecessary, Int. Num. Meth. Engn. 9 (1975) 469-507

10. Barsoum, R.S.: Triangular quarter-point elements as elastic and perfectly plastic tip elements, Int. J.Num. Meth. Engn. 11 (1977) 85-98

11. Parks, D.M.: The virtual crack extension for non-linear material behavior, Comp. Meth. in Appl. Mech. & Engn. 12 (1977) 353-364

12. Hollstein, T.: Experimentelle Untersuchungen zum Verhalten von Rissen bei elasto-plastischen Werkstoffverformungen, IWM-Bericht W 1/82, Fraunhofer-Institut für Werkstoffmechanik, Freiburg, FRG (1982)

13. Schmitt, W.: Anwendung der Methode der finiten Elemente in der Bruchmechanik unter besonderer Berücksichtigung dreidimensionaler und elastisch-plastischer Probleme, IWM-Bericht W 2/82, Fraunhofer-Institut für Werkstoffmechanik, Freiburg, FRG (1982)

14. Hollstein, T., Schmitt, W. and Blauel, J.G.: Numerical analysis of ductile fracture experiments using single-edge-notched tension specimens, Journal of Testing and Evaluation, JTEVA, 11 (1983) 174-181

15. Hollstein, T., Blauel, J.G. and Voss, B.: On the determination of elastic-plastic material parameters - a comparison of different test methods, ASTM STP 856 (1984) 104-116

16. Voss, B. and Mayville, R.A.: On the use of partial unloading compliance method for the determination of J-R curves and $J_{Ic}$, ASTM STP 856 (1984) 117-130

17. Schmitt, W. and Voss,, B.: Numerical simulation of post-yield fracture mechanics experiments as a basis for the transferability to components, Nuclear Engineering and Design 81 (1984) 247-255

18. Schmitt, W., Siegele, D., Kordisch, H. and Baudendistel, E.: Transferability of results of small scale experiments to real structures, Nuclear Engineering and Design 81 (1984) 247-255

19. Schmitt, W.: Three-dimensional finite element simulation of post-yield fracture experiments, in: Computational Fracture Mechanics-Nonlinear and 3-D Problems, PVP-Vol. 85, AMD-Vol. 61 (1984) 119-131

20. Schmitt, W., Blauel,, J.G., Hodulak, L. und Janski, J.: Rißfortschrittsberechnungen (3D-FE) für die Stutzenkante des RDB bei Thermoschock und erster Vergleich mit Versuchsergebnissen, HDR-Sicherheitsprogramm, 8. Statusbericht 1984

# Problems of Numerical CTOD Anlayses

H. Amstutz, T.Seeger

FG Werkstoffmechanik, Technische Hochschule Darmstadt, Petersenstr. 13, D-6100 Darmstadt, F.R.G.

Abstract:

The interest of fracture mechanics research in CTOD (crack-tip opening displacement) results from its close relationship to the asymptotic strain field. In theoretical investigations a realistic CTOD may therefore depend on a representative strain field in the near-tip region.

In order to find out the requirements of an adequate approximation of the near-tip strain field using FEM (Finite Element Method) a set of numerical analyses were performed. The results show that a calculated mesh refinement and an efficient load incrementation regarding constant as well as variable stiffness reveals a definite convergence of the numerical CTOD-values.

Applying constant-strain triangular elements (CST-elements) and isoparametrical 8-node-elements the obtained results are in a good agreement with DUGDALE- and HRR-field-solutions.

## 1. Introduction

Using finite elements (FE) often a good approximation of the stress-strain states is only obtained for areas in some distance from the crack-tip. Whereas the behaviour of the cracked structure is determined by the situation close to the crack-tip.

In CTOD-concept the asymptotic strain field described by the integrated quantity of the resulting crack opening is considered to be the controlling parameter (Fig. 1). Therefore the success of the CTOD-concept will depend on a positive answer on the question wether a representative opening displacement in the near tip region can be determined.

In this paper first results are presented of investigations dealing with the requirements for an adequate numerical approximation of the near-tip stress- strain-fields in order to get realistic CTOD.

## 2. Fundamentals and conditions of the numerical solution

Object of the computations is a center-cracked tension (CCT-) specimen in plane stress with a crack length ratio of 0.5 (Fig.2). The elastic-plastic stress-strain relations follow a piece-wise linear elastic and elastic-plastic power law on the basis of isotropic hardening. The chosen hardening exponents of oo for elastic-perfectly-plastic material and 3.333 for distinct hardening refer to the upper and lower limits of the real material behaviour of metals (Fig. 2). .

As the crack surface displacements are interesting only in an extension of a few times dt it is possible to focus the attention to the near-tip region and furthermore to ignore the CCT-specimen and the special crack length. To denote the load level we can refer to the elastic values of K or J.

The condition of plane stress is also not an essential limitation of the results and the conclusions. In this context the plane stress case can be considered exemplary for crack-tip problems with other stress-strain states when discussing the dependences of convergence and accuracy of the numerical solution.

Two element-types of general importance are applied:

1. Constant-strain-triangular (CST-)elements of the program EPSYS
2. Isoparametric 8-node-elements of quadratic order of ADINA

The basic patterns of the FE-models for the two element-types can be seen in Fig. 3. For both types different stages of mesh refinement in the near-tip region are examined.

Using isoparametric 8-node-elements special node configurations with collapsing element-sides and coinciding nodes are known to induce singularities. Crack-tip singularities can be generated by a ring of triangular 8-node-elements around the tip each with three coinciding nodes at the crack-tip. The 1/r-singularity of the strain field for elastic-perfectly- plastic material or a $(1/r + 1/\sqrt{r})$-singularity will arise depending on wether the mid-side nodes of the element-sides adjacent to the crack-tip are kept at mid-side position (half-point element - HPE) or moved to a quarter distance from the tip (quarter point element - QPE) leaving the supernumerary crack-tip nodes unrestricted in any motion.

3. Numerical approximation of near-tip fields

Three main topics have to be taken into consideration discussing numerical approximations of elastic-plastic near-tip stress-strain fields:

1. Subdivision of structure

2. Incrementation of load

3. Convergence-criterion of the equilibrium condition

In this paper the last topic will be not discussed. A satisfying convergence of the iterative process is guaranteed by a strong criteria.

The design of the FE-model ist mainly determined by the necessity of mesh refinement in the near-tip region. The element-size should have a close relationship to the expected strain gradients and the geometry of crack opening taking into account the order of the element functions. Two aspects have to be kept in mind:

1. A sufficient approximation of the strain-field with respect to the accuracy of the opening displacements

2. A good representation of the opening profile by displaced nodes and element-edges to get a properly defined CTOD.

An estimation of the opening displacements, using the DUGDALE- or the HRR- solution, may be helpful.

The need for load incrementation results from the linearization in the analysis of the non-linear problem and the consequences arising in inhomogenious fields. In this case the incompatibilities of the stress increments from the linear step can considerably differ between elements belonging to the near-field or the far-field. Greater and largely different incompatibilities counter-act a definite tendency of convergence, and lead to a deficient solution.

The following figures and diagrams illustrate, how FE describe the crack opening, depending on different stages of input. They all refer to a small-scale-yielding-(SSY-)level with $K_I/K_{IY}$ of 1/15.

The opening profiles are plotted in their real shape without any size distortion. For this reason the opening contour of the DUGDALE-solution, which serves as a reference, does not show the well-known profile of Fig. 4.a refering to different scales in x- and y-direction. Using equal scales in both directions, we are faced with the uncommon shape of opening in Fig. 4.c.

As suggested by RICE, CTOD is defined as the opening distance given by the intercept of two $45^o$ lines from tip to crack surface.

### 3.1 Opening profiles with CST-elements

As the displacements of CST-elements vary linearly, and the opening contour in total can only be represented in polygonal form, one has to arrange a reasonable number of nodes in the relevant section to get a well-defined CTOD.

For a first approximation a subdivision of five elements in the significant part of the opening profile has been chosen, being denoted as structure E.I. Fig. 5 shows the crack opening for elastic-perfectly-plastic material, resulting from a single step load incrementation.

As the opening displacements are considerably smaller than estimated by DUGDALE-solution, the wanted subdivision of elements in the CTOD-range is missed. CTOD just reaches a quarter of DUGDALE-CTOD. Moreover, the shape of the opening contour resembles more the parabolic curve of elastic displacements.

To get an insight into the background of this poor approximation a sequence of local analyses on the basis of the line-integrals of J were carried out. The radius of circular integration-paths is altered in powers of ten. The relation between J and the radius can be seen in Fig. 6.

The extension of the range with nearly constant, path-independant J serves here as an indicator for a suitable local approximation. Marking the radius of the current CTOD position, the rate of convergence in the corresponding states can now be judged. In spite of the wide range of constant J, there remains a relatively large space (compared to the small CTOD) with a strong dropping of J, indicating an insufficent approximation of stress-strain states.

The lack of element subdivision in the significant part of the crack surface yields to the idea, that a more refined mesh could improve the numerical accuracy. To check this influence the mesh in the near-tip region was reduced by multiplying the radius of the nodes with a factor of 0.12, leading to structure E.II. As Fig.7 and Fig. 8 demonstrate, the improved FE-model causes no distinct effect on the opening contour and the local J. This shows that after reaching a certain degree of mesh refinement, a further subdivision of the structure cannot improve the solution.

A substantial advance of approximation can finally be achieved by load incrementation. Even a few load steps result in a decisive increase in size and shape of the opening profile (Fig. 9). The extension of the range with nearly constant J together with the counter-movement of the CTOD-position in Fig. 10 indicate a distinct tendency of convergence. Without reformation of stiffness, an incrementation of 20 load steps leads to a largely converged solution. A higher subdivision of load only affects the fields in an area that does not influence the opening distance. CTOD as an integrated quantity is relatively insensitive to local deficiencies in a zone around the crack-tip, which is small compared to CTOD.

A final parallel computation with high load incrementation, applying the finer mesh of structure E.II, again shows the little effect of further mesh refinement, and proves the advanced general convergence of the previous solution (Fig. 11 and Fig. 12).

For a quantitative comparison with the DUGDALE-solution the varying yield conditon of the uniaxial model and the MISES-criterion in plane stress has to be regarded. As RICE pointed out, an identical solution to the DUGDALE-solution can be obtained by applying the yield criterion of MOHR/TRESCA [3]. A corresponding FE-solution (Fig. 13) results in 15 % higher displacements, which is the expected difference between the MISES and MOHR/TRESCA yield condition in plane stress. There still remains a difference of 6 % between the FE- and DUGDALE-solution.

The influence of element subdivision and load incrementation, arising in the case of elastic-perfectly-plastic material, was found as well for hardening material, considering the smaller values of crack opening on mesh refinement (Fig. 14 and Fig. 15). As expected, the opening profile becomes more rounded compared to the squared contour, which is characteristic for elastic-perfectly-plastic material. The CTOD-values of the two structures presented in Fig. 14 agree well with the HRR-field-solutions, using the factors of SHIH [4]. However, again no further benefit by a more refined FE-mesh could be achieved. The CTOD also differs only slightly from SCHWALBE`s formula, being derived on an analogy to the mode-III-state [5].

3.2 Opening profiles with isoparametric 8-node-elements

Different from the CST-element, the quadratic functions of the 8-node-element can formally describe the opening contour parabolically, just within the crack-tip elements. Applying degenerated elements this property is usually not important, as the opening profile is mainly modelled by the displacements of the free crack-tip nodes (Fig. 16).

In spite of the flexibility of the 8-node-elements the size of the crack-tip elements is not arbitrary, because the integration points lie in a finite distance from the crack-tip. To induce yielding the element-size has to be reduced until the yield limit at least in the inner integration points is exceeded.

The area, in which the stress state lies beyond the bounds of the yield stress ($\sigma_{eq} > \sigma_Y$), is enclosed by the plastic radius (Fig. 17)

$$r_{pl} = \frac{1}{4\pi}\cdot\left(\frac{K_I}{\sigma_Y}\right)^2 \left(1 + \frac{3}{2}\sin^2\phi + \cos\phi\right) \qquad (1)$$

$$\phi = 0: \qquad r_{pl} = 0.159\cdot\left(\frac{K_I}{\sigma_Y}\right)^2 \qquad (2)$$

Inserting

$$\delta_t = \frac{K_I}{E\sigma_Y} \doteq \text{CTOD} \qquad \text{(DUGDALE)} \qquad (3)$$

$$r_{pl} = 111.3\cdot\text{CTOD} \qquad (4)$$

is obtained with $\sigma_Y/E = 0.00143$ for the given material.

The distance between the crack-tip and the inner integration points (Gauss points) is in the case of HPE (1/r-singularity) about 1/5 of the element-side and for QPE ($(1/r + 1/\sqrt{r})$-singularity) about 1/25 of the element-side. Equating plastic radius and Gauss point distance, we get an upper element-size for the two elements of

$$\Delta_t = 0.795 \cdot \left(\frac{K_I}{\sigma_Y}\right)^2 \qquad \text{(HPE)} \qquad (5)$$

$$\Delta_t = 3.975 \cdot \left(\frac{K_I}{\sigma_Y}\right)^2 \qquad \text{(QPE)} \qquad (6)$$

or

$$\Delta_t \cong 500 \cdot \text{CTOD} \qquad \text{(HPE)} \qquad (7)$$

$$\Delta_t \cong 2780 \cdot \text{CTOD} \qquad \text{(QPE)} \qquad (8)$$

respectively

$$\Delta_t \cong 300 \cdot \text{CTOD} \qquad \text{(HPE)} \qquad (9)$$

$$\Delta_t \cong 1500 \cdot \text{CTOD} \qquad \text{(QPE)} \qquad (10)$$

when a fully plastified integration area is striven for.

The first ADINA-analyses for the two node configurations were performed on the basis of a mesh with a crack-tip element-size of 300 · CTOD (mesh A.I). The corresponding crack openings refering to elastic-perfectly-plastic material can be seen in Fig. 18 and Fig. 19.

Considering HPE an incremenation of a single load step leads to a well-shaped opening profile with a CTOD-value of nearly 85 % of the correct one. But this positive result seems to be more accidental because the solution for a finer mesh as shown in the next section converges to a smaller crack opening.

As stiffness is reformed after each load step just a load sequence of five steps is required for a high degree of convergence in respect of load incrementation. Using constant stiffness like the analysis with CST-elements, a comparable convergence would again demand twenty load steps. Accidentally, CTOD converges on the level of DUGDALE-solution, which exceeds the value valid for MISES yield function by 15%.

Though the Gauss points for QPE compared to HPE lie five times nearer to the crack-tip, there is no profit (Fig. 19). In the case of a single load step some strange overlap of the free crack-tip nodes and negative displacements in both directions are obtained. The value of CTOD is a quarter of DUGDALE-CTOD.

Subdividing the load the opening displacements grow and the surface-overlap disappears, but the tendency of the free nodes to move behind the crack-tip position remain like the previous solution. In spite of continous reformation of stiffness, a largely converged solution with 75 % of the nominal CTOD-value needs twenty load steps.

The deviation in shape and size of the expected opening contour may be caused by inconstancies, arising at the interface between elastic and plastic areas of partially yielded elements. To ensure fully plastified elements at the crack-tip, the element-side adjacent to the tip should have a maximum of 100·CTOD. Accordingly, an advanced mesh with a crack-tip element around 40·CTOD and a distance of the inner Gauss point of 8·CTOD of HPE or 1.6·CTOD of QPE was used.

The five-step load incrementation in Fig. 20 refering to HPE delivers now a CTOD, which is practically equal the estimated one, regarding the used yield function. Also the shape of the crack surface contour deviates only a little from the nearly rectangular opening profile, being characteristic for elastic-perfectly-plastic material. A smaller crack-tip element has no further effect on the opening distance, but reduces the irregular drift of the free nodes behind the crack-tip.

Now turning to QPE, the problem of reproduction of the 1/r-singularity becomes obvious (see Fig. 21). A single load step results in nearly zero opening displacements for all the free nodes at the crack-tip. Even for an incrementation of five load steps an overlap of the crack surface is computed. Curiously, a reasonable opening in size and shape can yet finally be achieved, but dividing the load into 10 to 20 steps and reforming the stiffness at each step. In the latter case CTOD differs only for about 3 % from the nominal value.

The singularities of the degenerated elements should not be over-valued in the question of element-size and approximation of the near-tip fields. Apart from the fact, that the type of singularity depends on the hardening behaviour of the material, and therefore the 1/r-singularity of HPE agree exactly only with elastic-perfectly-plastic material, there is no continuous description of compatible elastic-plastic states within the whole element. The formal realization of a certain singularity in the element functions is no sufficient criterion for the resulting accuracy in the elastic-plastic stress-strain-states - not in the special case, even less for differing material behaviour.

To describe the complex strain singularity of a hardening material it is necessary to move the integration points into the CTOD-controlling range of the near-field. How far the mixed singularity of QPE can have a practical importance is quite unknown. It seems that they bring no advantage in elastic-plastic analysis.

The investigations regarding the conditions of convergence for hardening material could not be finished up to now. The latest analysis using HPE refer to a mesh with a crack-tip element-size of only 10•CTOD (structure A.III: inner Gauss point distance 2•CTOD). However it turned out that the mesh is still not fine enough to reach a satisfying convergence. Compared to HRR-field- and SCHWALBE-solution there are differences of 2O %to 15 % (Fig. 22).

The deficiencies of the solution becomes also visible in the lack of inclination of the opening contour, which should be particularly marked for the given hardening exponent. It is supposed that current analysis with crack-tip elements around the size of CTOD will finally bring a general convergence.

The examination of QPE with respect to the potentialities of the $(1/r + 1/\sqrt{r})$-singularity to describe the influence of hardening was limited to an analysis of structure A.II (Fig. 23). In view of the unsatisfactory results of this elements, no further effort seems to be useful. It appears, that no advantage can be expected in comparison with HPE.

## 4. Summary

An adequate FE-approximation of the near-tip fields, and a representative CTOD requires a calculated subdivision of structure, and a reasonable incrementation of load regarding the mechanical situation at the crack-tip and the realization in the FE-formulation.

The investigations lead to the conclusion, that in the considered SSY-range the crack-tip element of CST should have an element-side about 1/10 · CTOD, that of 8-node-HPE in the general case with hardening material in the order of CTOD and in the special case of elastic-perfectly-plastic material up to 40 · CTOD. The results of the 8-node-QPE show that the $(1/r + 1/\sqrt{r})$-singularity is not well suited to approximate the $1/r^{(1/(n+1))}$-singularity of the elastic- plastic strain field.

A great influence on convergence has the incrementation of load in connection with constant or variable stiffness. The results indicate, that without reformation of stiffness an incrementation of 20 load steps may be necessary while using variable stiffness after each load step an incrementation of five load steps will be quite sufficient.

References

[1] Henshell, R. D., Shaw, K. G.
Crack Tip Finite Elements Are Unnecessary
Int. J. Num. Meth. Engn., 9 (1975), 496-507

[2] Barsoum, R. S.
Triangular Quarter-Point Elements as Elastic and Perfectly-Plastic Crack Tip Elements
Int. J. Num. Meth. Engn., 11 (1977), 85-98

[3] Drucker, D. C., Rice, J. R.
Plastic deformation in brittle and ductile fracture
Eng. Fract. Mech. J. (1970), 577-602

[4] Shih, C. F.
Relationships between the J-Integral and Crack Opening Displacement for Stationary and Extending Cracks
Rep No 79CRDO75 April 1979, General Electric Company, Corporate Research and Development Schenectady, New York

[5] Schwalbe, K. H.
Crack tip opening displacement for work hardening materials
Rep GKSS 84/E/26, GKSS-Forschungszentrum Geesthacht GmbH, Geesthacht 1984

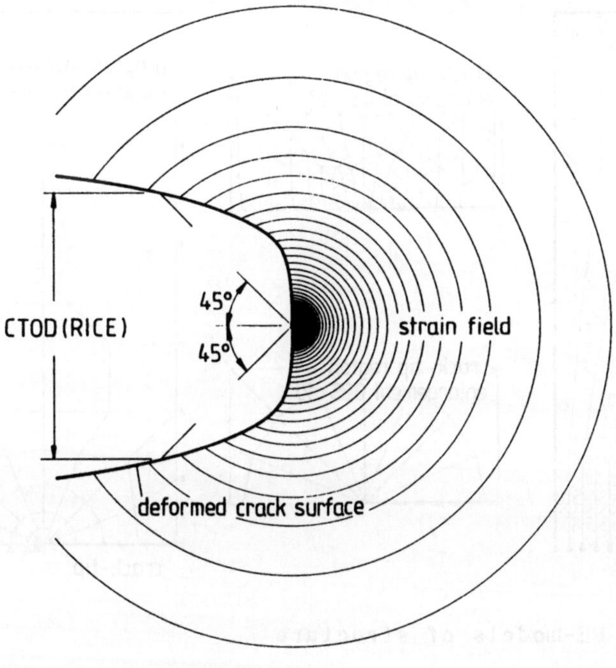


Fig. 1 CTOD-Concept

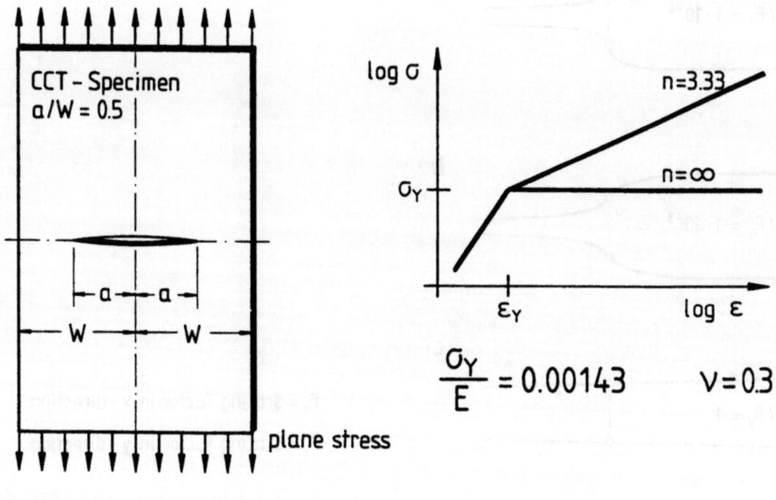


Fig. 2 Geometry and material of FE-Analysis

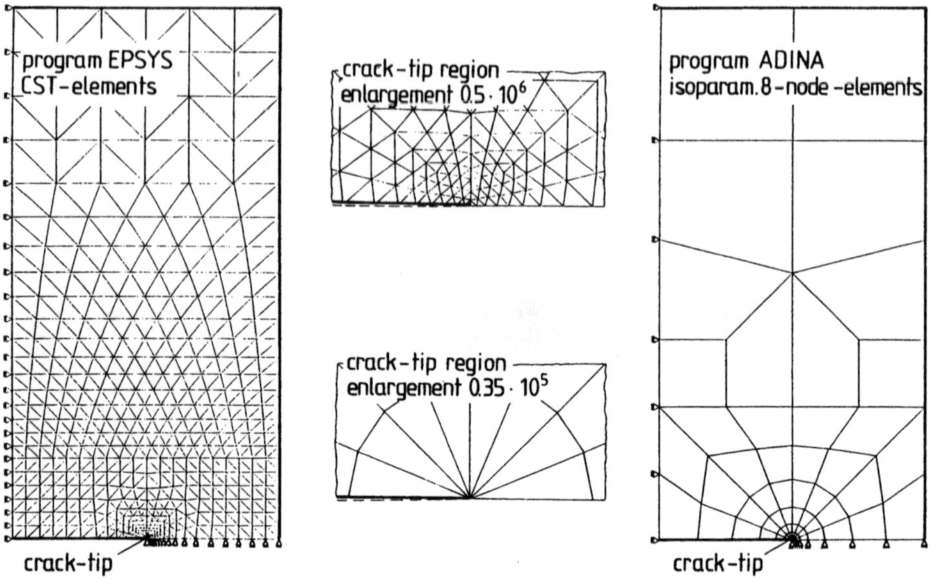


Fig. 3 FE-models of structure

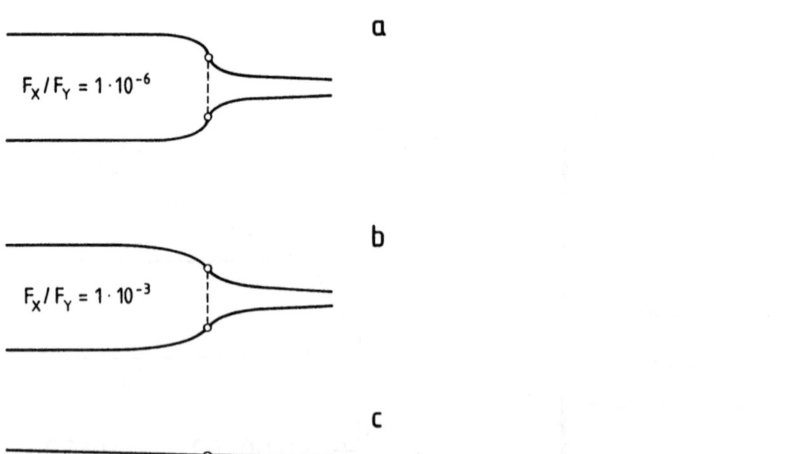


Fig. 4 Opening profiles of DUGDALE-solution in different scaling

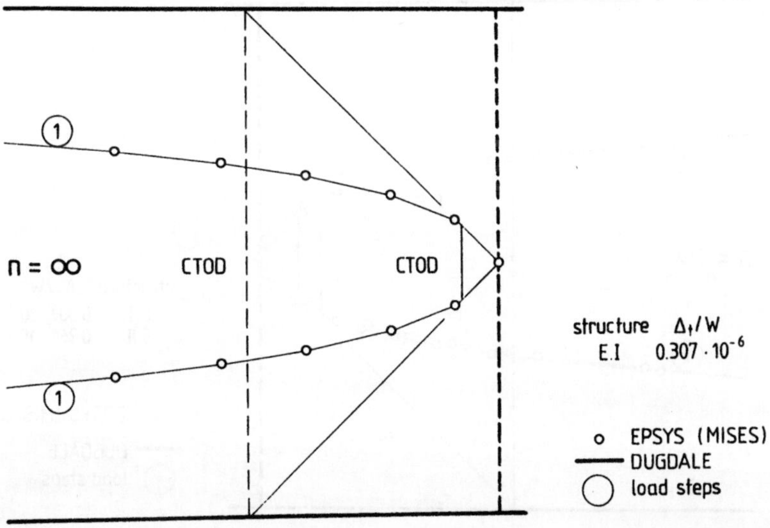


Fig. 5 Crack opening profile structure E.I (EPSYS) - n = ∞

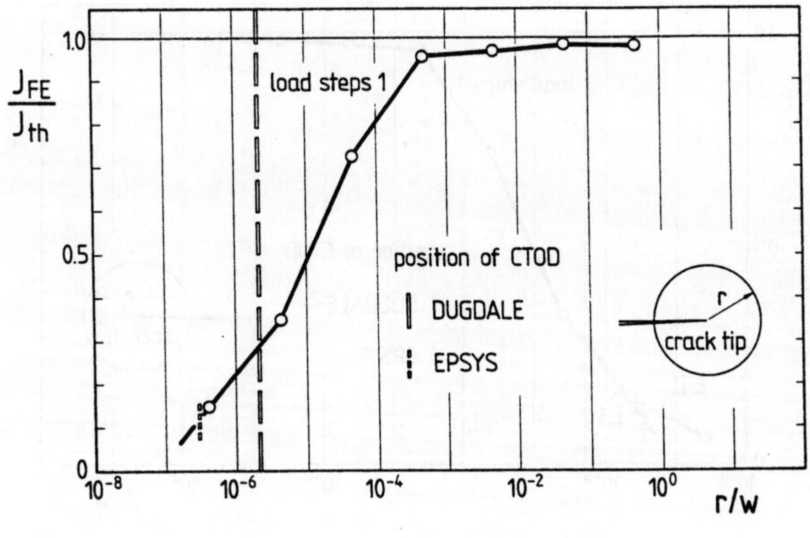


Fig. 6 J-integral structure E.I (EPSYS) - n = ∞

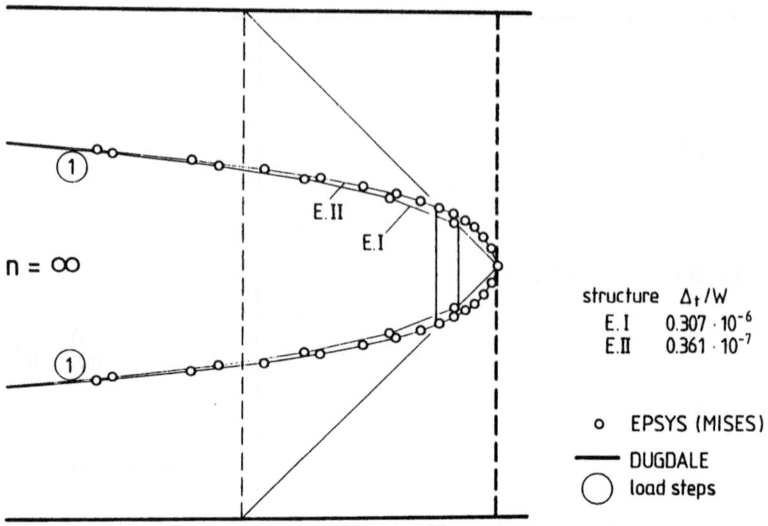


Fig. 7 Crack opening profiles struct. E.I, E.II (EPSYS) - n=∞

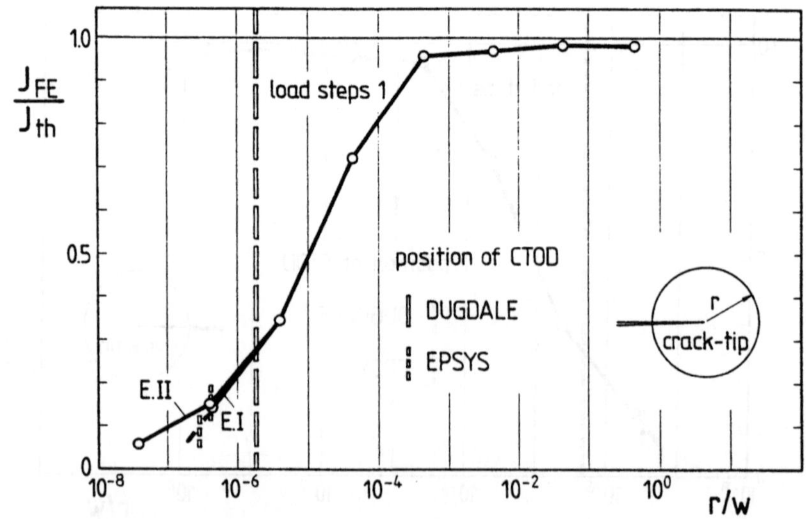


Fig. 8 J-integral structure E.I, E.II (EPSYS) - n = ∞

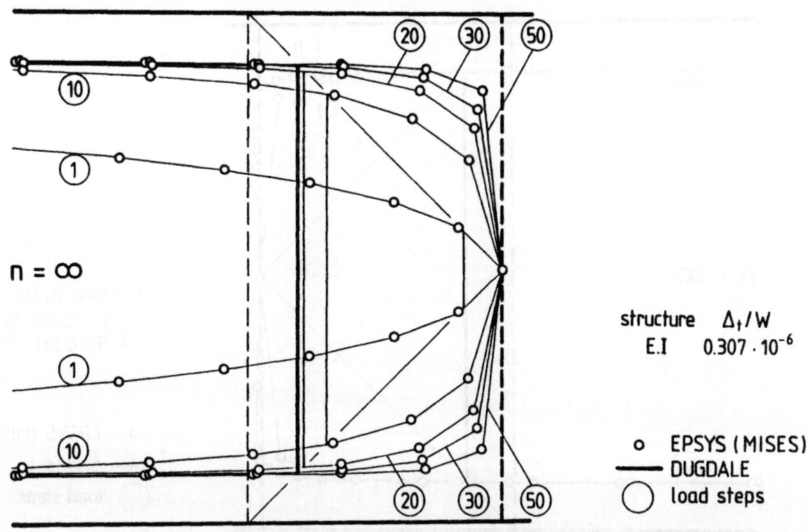


Fig. 9 Crack opening profiles for different load incrementation structure E.I (EPSYS) - n = ∞

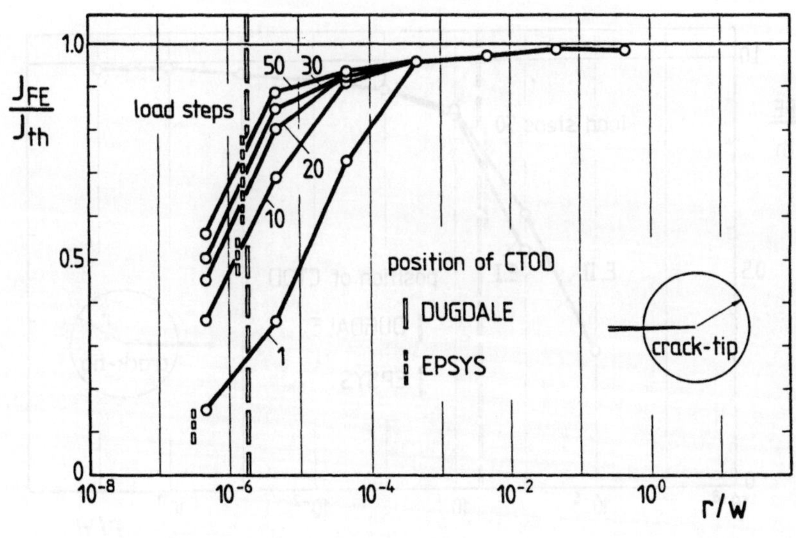


Fig. 10 J-integral E.I for different load incrementation structure (EPSYS) - n = ∞

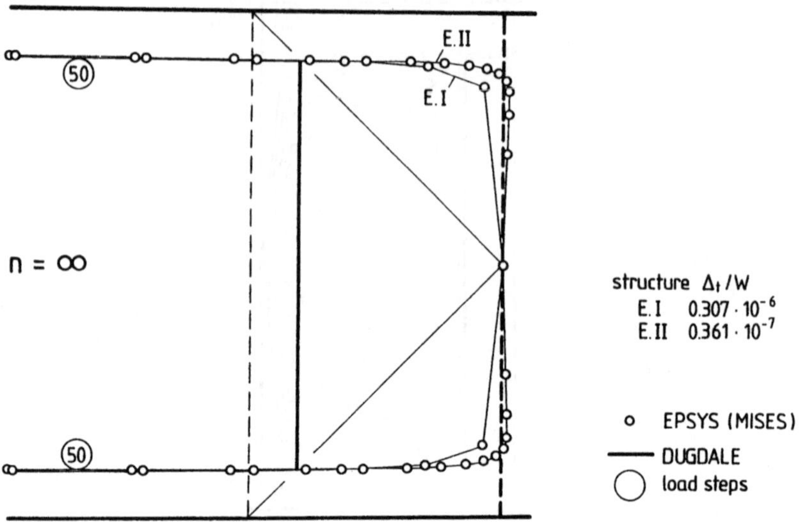


Fig. 11 Crack opening profiles for 50 load steps structure E.I, E.II (EPSYS) - n = $\infty$

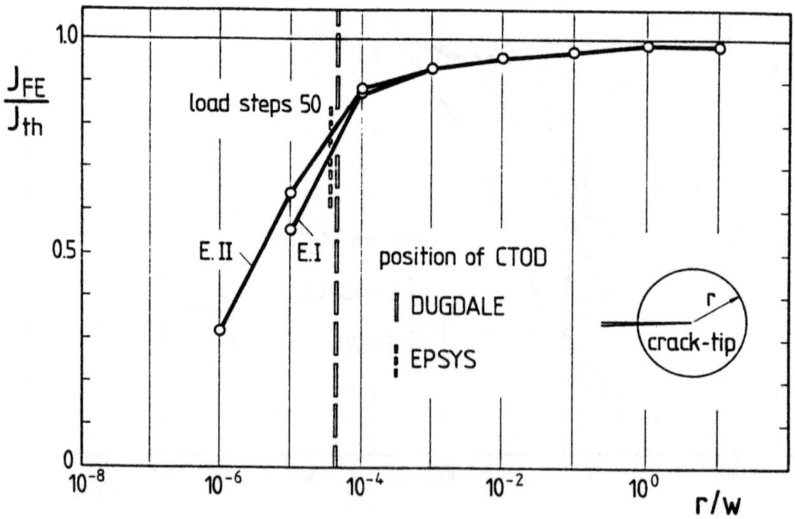


Fig. 12 J-integral for 50 load steps structure E.I, E.II (EPSYS) - n = $\infty$

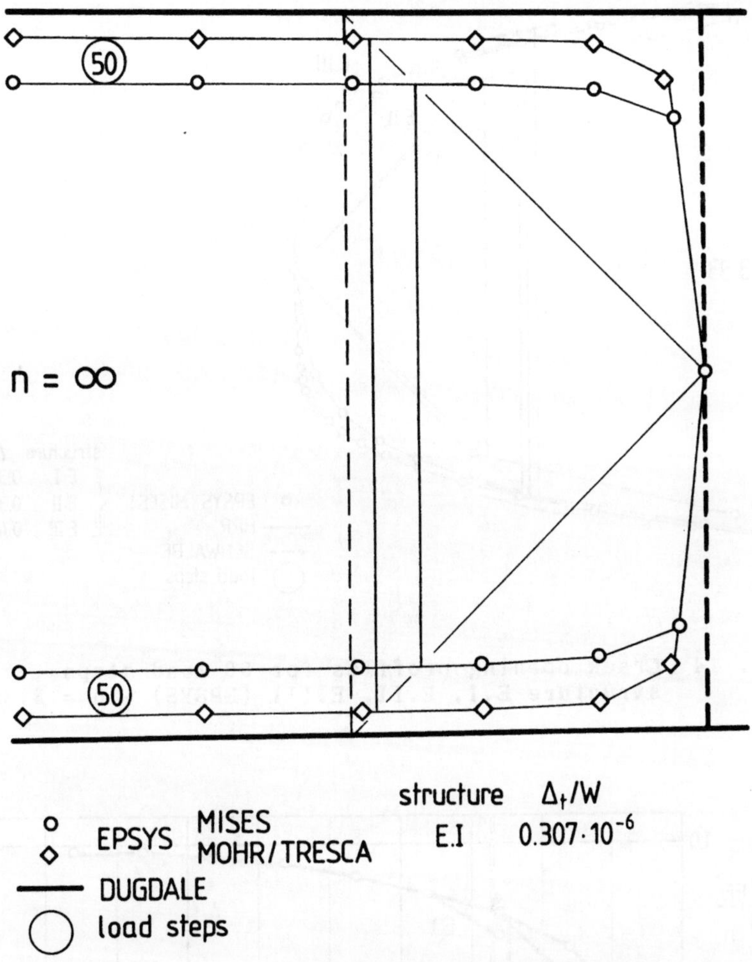


Fig. 13 Crack opening profiles with MISES or MOHR/TRESCA yield function for 50 load steps struct. E.I (EPSYS) - n=∞

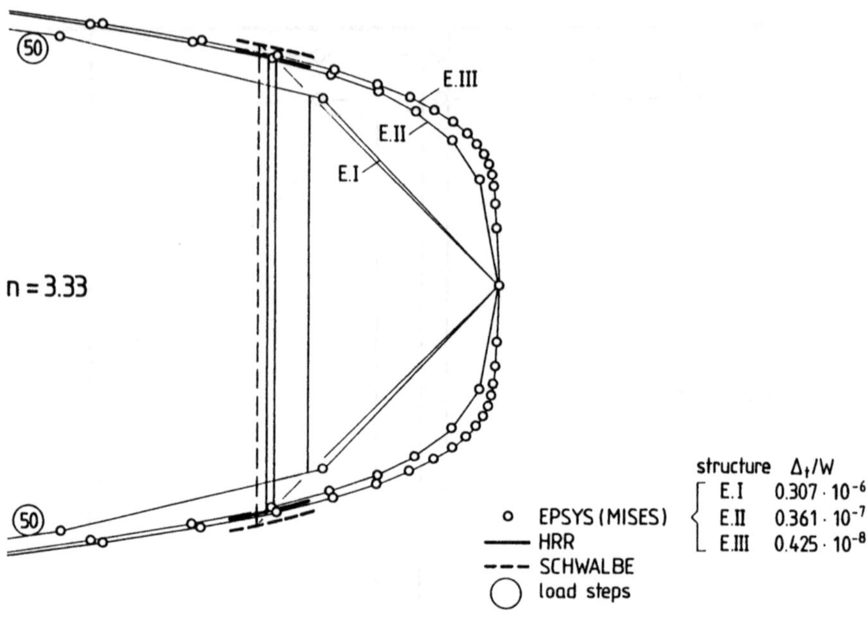


Fig. 14 Crack opening profiles for 50 load steps structure E.I, E.II, E.III (EPSYS) - n = 3.33

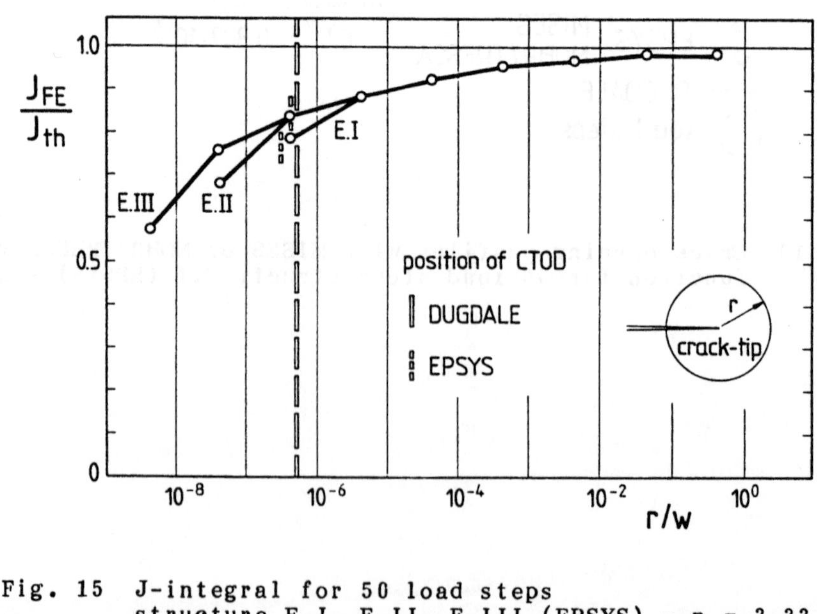


Fig. 15 J-integral for 50 load steps structure E.I, E.II, E.III (EPSYS) - n = 3.33

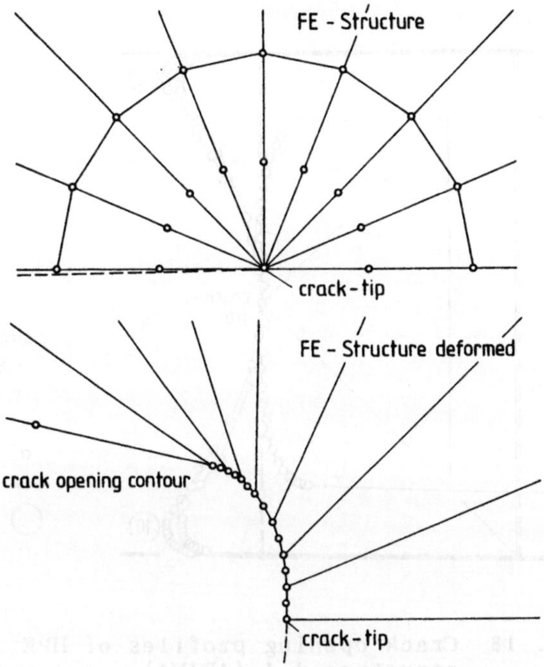


Fig. 16 Node configuration of isoparametric 8-node-elements before and after loading

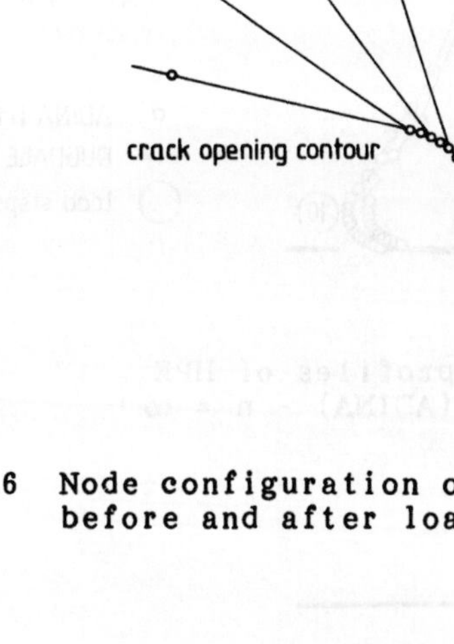


$$\sigma_x = \frac{K_I}{\sqrt{2\pi r}} \cos\frac{\varphi}{2}\left(1 - \sin\frac{\varphi}{2}\sin\frac{3\varphi}{2}\right)$$

$$\sigma_y = \frac{K_I}{\sqrt{2\pi r}} \cos\frac{\varphi}{2}\left(1 + \sin\frac{\varphi}{2}\sin\frac{3\varphi}{2}\right)$$

$$\tau_{xy} = \frac{K_I}{\sqrt{2\pi r}} \sin\frac{\varphi}{2}\cos\frac{\varphi}{2}\cos\frac{3\varphi}{2}$$

MISES yield function - plane stress

$$\sigma_{eq} = \frac{K_I}{\sqrt{4\pi r}} \sqrt{1 + \frac{3}{2}\sin^2\varphi + \cos\varphi}$$

$$\sigma_{eq} \overset{!}{=} \sigma_Y$$

$$r_0 = \frac{1}{4\pi} \cdot \left(\frac{K_I}{\sigma_Y}\right)^2 \left(1 + \frac{3}{2}\sin^2\varphi + \cos\varphi\right)$$

Fig. 17 Plastic radius (plane stress state)

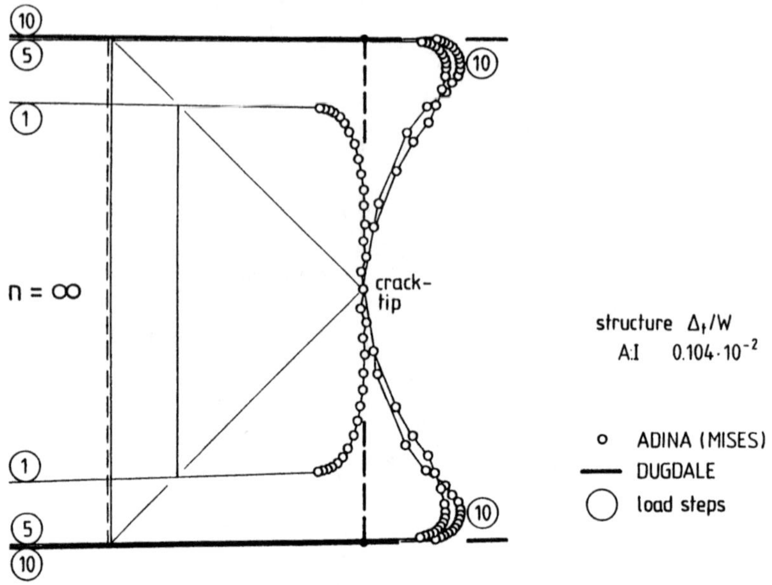


Fig. 18 Crack opening profiles of HPE structure A.I (ADINA) - n = ∞

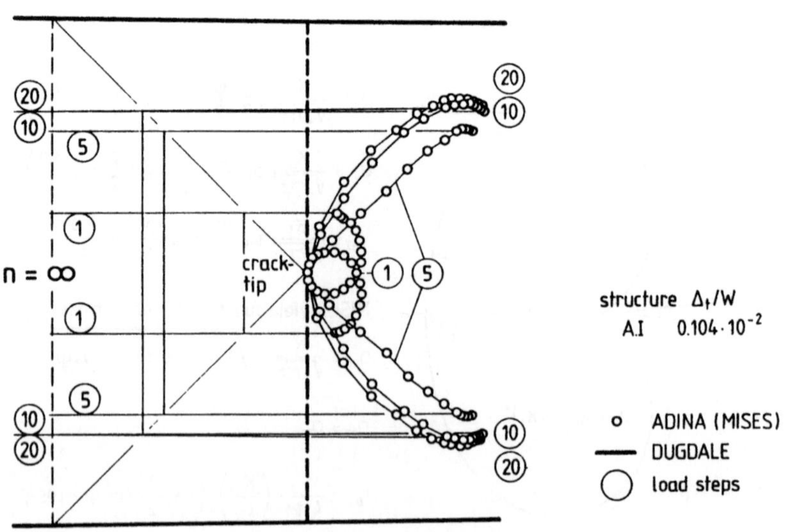


Fig. 19 Crack opening profiles of QPE structure A.I (ADINA) - n = ∞

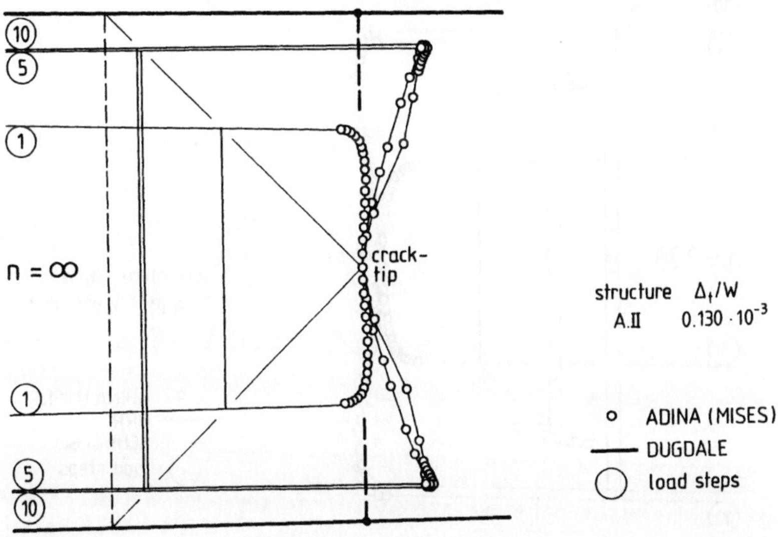


Fig. 20 Crack opening profiles of HPE structure A.II (ADINA) - n = ∞

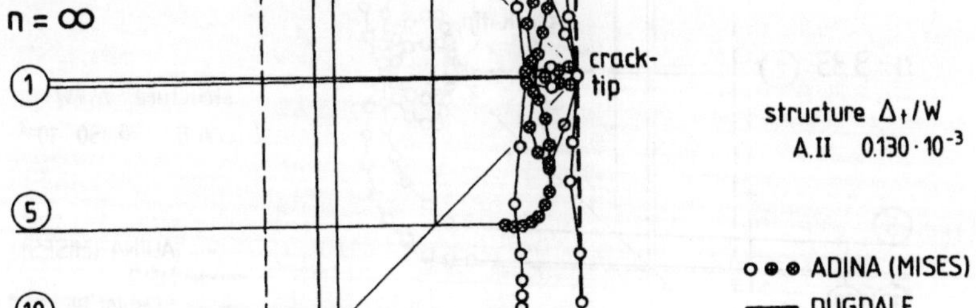


Fig. 21 Crack opening profiles of QPE structure A.II (ADINA) - n = ∞

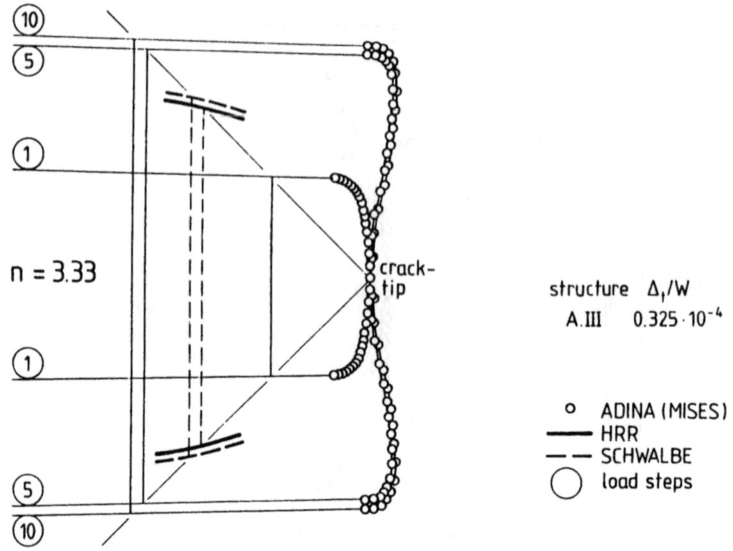


Fig. 22 Crack opening profiles of HPE structure A.III (ADINA) - n = 3.33

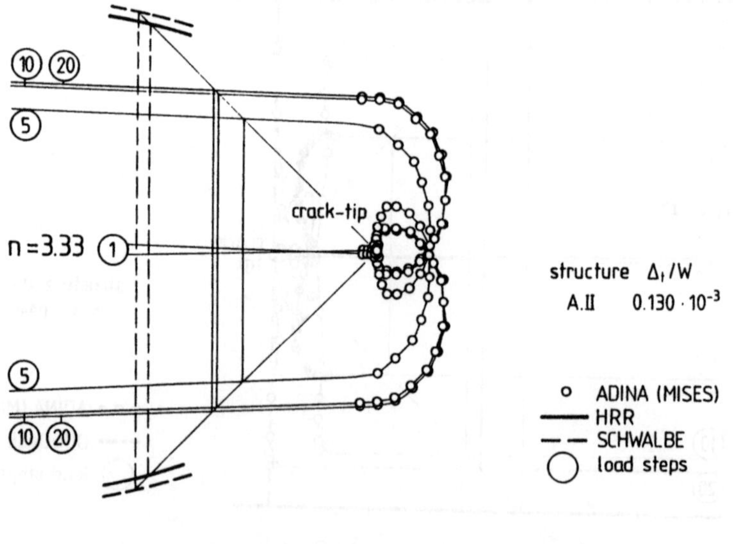


Fig. 23 Crack opening profiles of QPE structure A.II (ADINA) - n = 3.33

# Theoretical Calculation of CTOD Using a Dugdale Model Including Strain Hardening for Surface Cracks

C. Mattheck and F. Görner

Kernforschungszentrum Karlsruhe, Institut für Reaktorbauelemente, Arbeitsgruppe Zuverlässigkeit und Schadenskunde im Maschinenbau
D-7500 Karlsruhe, W-Germany

Abstract:

A generalized Dugdale model is used to predict ligament yielding and ligament rupture of semi-elliptical surface cracks. Strain hardening effects are regarded approximately. The weight function method is used to calculate the stress intensity factors due to the plastic stresses acting in the yield zone.
The CTOD is calculated at the deepest point of the semi-elliptical surface crack. Good agreement with experiments is shown.

Introduction:

The original Dugdale model /1/ is a description of plastic effects at the crack tip by combination of two elastic fracture mechanical solutions.
Newman /2/ regarded strain hardening effects with an assumed strain distribution in the yield zone. Hoffmann and Gross /3/ assumed linear decrease of plastic stresses from maximum stress to yield stress in the plastic zone. As well in /2/ as in /3/ the crack opening displacement field has been calculated.
The success of the Dugdale model has encouraged us to generalize the original Dugdale model to three-dimensional crack problems /4,5,6/. In the present paper the crack tip opening displacement will be calculated as the final result of this generalized Dugdale model.

Analysis and Results:

Fig.1 shows a semi-elliptical surface crack with a plastic zone which is assumed to be also semi-elliptical.
The dimensions of the plastic zone along the semi-axis of the crack, $d_A$ and $d_B$, are given as in the original Dugdale model from the requirement of vanishing stress singularity in points A and B. In other words the stress intensity factors $K_r$ due to the externally applied stresses $\sigma$ have to be equal to the stress intensity factors $K_{pl}$ due to the plastic loads $\sigma_{pl}$. Following /7/ this stress intensity factors are calculated in weighted average using a weight function derived in /8/:

$$\overline{K}_A = \frac{H}{\overline{K_r^A}} \int_S \sigma_{new}(x,y) \frac{\partial u_r}{\partial S_A} dS \tag{1}$$

$$\overline{K}_B = \frac{H}{\overline{K_r^B}} \int_S \sigma_{new}(x,y) \frac{\partial u_r}{\partial S_B} dS \tag{2}$$

$dS_A$ and $dS_B$ are explained in Fig.2 H=E for plane stress and $H=E/(1-\nu^2)$ for plane strain. The index r means the reference loading case as described in /8/ from which the weight function is derived. In the present case the reference stress intensity factor $K_r$ is given in /9/ by Newman and Raju. $K_{pl}$ will be calculated by eq.(1,2) with $\sigma_{new}=\sigma_{pl}$, where $\sigma_{pl}$ is the distribution of plastic stresses in the yield zone. $\sigma_{pl}(x,y)$ will be specified later. The equations for the determination of $d_A$, $d_B$ are now

$$\overline{K_r^A} = \overline{K_{pl}^A} \tag{3}$$

$$\overline{K_r^B} = \overline{K_{pl}^B} \tag{4}$$

From equations (3,4) the dimensions of the plastic zone related to an externally applied stress are computed iteratively with the computer program EASY /10/ as described in /6/ in a more detailed way.
It has been shown in /6,11/ that for the prediction of the state of a fully yielded ligament ($d_A = t-a_o$ in Fig. 1) the strain hardening does not play an important role. The assumption of constant plastic stresses in the plastic zone

$$\sigma_{pl}(x,y)=1.15\cdot\sigma_y=\text{const.} \tag{5}$$

gives very good agreement with experimental results as Fig.3 shows for the steel 90 MnV8 (German designation 1.2842).
In order to predict the ligament rupture the strain hardening of the material has to be taken into account. In /6/ following distribution of plastic stresses in the plastic zone was assumed (Fig.4)

$$\sigma(x,y) = \sigma_u^{true}+\left[\sigma_y-\sigma_u^{true}+\frac{2}{\pi}(\sigma_u^{true}-\sigma_y)\arctan\left(\frac{x}{y}\right)\right]\cdot\sqrt{\frac{(x-x_i)^2+(y-y_i)^2}{(x_o-x_i)^2+(y_o-y_i)^2}}$$

with the intersection points $(x_i, y_i)$ and $(x_o, y_o)$ (6)

$$x_i = \sqrt{\frac{1}{\left(\frac{1}{a_o}\right)^2+\left(\frac{x}{y}\right)^2\left(\frac{1}{c_o}\right)^2}} \tag{7}$$

$$y_i = c_o\sqrt{1-\left(\frac{x_i}{a_o}\right)^2} \tag{8}$$

and

$$x_o = \sqrt{\frac{1}{\left(\frac{1}{a}\right)^2+\left(\frac{y}{x}\right)^2\left(\frac{1}{c}\right)^2}} \tag{9}$$

$$y_o = c\sqrt{1-\left(\frac{x_o}{a}\right)^2}\ . \tag{10}$$

with $\sigma_u^{true}$ being the true ultimate tensile strength. This includes the assumption that the ligament is fully hardened and that a linear decrease of plastic stresses exists along the y-axis and along the front of the fictitious crack.

The unknown symbols in eq. (6) are explained in Fig.5.
Fig.6 shows the comparison between experimentally measured and theoretically predicted external loads applied at the ligament rupture. Good agreement is shown also with the results by Hasegawa et al. /12/ which are based on FE-calculations and fitting of one parameter to experimental results.
These previous considerations show alltogether that the generalized 3 D-Dugdale model is a description which is able to characterize the experimental situation in good quality.
Now the crack opening displacement at the deepest point will be calculated. The fictitious crack in Fig.1 has the semi-axis a and c. It is opened by the externally applied stress and closed partially by the plastic stresses. The resulting crack opening displacement field is

$$\delta(x) = \delta_r(x) - \delta_{pl}(x) \tag{11}$$

where $\delta_r$ is the crack opening due to external tension and $\delta_{pl}$ is the crack closure due to plastic loads. $\delta_r$ is calculated simply by use of the approximation by Petroski and Achenbach /13/ extended in /8/ to semi-elliptical cracks:

$$u_r(x,a) = \frac{\sigma_r}{H\sqrt{2}} \left(4F(a)\sqrt{a}\,\sqrt{a-x} + G(a)\,\frac{(a-x)^{3/2}}{\sqrt{a}}\right) \tag{12}$$

where F (a) is given by the known reference stress intensity factor /9/

$$K_r = \sigma_r \sqrt{\pi a}\; F(a) \tag{13}$$

and G (a) follows from the requirement of selfconsistency of the weight function method as outlined for example in /4,5,6,13/. $\delta_{pl}$ is unknown up to now. In order to calculate it, the equation for the Rice-integral

$$J = \sigma_f \cdot \delta \tag{14}$$

is used, where $\sigma_f$ is some constant flow stress. Because the Dugdale model is based on linear theory it is

$$J = \frac{K_{pl}^2}{H} \tag{15}$$

Combination of eqs. (14,15) gives

$$\delta_{pl} = \frac{K_{pl}^2}{H\,\sigma_f} \tag{16}$$

with eqs. (11,16) the CTOD is

$$\delta = \delta_r - \frac{K_{pl}^2}{H\,\sigma_f} \tag{17}$$

At the deepest point of the semi-elliptical crack eq. (17) reads as

$$\delta_A = \delta_r^A - \frac{\overline{K_{pl}^A}^{\,2}}{H\,\sigma_f} \tag{18}$$

Fig.7 shows the theoretically calculated CTOD-values for the applied stress at ligament rupture for different ratios a/c.
The difference of $\delta_A$ at ligament rupture is surprising small for different ratios a/c. Nevertheless from this Fig.7 cannot be concluded that a critical $\delta_A$ for ligament rupture exists because the value decreases with increasing crack depth and $\delta_A$ is not a constant value at ligament rupture.

## 3. Conclusions

A generalized Dugdale model which describes the behaviour of semi-eliptical surface cracks with respect to ligament yielding and ligament rupture is in good agreement with experimental measurements.
CTOD-calculations show the same qualitative behaviour as the measured crack mouth opening displacements.

Literature:

/1/ Dugdale, D.
Yielding of Steel Containing Hits
J.Mech.Phys.Sol. 8(1960)100

/2/ Newman, J.C.
Fracture of Cracked Plates Under Plane Stress
Eng.Fract.Mech. 1(1968)137

/3/ Hoffmann, M.
Ein erweitertes Dugdale Modell
Studienarbeit am Institut für Mechanik (Prof.Gross)
TH Darmstadt, W-Germany

/4/ Mattheck, C., Morawietz, P., Munz, D.
Ligament instability of Semi-elliptical Surface Cracks in Plates and Pipes
SMIRT 1983, Chicago USA, Paper-No. G/F4/6

/5/ Mattheck, C., Morawietz, P., Munz, D.
Ligament Yielding of a Plate with Semi-elliptical Surface Cracks Under Uniform Tension
J. Press.Ves.Piping 16(1984)131

/6/ Mattheck, C., Görner, F.
Leak Prediction by Use of a Generalized Dugdale Model for Semi-elliptical Surface Cracks in Plates
Proc. of 5th European Conf.Fract.,
Lissabon 1984

/7/ Cruse, T., Besuner, P.
Residual Life Prediction for Surface Cracks in Complex Structural Details
Journ. of Aircraft 12(1975)369

/8/ Mattheck, C., Morawietz, P., Munz, D.
Stress Intensity Factors at the Surface and the
Deepest Point of a Semi-elliptical Surface Crack in Plates
Under Stress Gradients
Int.J.Fract. 23(1983) 201

/9/ Newman, J.,Raju, I.
An Empirical Stress Intensity Factor Equation for
the Surface Crack
Eng.Fract.Mech. 15(1981) 185

/10/ Mattheck, C., Munz, D.
EASY- A Fracture Mechanics Computer Program
on the Basis of the Weight Function Method
Nuclear Research Center
Karlsruhe, West-Germany

/11/ Mattheck, C., Gross, D.
A Dugdale Model Including Strain Hardening for
Through Wall and Surface Cracks
SMIRT 1985, Bruxelles, Paper-NO. G4/9

/12/ Hasegawa, K., Sakata, S.
Prediction of Fracture Tolerances for Stainless Steel
Pipes with Circumferential Cracks
ASME 4th PVP Congress, Session 24

/13/ Petroski, J., Achenbach, J.
Computation of a Weight Function from a Stress
Intensity Factor,
Eng.Fract.Mech. 10(1978) 257

/14/ Göring, J.
Versagensverhalten von Flachzugproben mit einem halb-
elliptischen Riß aus einem Werkzeugstahl
Diplomarbeit am Institut für Zuverlässigkeit und
Schadenskunde im Maschinenbau,
Universität Karlsruhe 1983

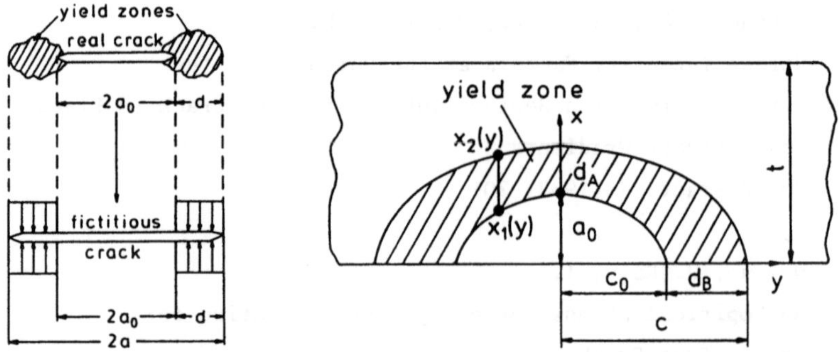


Fig.1: Sketch of the transition from original plane Dugdale model to semi-elliptical surface crack with plastic zone

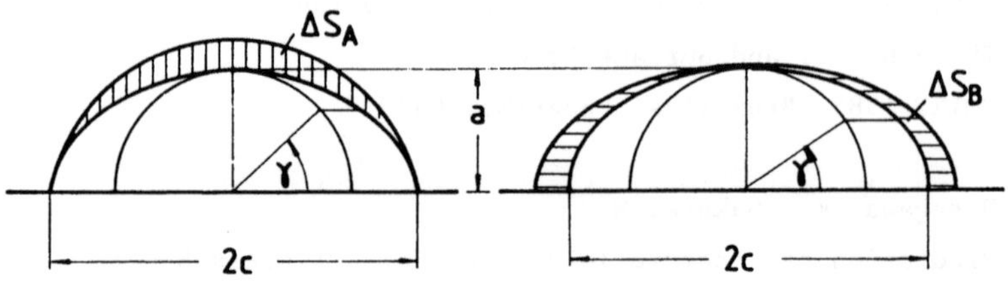


Fig.2: Degrees of freedom used for calculation of stress intensity factors in weighted average

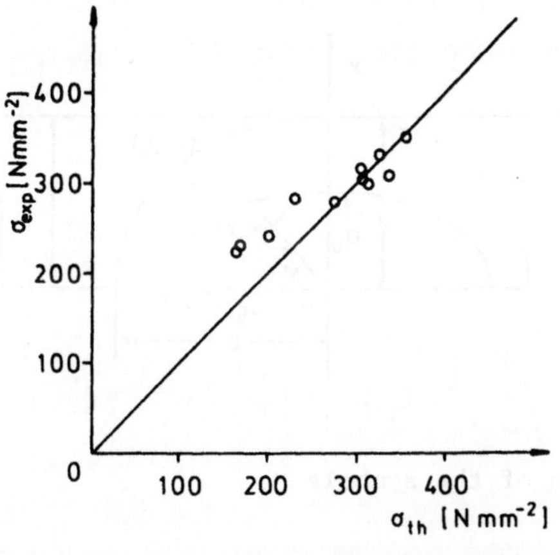


Fig. 3 Comparison of measured and theoretically predicted external stresses at ligament yielding ($a_o/c_o$= 0.24 up to 1.03, $a_o/t$= 0.28 up to 0.81 material: steel 90 MnV8 (1.2842)

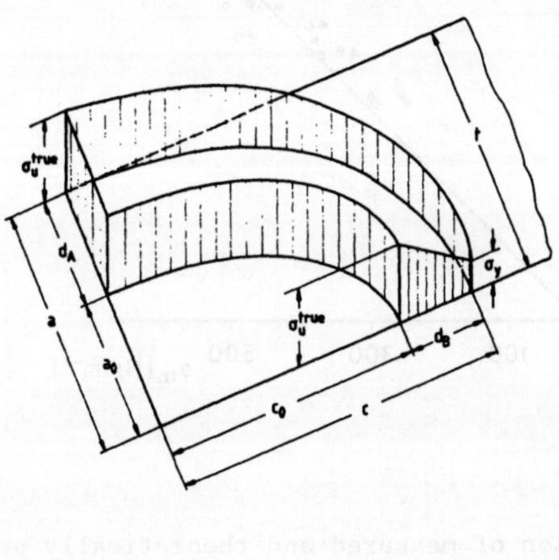


Fig. 4: Sketch of the plastic stress distribution at ligament rupture

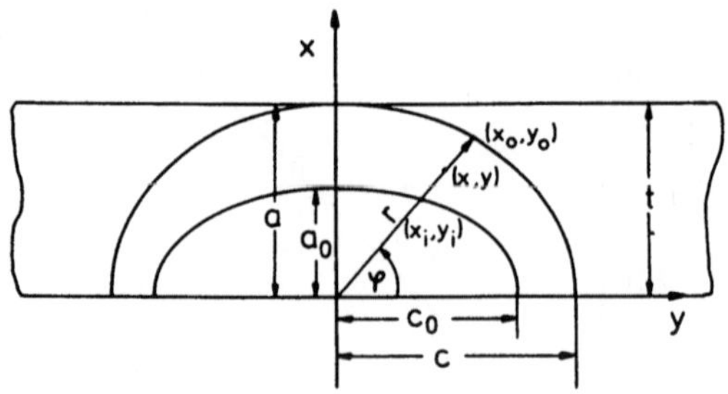


Fig. 5: Explication of the symbols

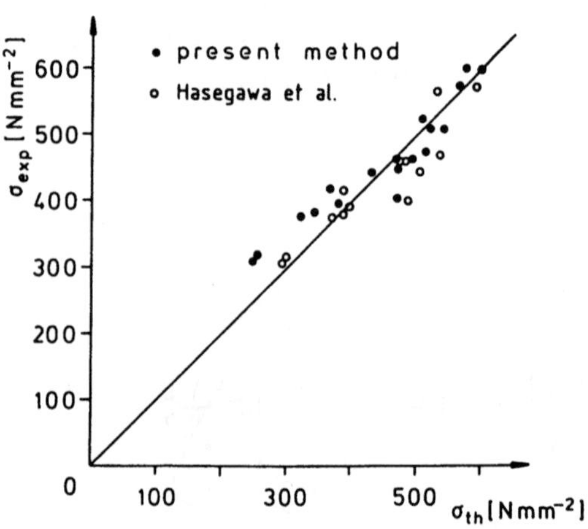


Fig. 6: Comparison of measured and theoretically predicted external stresses at ligament rupture

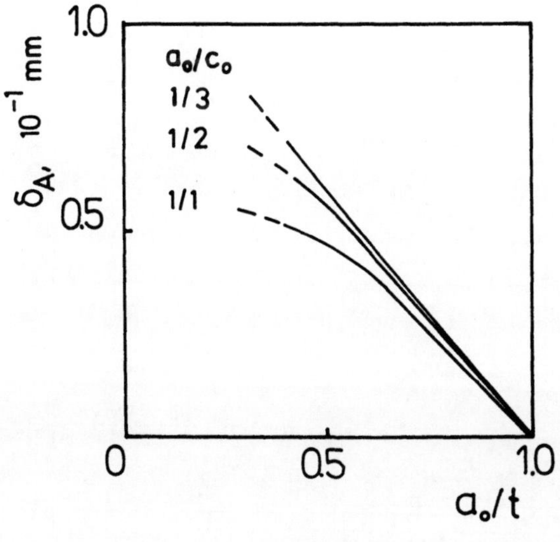


Fig. 7: Calculated CTOD-values just before ligament rupture

# Dugdale Solutions for Strain Hardening Materials

M. Hoffmann, T. Seeger

FG Werkstoffmechanik, Technische Hochschule Darmstadt
Petersenstraße 13, D-6100 Darmstadt, F.R.G.
phone: 06151/165338

Abstract:
A short survey of the field of application of Dugdale's model is given.

Two Dugdale type solutions for strain hardening materials are presented and compared to Finite-Element results and to the HRR-field.

The neccessary input for a Dugdale model including strain hardening - the shape of the stress distribution within the plastic zone - is taken for the first one from Finite-Element calculation.

The second one is based on a stress distribution assumption.

An assessment of Dugdale models for strain hardening materials closes the paper.

## 1. Introduction

The advantage of the Dugdale model [1] consists in replacing extensive nonlinear analyses by elastic ones. Dugdale solutions are adopted as well in static as in cyclic fracture mechanics, Fig. 1.

In static fracture mechanics they are used to study stability problems [2], to calculate crack tip opening displacements, e.g. [3], or to find extrapolation formulas which relate measured crack surface displacements to the crack tip opening displacement.

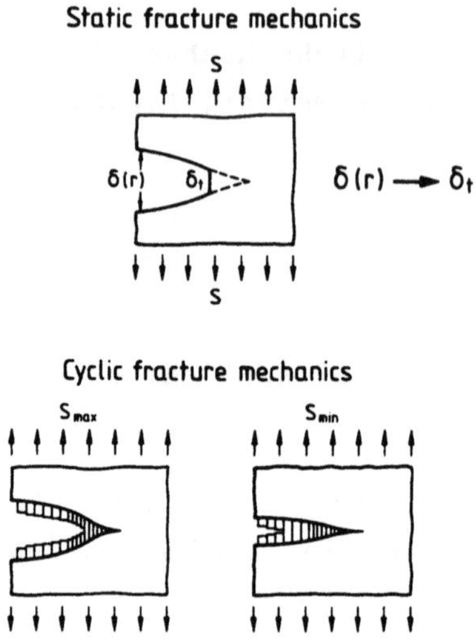


Fig. 1. Field of application of Dugdale solutions

The usefulness of Dugdale solutions for predicting fatigue crack propagation has been shown by Führing/Seeger [4,5] and Newman [6]. In this field it is important to know the complete crack surface displacements in order to allow consideration of contact problems.

The original model has been developed on the assumption of an elastic perfectly-plastic material. For this case the stress distribution within the plastic zone is known; the stress is equal to the yield stress for plane stress conditions. It can be shown that for Tresca's yield criterion the Dugdale model is identical with the exact solution [7].

In the last 20 years there have been several papers dealing with Dugdale solutions for strain hardening materials [8-14]. All of them are based on an assumption concerning the stress distribution within the plastic zone. In cooperation with Prof. Gross of the Technical University of Darmstadt the authors investigated a linear stress distribution assumption [13,14]. Although this assumption may describe strain hardening effects, and, e.g., can be used to study stability problems [2], the results were unsatisfying because an overlapping of the fictitious crack surfaces within the plastic zone was observed at high load levels.

On this background the questions arose: "Does a modified Dugdale model lead at all to the correct crack surface displacements and net section stresses if the actual stress distribution within the plastic zone is known? And if so, is it possible to describe the shape of the stress distribution?"

## 2. Dugdale solutions based on the exact stress distribution

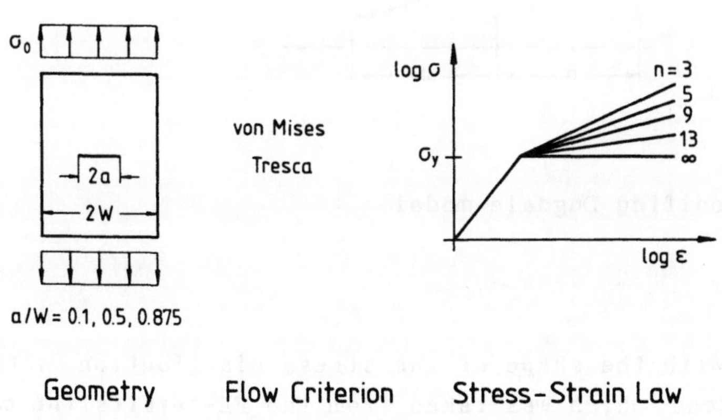


Fig. 2. Parameter of the FE-calculations (plane stress)

In order to answer the outlined questions FE-results of center cracked panels with different crack lengths were investigated,

Fig. 2. The FE-calculations were carried out for a piecewise power law with different strain hardening exponents.

$$\varepsilon = \begin{cases} \sigma/E & \sigma \leqslant \sigma_Y \\ \sigma_Y/E \cdot (\sigma/\sigma_Y)^n & \sigma > \sigma_Y \end{cases} \quad (1)$$

The elastic solution for an infinite periodic array of cracks served as basis for the Dugdale model, Fig. 3.

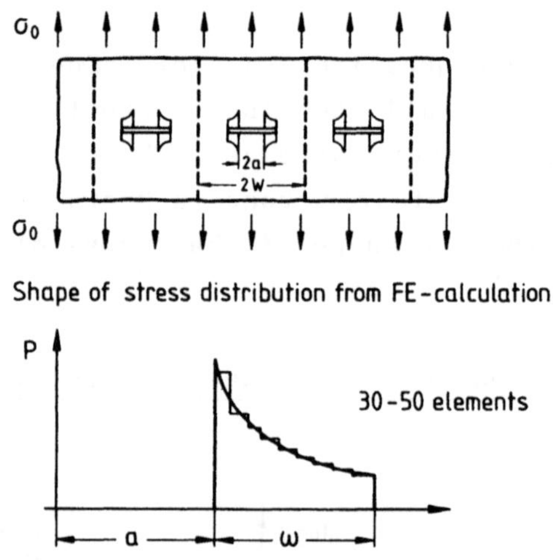


Fig. 3. Modified Dugdale model

Together with the shape of the stress distribution within the plastic zone which was taken from the FE-results the modified Dugdale model enables the calculation of

- o the plastic zone size
- o crack surface displacements

o net section stresses

o and the J-integral

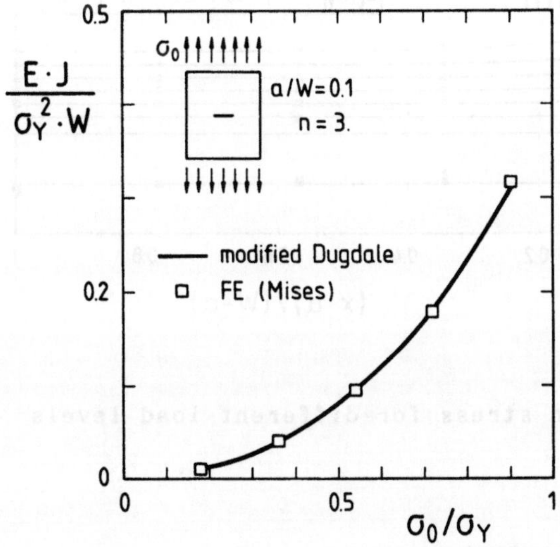


Fig. 4. J-Integral versus applied load

If the J-integral for a crack length ratio of 0.1 and a hardening exponent of 3 is considered, Fig. 4, the Dugdale solution agrees with the FE-results, even for loading near general yielding.

In Fig. 5 the net section stresses are shown for different load levels. The deviations, Dugdale related to FE, shown in the lower diagram are less than 5 per cent for all load levels.

So far, the modified Dugdale model delivers the same results as the FE-calculations.

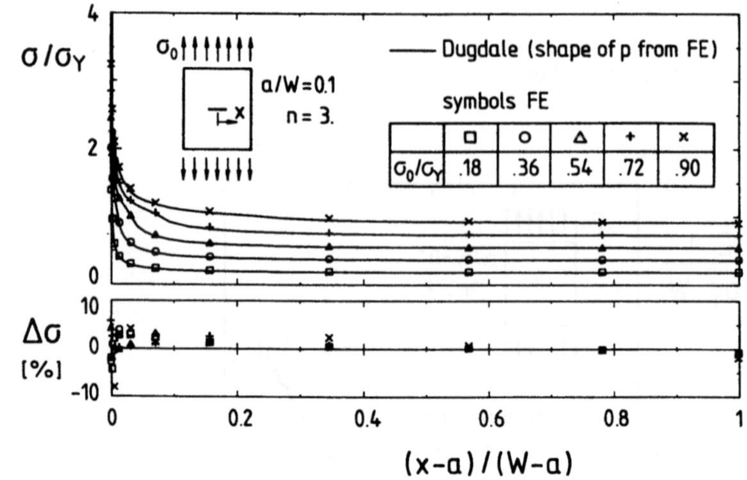


Fig. 5. Net section stress for different load levels

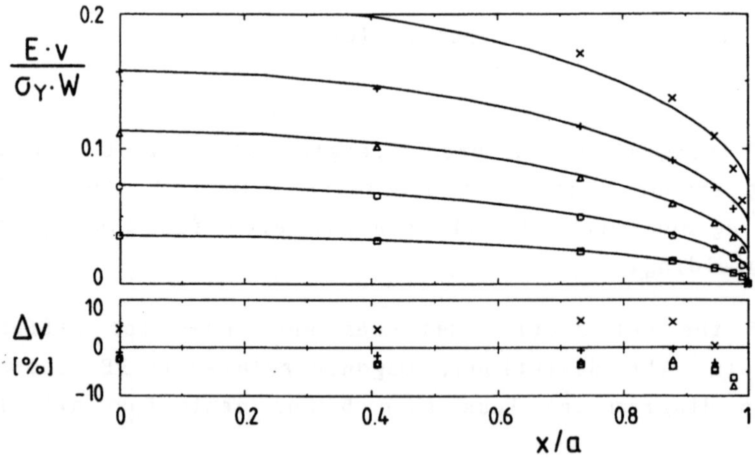


Fig. 6. Crack surface displacements for different load levels (legend see Fig. 5)

Considering the crack surface displacements, Fig. 6, leads to a different result. Although the exact shape of the stress distribution has been applied to the Dugdale model considerably larger displacements than the FE-calculations are found in the very near crack tip region. However, for 95 per cent of the crack length the Dugdale solution agrees with the FE-results.

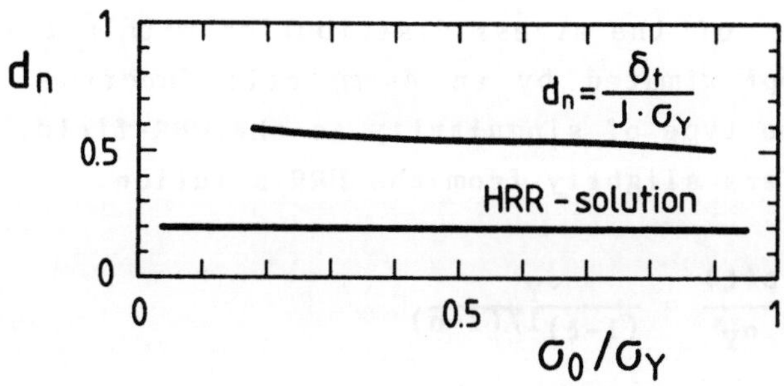


Fig. 7. $d_n$ versus applied load

The near crack tip behaviour is illustrated clearer in Fig. 7. There, the value $d_n$ defined as the ratio of the crack tip opening displacement related to the J-integral times the yield stress is plotted versus the applied stress. The Dugdale value is about a factor 2.5 larger than the HRR-value obtained from reference 15.

Up to this point it can be summarized that even if the exact stress distribution is known the Dugdale model doesn't enable correct predictions of $\delta_t$ in case of strain hardening.

However, it is encouraging that the Dugdale $d_n$ remains approximately constant during loading and thus there is always an unequivocal correlation between Dugdale and HRR-field solutions merely depending on the amount of strain hardening.

Investigations of other crack length ratios and hardening exponents delivered similar results and encouraged for further work on Dugdale solutions. In order to obtain analytical solutions the relatively simple case of small scale yielding was studied.

## 3. Dugdale solutions based on a stress distribution assumption

The evaluation of the FE-results (piecewise power law) showed that the shape of the stress distribution within the plastic zone can be approximated by an asymptotic function, Fig. 8, having the same type of singularity as the HRR-field. Only the amplitude differs slightly from the HRR-solution.

$$\bar{p}(\xi) = \frac{p(\xi)}{\sigma_Y} = \frac{\sigma_Y}{(1-\xi)^{1/(1+n)}} \qquad (2)$$

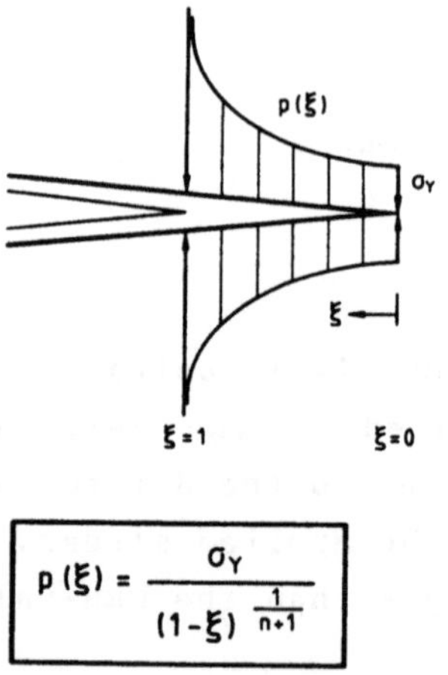


Fig. 8. Stress distribution assumption

For small scale yielding conditions the formulas for the plastic zone size and the crack surface displacements read:

$$\bar{\omega} = \frac{\omega}{(K/\sigma_Y)^2} = \frac{\pi/2}{\left(\int_0^1 \frac{\bar{p}(\xi)}{\xi}\, d\xi\right)^2} \tag{3}$$

$$\frac{v}{(K^2/(E\cdot\sigma_Y)} = \sqrt{\frac{8}{\pi}}\sqrt{\bar{\omega}\cdot s} - \frac{2}{\pi}\cdot\bar{\omega}\int_0^1 \bar{p}(\xi)\cdot \ln\left|\frac{\sqrt{s}+\sqrt{\xi}}{\sqrt{s}-\sqrt{\xi}}\right| d\xi \tag{4}$$

Where s denotes the distance from the fictitious crack tip related to the plastic zone size $\omega$. The integrals must be solved numerically (see Appendix).

Considering the special case s equal 1, the crack tip displacement, leads to a quite interesting fact: the crack tip displacement $v_a$ can be expressed in closed form in terms of the hardening exponent and the plastic zone size (see Appendix).

$$\frac{v_a}{(K^2/(E\cdot\sigma_Y)} = \frac{n-1}{n}\cdot\sqrt{\frac{2}{\pi}\cdot\bar{\omega}} \tag{5}$$

Therefore, if the integral in Eq. (3) has been solved the crack tip displacement can be directly calculated.

Before the influence of strain hardening on plastic zone size and crack tip displacement will be discussed the crack surface profile is examined. The question is whether there is still an overlapping within the plastic zone as with the assumption of a linear stress distribution [13,14].

In Fig. 9 the crack surface profile is depicted for different hardening exponents, n equal infinity represents the solution for elastic perfectly-plastic material. In contrary to a linear stress distribution there is no overlapping of the fictitious crack surfaces.

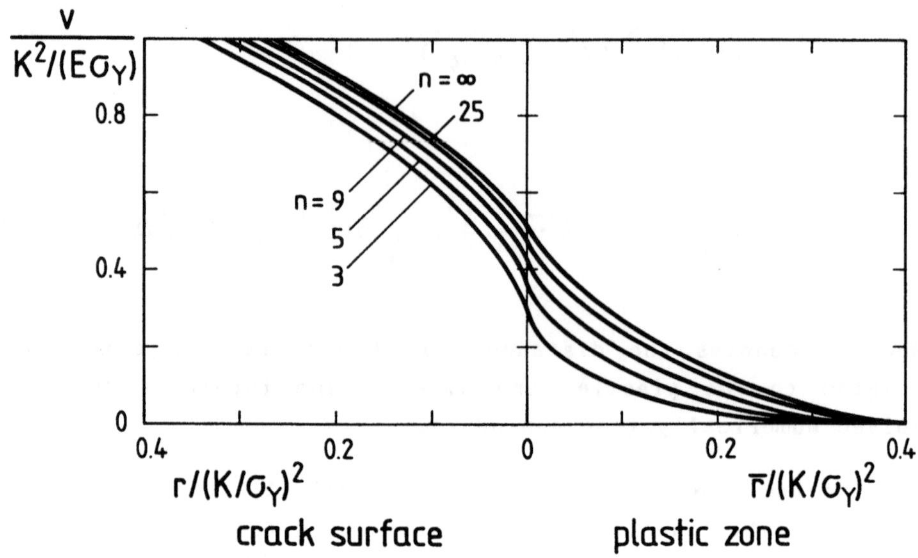


Fig. 9. Crack surface profile for different hardening exponents

Returning to the plastic zone size: The numerical solution of the integral in Eq. 3 has been approximated by a simple function.

$$\frac{\omega}{(K/\sigma_Y)^2} = \frac{\pi}{8}\cdot\frac{n}{n+\pi^2/4-1} \tag{6}$$

For n equal infinity, elastic perfectly-plastic material, the original Dugdale solution is obtained.

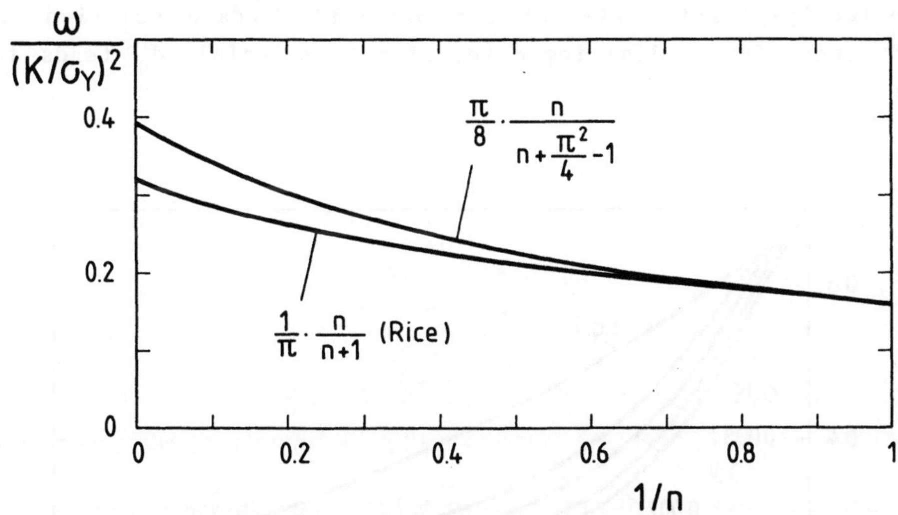


Fig. 10. Plastic zone size as a function of the hardening exponent

For comparison the solution of Rice, developed from Mode III considerations, is additionally depicted in Fig.10. In the limiting case of elastic material both solutions tend to the same result.

As mentioned previously, if we have found a closed solution for the plastic zone size $\omega$ we can write the crack tip displacement $v_a$ in closed form, too.

$$d_n = \frac{2\ v_a}{(K^2/(E\cdot\sigma_Y)} = \frac{n-1}{n}\cdot\sqrt{\frac{n}{n+\pi^2/4-1}} \tag{7}$$

Here, the formula is given for twice $v_a$. For small scale yielding this expression is equal to the quantity $d_n$ known from the HRR-field solutions. Again for elastic perfectly-plastic material the original Dugdale solution is obtained. In the limiting case, elastic material, $d_n$ tends to zero.

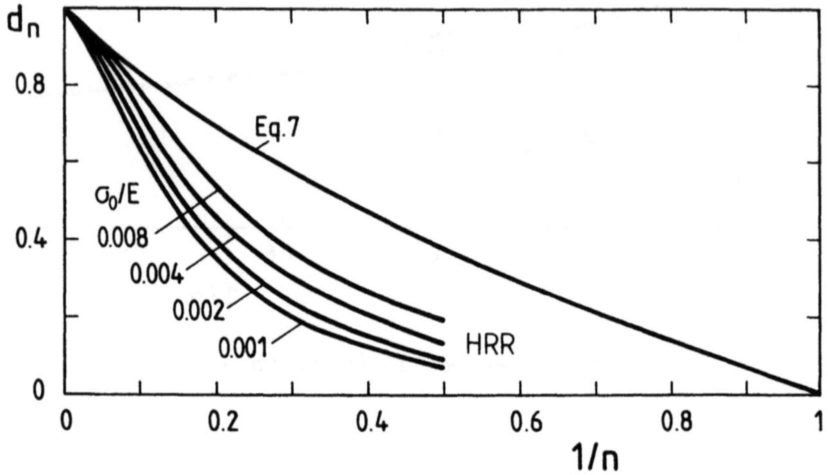


Fig. 11. $d_n$ obtained from modified Dugdale and HRR-solution

The comparison with the HRR-solution, Fig. 11, shows the same results as already seen previously. For a small amount of hardening HRR-field and the modified Dugdale model predict nearly the same crack tip opening displacements. With increasing hardening the differences between both solutions increase, too.

Remembering that the Dugdale $d_n$ remains approximately constant during loading this diagram can be used to correlate crack tip opening displacements obtained from Dugdale calculations to those from FE-calculations or HRR-field solutions.

Finally the modified Dugdale solution is compared in terms of the complete crack surface displacements to Finite-Element calculations of a center crack panel, Fig.12. The applied stress is one thirtieth of the yield stress, thus small scale yielding conditions are met. The hardening exponent amounts to 3.33.

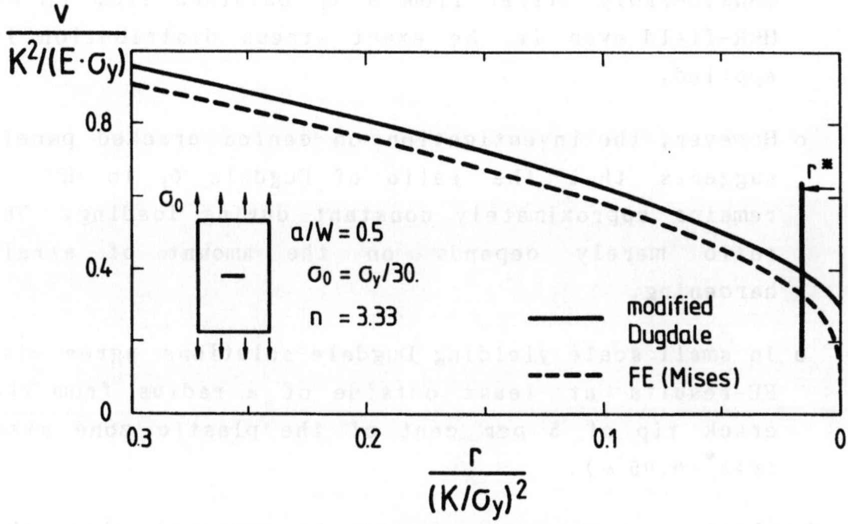


Fig. 12. Crack surface displacements of a center crack panel.

For this case the related plastic zone size $\bar{\omega}$ is approximately 0.3. Taking the plastic zone size as reference it can be stated that the modified Dugdale solution differs considerably from FE-results only in a range of 5 per cent of the plastic zone size. Crack surface displacements distant from the crack tip are well predicted by the Dugdale model.

# 4. General discussion

## 4.1 Assessment of Dugdale models for strain hardening materials

Summarizing it can be stated

- For strain hardening materials the Dugdale $\delta_t$ can considerably differ from a $\delta_t$ obtained from FE or HRR-field even if the exact stress distribution is applied.
- However, the investigation on center cracked panels suggests that the ratio of Dugdale $\delta_t$ to HRR $\delta_t$ remains approximately constant during loading. The ratio merely depends on the amount of strain hardening.
- In small scale yielding Dugdale solutions agree with FE-results at least outside of a radius from the crack tip of 5 per cent of the plastic zone size ($r>r^*=0.05\ \omega$).

With the last two statements it can be expected that the modified Dugdale model is suitable for:

- finding extrapolation formulas to correlate a measured crack surface displacement to the wanted crack tip opening displacement ($\delta(r) \rightarrow \delta_t$).
- and for considering contact problems in fatigue crack propagation.

For applying the Dugdale model is must be ensured that, for the first point, the radius r is in the range where the Dugdale model delivers reliable results, and, for the second point, that the contact length is considerably larger than $r^*$.

## 4.2 Large scale yielding solutions

The solutions of the modified Dugdale model given in the previous chapter for small scale yielding conditions can be extended to large scale yielding in the following manner. As example serves the infinite plate.

The original Dugdale solution for the plastic zone size is modified according to a proposal of Newman [8].

$$\omega = A \cdot a \cdot \left[ \sec\left(\frac{\pi}{2} \frac{\sigma_0}{B \cdot \sigma_Y}\right) - 1 \right] \tag{8}$$

The two constants A and B are calculated with the conditions that

- o the small scale yielding solution of chapter 3 should be satisfied
- o the structural limit behaviour is included.

Starting with the latter point, the applied stress $\sigma_{0,Y}$ producing general yielding can be calculated by

$$\sigma_{0,Y} = \int_0^1 p(\xi)\, d\xi = \frac{n+1}{n}\, \sigma_Y \tag{9}$$

For this load level the plastic zone size $\omega$ must tend towards infinity, i.e.

$$\frac{\sigma_{0,Y}}{B \cdot \sigma_Y} = 1 \tag{10}$$

Together with Eq. (9) the constant B reads

$$B = \frac{n+1}{n} \tag{11}$$

Substituting Eq. (11) in Eq. (8) and considering small scale yielding ($\omega \ll a$, $\sigma_0 \ll \sigma_Y$) leads to

$$\frac{1}{1+\frac{\omega}{A \cdot a}} = \cos\left[\frac{\pi}{2}\,\frac{n \cdot \sigma_0}{(n+1)\sigma_Y}\right] \tag{12}$$

$$1 - \frac{\omega}{A \cdot a} = 1 - \frac{1}{2}\cdot\left(\frac{\pi}{2}\cdot\frac{n}{n+1}\cdot\frac{\sigma_0}{\sigma_Y}\right)^2 \tag{13}$$

$$\frac{\omega}{(\sigma_0/\sigma_Y)^2 \cdot \pi a} = A \cdot \frac{\pi}{8} \cdot \frac{n^2}{(1+n)^2} \tag{14}$$

Finally, by comparison with Eq. (6) A is obtained to

$$A = \frac{(1+n)^2}{n(n+\pi^2/4-1)} \tag{15}$$

Eq. (8) now reads

$$\omega = \frac{(1+n)^2}{n(n+\pi^2/4-1)} \cdot a \cdot \left[\sec\left(\frac{\pi}{2}\cdot\frac{n}{n+1}\cdot\frac{\sigma_0}{\sigma_Y}\right) - 1\right] \tag{16}$$

The formula for the crack tip displacement is derived in an analogous way.

Assumption:

$$\frac{E \cdot v_a}{\sigma_Y \cdot a} = C \cdot \frac{4}{\pi} \cdot \ln\left(1+\frac{\omega}{a}\right) \tag{17}$$

Result:

$$\frac{E\cdot v_a}{\sigma_Y\cdot a} = \frac{4}{\pi}\sqrt{\frac{n+\pi^2/4-1}{n}}\cdot\frac{n-1}{n}\cdot\ln(1+\frac{\omega}{a}) \qquad (18)$$

With the outlined procedure the different Dugdale solutions for specific structures [16,17], developed for elastic perfectly-plastic material, can be extended to strain hardening materials.

## 4.3 Logarithmic trilinear stress-strain curve

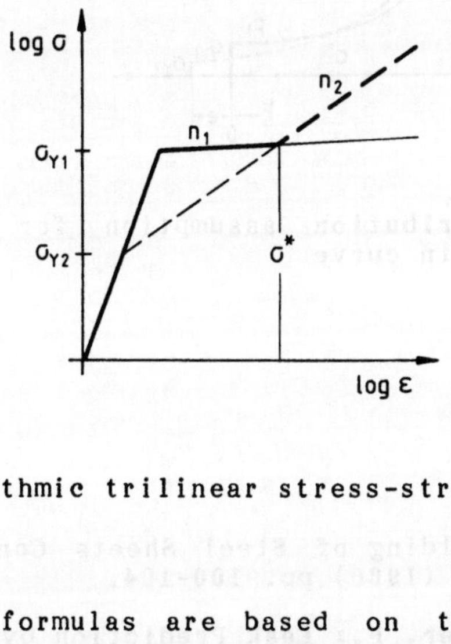


Fig. 13. Logarithmic trilinear stress-strain curve

The presented formulas are based on the assumption of a piecewise power law. If the stress-strain law of the material under consideration is better approximated by a logarithmic trilinear curve, Fig. 13, the stress distribution assumption, Eq. (2), should be replaced as depicted in Fig. 14 with

$$p_1(\xi) = \frac{\sigma_{Y1}}{(1-\xi)^{1/(1+n_1)}} \qquad (19)$$

$$p_2(\xi) = \frac{\sigma_{Y2}}{(1-\xi-e)^{1/(1+n_2)}} \quad (20)$$

$$e = \left(\frac{\sigma_{Y1}}{\sigma^*}\right)^{1+n1} - \left(\frac{\sigma_{Y2}}{\sigma^*}\right)^{1+n2} \quad (21)$$

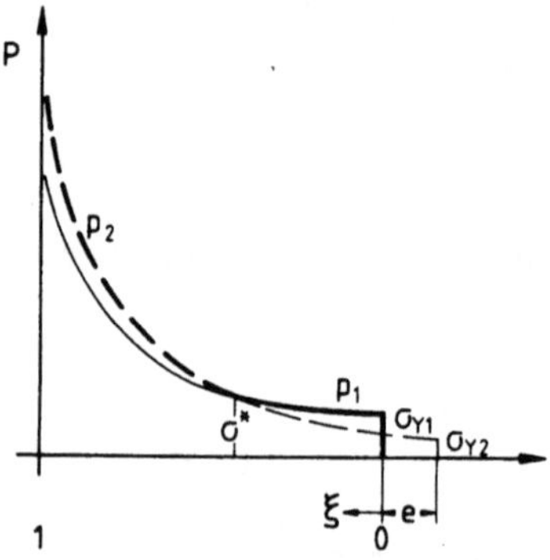


Fig. 14. Stress distribution assumption for a logarithmic trilinear stress-strain curve

## References

1. Dugdale, D.S.: Yielding of Steel Sheets Containing Slits. J. Mech. Phys. Sol. 8 (1960) pp. 100-104.

2. Mattheck, C.; Görner, F.: Leak Prediction by Use of a Generalized Dugdale Model for Semi-Elliptical Surface Flaws in Plates under Tension Loading. Proc. 5th Europ. Conf. on Frac., Ed.: I. Faria, Lisbon, Portugal, 1984 .

3. Newman, J.C.: The Crack Tip Opening Displacement ($v_R$) Resistance Curve Method. Workshop on the CTOD-Methodology, Geesthacht, F.R.G., 1985.

4. Führing, H.: Berechnung von elastisch-plastischen Beanspruchungsabläufen in Dugdale-Rißscheiben mit Rißuferkontakt auf der Grundlage nichtlinearer Schwingbruchmechanik. Heft 30, Veröffentlichungen des Instituts für Statik und Stahlbau der TH Darmstadt, 1977.

5. Führing, H.; Seeger, T.: Dugdale Crack Closure Analysis of Fatigue Cracks under Constant Amplitude Loading. Engng. Fract. Mech. 11 (1979), pp. 99-122.

6. Newman, J.C.: A Crack-Closure Model for Predicting Fatigue Crack Growth under Aircraft Spectrum Loading. Methods and Models for Predicting Fatigue Crack Growth under Random Loading, ASTM STP 748, J.B. Chang and C.M. Hudson, Eds., 1981, pp. 53-84

7. Drucker; Rice, J.R.: Plastic Deformation in Brittle and Ductile Fracture. Engineering Fracture Mechanics, 1 (1970) pp. 577-602.

8. Newman, J.C.: Fracture of Cracked Plates under Plane Stress. Eng. Frac. Mech. 1 (1968) pp. 137-154.

9. Andersson, H. ; Bergkwist, H.: Analysis of a Non-Linear Crack-Modell. J. Mech. Phys. Solids 18 (1970) pp. 1-20.

10. Bergkwist, H.; Andersson, H.: Plastic Deformation at a Stably Growing Crack Tip. Int. Journ. of Fracture Mech. 8 (1972) pp. 139-156.

11. Isherwood, D.P.; Williams, J.G.: The Effect of Stress-Strain Properties on Notched Tensile Failure in Plane Stress. Eng. Frac. Mech. 2 (1970) pp. 19-35.

12. Theocaris, P.S.; Gdoutos, E.E.: The modified Dugdale-Barenblatt model adapted to various fracture configurations in metals. Int. Journ. of Fracture 10 (1974) pp. 549-564.

13. Hoffmann, M.: Ein erweitertes Dugdale-Modell. Studienarbeit im Fachbereich Mechanik der Technischen Hochschule Darmstadt 1980.

14. Schwalbe, K.H.; Hellmann, D.; Seeger, T.; Amstutz, H.; Cornec, A.; Gross, D.; Hoffmann, M.: Entwicklung eines modifizierten COD-Konzepts. FD-9/1981, FG Werkstoffmechanik TH Darmstadt (1981).

15. Shih, C.F.: Relationships Between the J-Integral and Crack Opening Displacement for Stationary and Extending Cracks. General Electric Technical Information Series No. 79CRD075 April 1979.

16. Gray, T.G.F.: A Closed Form Approach to the Assessment of Practical Crack Propagation Problems. Fracture Mechanics in Engineering Practice/Conference of Stress Analysis 1976, Sheffield, England pp. 123-147.

17. Hayes, D.J.; Williams, J.G.: A Practical Method for Determining Dugdale Model Solutions for Cracked Bodies of Arbitrary Shape. Int. Journal of Fracture Mech. 8 (1972) pp. 239-256.

## Appendix: Numerical integration of the small scale yielding formulas

In order to avoid numerical problems the integrals are splitted in

- o integrals containing singularities which can be solved analytically
- o integrals with finite integrand which have to be solved numerically

### A.1 Plastic zone size

The integrand in Eq. (3) is singular for $\xi=0$ and $\xi=1$.

$$\int_0^1 \frac{\bar{p}(\xi)}{\xi}\,d\xi = \int_0^{1/2} \frac{\bar{p}(\xi)-\bar{p}(0)}{\xi}\,d\xi + \int_0^{1/2} \bar{p}(0)\,\frac{1}{\sqrt{\xi}}\,d\xi + \int_{1/2}^{1} \bar{p}(\xi)\cdot\left(\frac{1}{\sqrt{\xi}}-1\right)d\xi + \int_{1/2}^{1} \bar{p}(\xi)\,d\xi \tag{A1}$$

$$\int_0^1 \frac{\bar{p}(\xi)}{\xi}\,d\xi = \int_0^{1/2} \frac{\bar{p}(\xi)-1}{\xi}\,d\xi + \int_{1/2}^{1} \frac{(1-\xi)\bar{p}(\xi)}{\sqrt{\xi}}\,d\xi + \sqrt{2} + \frac{1+n}{n\cdot 2^{n/(1+n)}} \tag{A2}$$

### A.2 Crack surface displacements

Applying integration by parts $\int u\cdot dv = uv - \int v\cdot du$ to the integral in Eq. (4) with $dv = \bar{p}(\xi)$ leads to

$$\int_0^1 \bar{p}(\xi)\cdot\ln\left|\frac{\sqrt{s}+\sqrt{\xi}}{\sqrt{s}-\sqrt{\xi}}\right| d\xi = \frac{1+n}{n}\cdot\sqrt{s}\int_0^1 \frac{(1-\xi)^{n/(1+n)}}{\sqrt{\xi}\cdot(s-\xi)}\,d\xi \tag{A3}$$

considering the crack tip, $s=1$, the right-hand integral is equal to the integral in Eq. (3). Therefore, the crack tip displacement can be expressed in terms of the hardening exponent n and the related plastic zone size $\bar{\omega}$ (Eq. (5)).

For calculation of crack surface displacements ($s \neq 1$) the integral, Eq. (A3), is splitted analogous to the previous section ( $\beta = n/(1+n)$ ).

Case $0<s<1$

$$\frac{\sqrt{s}}{\beta}\int_0^1 \frac{(1-\xi)^\beta}{\xi(s-\xi)}\,d\xi = \frac{\sqrt{s}}{\beta}\int_0^{s/2} \frac{(1-\xi)^\beta-1}{\xi\cdot(s-\xi)}\,d\xi + \frac{\sqrt{s}}{\beta}\int_{s/2}^1 \frac{(1-\xi)^\beta-(1-s)^\beta}{\xi\cdot(s-\xi)}\,d\xi$$
$$+ \frac{1}{\beta}\cdot(1-(1-s)^\beta)\cdot\ln\frac{\sqrt{2}+1}{\sqrt{2}-1} + (1-s)^\beta\cdot\ln\frac{1+\sqrt{s}}{1-\sqrt{s}} \tag{A4}$$

Case $s>1$

$$\frac{\sqrt{s}}{\beta}\,\frac{(1-\xi)^\beta}{\sqrt{\xi}(s-\xi)}\,d\xi = \frac{\sqrt{s}}{\beta}\left[\int_0^{1/2}\left(\frac{(1-\xi)^\beta}{(s-\xi)\sqrt{\xi}} - \frac{1}{\sqrt{s}\,\xi}\,d\xi + \int_{1/2}^1 \frac{(1-\xi)^\beta}{\sqrt{\xi}(s-\xi)}\,d\xi\right] + \frac{1}{\beta}\sqrt{\frac{2}{s}} \tag{A5}$$

# Experimental Techniques

# COD Concepts in R-Curve Testing

D. E. McCabe

Westinghouse R&D
1310 Beulah Road
Pittsburgh, PA

Summary

This report records the information that has been developed over many years of research on the effort to express the R-curve behavior of materials in a crack opening displacement (COD) format. In earlier tests, hinge rotation points were determined empirically and COD was inferred at the updated physical crack tip location ($\delta_t$ at $a_o+\Delta a_p$). Specimen geometry effects not seen in $J_R$-Curve or $K_R$-Curve representations had shown up at displacement levels of appreciable plastic deformation. The more recent suggestion of a COD measurement at the initial fatigue crack tip location has been shown to correlate well with modified J, and $J_M$ R-Curves. The commonality was demonstrated for a variety of cases where plane stress R-Curves were shown to be geometry independent, and where unique circumstances altered the basic geometry independent nature of R-Curve.

## Section I

### R-Curve in Thin Sheet and Plate Materials 1972-1974

The purpose of this first section is to describe how a test method originally designed to develop R-Curves on high-strength materials was modified through a COD based methodology to accommodate determinations on structural materials.[1,2] The scheme was to measure opening displacement at a current stably propagated crack tip by inference using plastic hinge, (crack opening stretch, COS). The procedure was based on the crack-line wedge loaded technique shown conceptually in Figure 1.1. The specimen was loaded along the crack plane using a wedge and split pin as shown and all information necessary to make elastic-plastic COS computations was obtained from the measurement of displacement at two locations along the crack plane, Figure 1.2. Both crack lengths and applied wedge forces were computed using previously determined compliance calibration curves.[3] The specimen was of compact tension proportionality and thickness was set according to the as-received material thickness for application. The dual

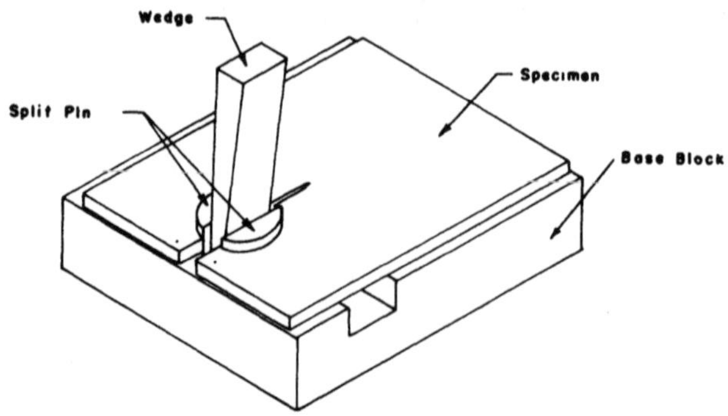


Figure 1.1 - Crack line loaded specimen with displacement controlled wedge loading.

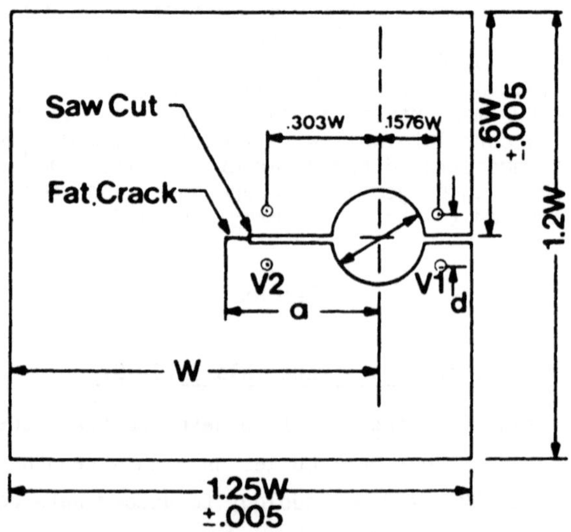


Figure 1.2 - Crack-line wedge-loaded specimen.

measurement points also allowed for location of the hinge rotation points as the test proceeded.

This COS methodology preceeded the more formal elastic-plastic $J_R$-Curve technology of the deformation theory J-integral. The COS practice had been predicated on many developments in CTOD testing in the U.K.,[4-7] and the somewhat related R-Curve techniques were developed later in the U.S. by R. H. Heyer and the present author. For structural materials, the hinge ratio in compact specimens had been shown to be between 0.4 and

0.5. In physical terms, hinging means that all displacements along the crack plane tend to be a linear function of the distance from the hinge point. See Figure 1.3.

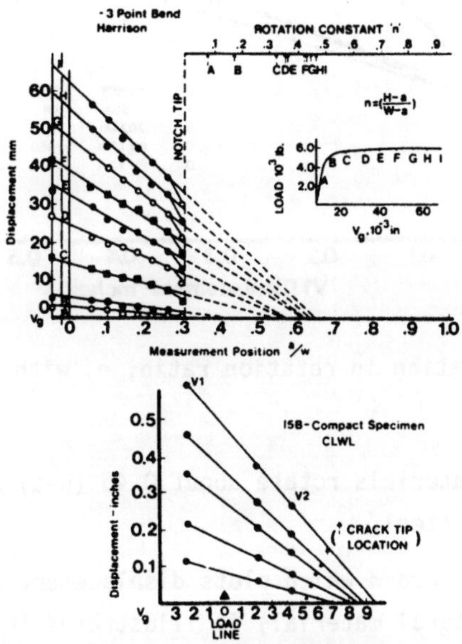


Figure 1.3 - Displacement versus measurement position for loading increments.

The upper plot comes from some early work of Harrison,[8] on bend bar specimens of a structural steel. The nature of hinging with linearity of displacements as shown in Figure 1.3 is clear. No stable crack growth was acknowledged in much of the early work, however. Hinge ratio, designated n in this case, plateaus during events G through I at about 0.45. The bottom plot for a compact specimen represents hinge action in a much larger wedge loaded specimen, again made of a structural steel. In this case, the work had been performed for R-Curve development and slow-stable crack growth was denoted via crack tip indication. By incorporating growth through updated crack size, a more comprehensive viewpoint on how hinge ratio, n, varies with material work hardening is shown in Figure 1.4.

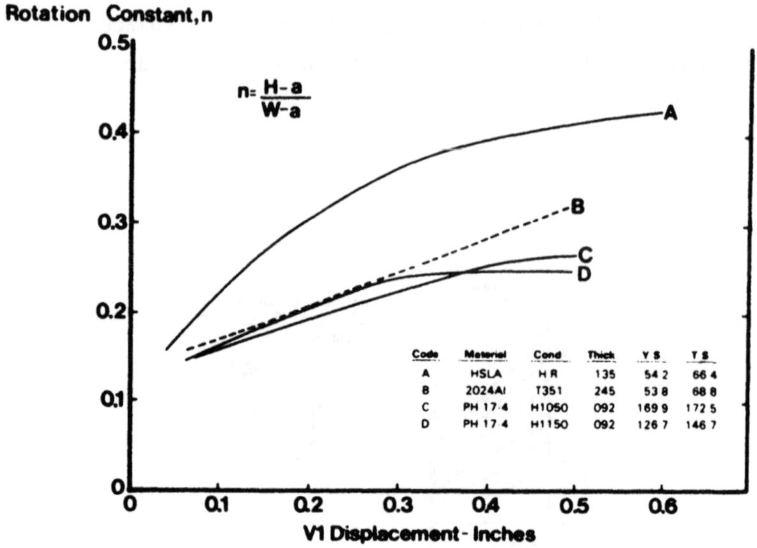


Figure 1.4 - Variation in rotation ratio, n, with materials.

An assumption that all materials rotate about 0.45 (W-a) was found to be true only for structural steel.

A double compliance test record which plots displacement at location V1 against V2, (for a structural material) is illustrated in Figure 1.5. Physical crack lengths were determined at each unload point from the linear unloading slopes. Hinge point, H, is located from linear projections of $V_1$ and $V_2$ displacements and CTOD (COS) is then determined as outlined in Figure 1.6.

Prior to the onset of fully plastic deformation and genuine hinge rotation, fracture toughness was determined by $K_R$ of ASTM method E561, where elastic plastic effects were computed using plastic zone adjusted crack sizes. At general yield, indicated by a strain gage mounted at the back edge of the specimen, all subsequent $K_R$ toughness values were calculated by:

$$K_R{}^2 = mE\ (COS)\ \sigma_{ys} \tag{1.1}$$

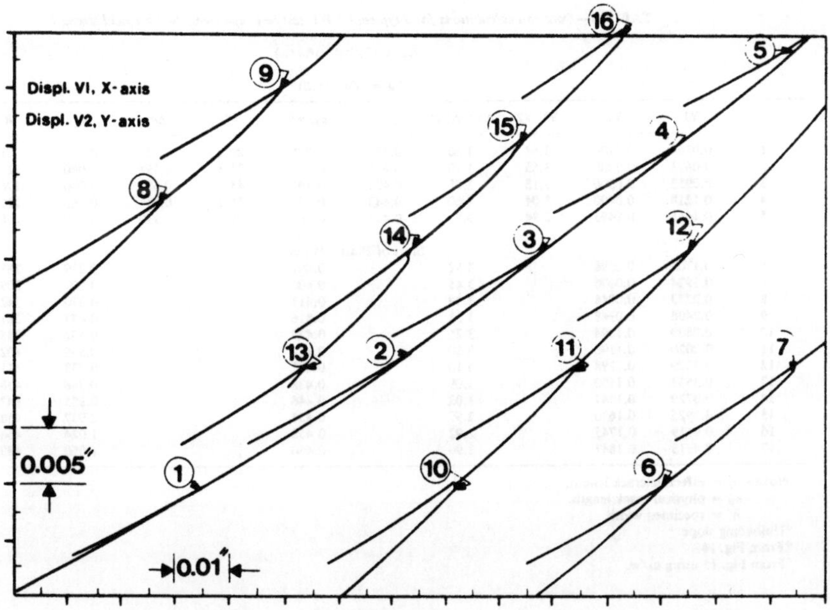


Figure 1.5 - Test record for pipe steel, 74.7 Ksi yield strength.

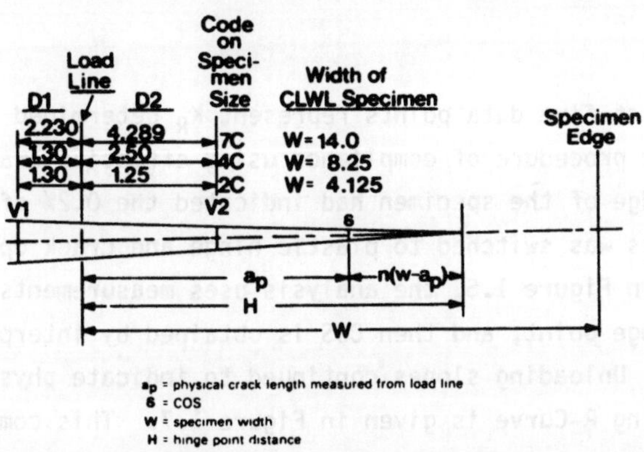


Figure 1.6 Plastic hinge point model for wedge loaded specimens. Dimensions shown for three specimen sizes.

where:

m is a proportionality constant

$1 \le m \le 2.5$

COS = Crack opening stretch at the physical crack tip position. In almost all of the R-Curve work reported at the time, m = 1 had been established as the correct plane stress proportionality constant. The following table gives an example calculation where raw data extracted from Figure 1.5 had been analyzed as listed:

TABLE 1—*Data and calculations for a typical CLWL test line pipe steel, 74.7 ksi yield strength.*

$$K_R = (P/BW^{1/2})f(a/w)$$

$$K_R = (E\sigma_y \text{ COS})^{1/2}$$

| | V1 | V2 | V1/V2 | ΔV1/ΔV2[a] | $a_e/w$[b] | $a_p/w$[b] | P[c] | $\Delta a_e$ | $\Delta a_p$ | $K_R$ | COS |
|---|---|---|---|---|---|---|---|---|---|---|---|
| 1 | 0.0364 | 0.0100 | 3.64 | 3.80 | 0.394 | 0.385 | 20.2 | 0.131 | 0 | 71.2 | |
| 2 | 0.0673 | 0.0200 | 3.65 | 3.70 | 0.413 | 0.390 | 34.5 | 0.398 | 0.080 | 127.8 | |
| 3 | 0.0953 | 0.0300 | 3.18 | 3.68 | 0.429 | 0.391 | 45.7 | 0.626 | 0.096 | 176.4 | |
| 4 | 0.1218 | 0.0400 | 3.04 | 3.60 | 0.443 | 0.396 | 55.1 | 0.816 | 0.165 | 220.6 | |
| 5 | 0.1455 | 0.0495 | 2.94 | 3.55 | 0.455 | 0.400 | 62.5 | 0.990 | 0.210 | 258.4 | |
| | | | | | Start of Plastic Hinge | | | | | | |
| 6 | 0.1705 | 0.0598 | | 3.52 | | 0.402 | | | 0.239 | 288.1 | 0.037 |
| 7 | 0.1934 | 0.0696 | | 3.45 | | 0.406 | | | 0.307 | 309.8 | 0.0428 |
| 8 | 0.2273 | 0.0846 | | 3.39 | | 0.411 | | | 0.370 | 342.6 | 0.0523 |
| 9 | 0.2496 | 0.0948 | | 3.33 | | 0.416 | | | 0.437 | 361.3 | 0.0582 |
| 10 | 0.2809 | 0.1094 | | 3.25 | | 0.423 | | | 0.532 | 385.8 | 0.0664 |
| 11 | 0.3020 | 0.1196 | | 3.20 | | 0.427 | | | 0.595 | 402.0 | 0.0721 |
| 12 | 0.3229 | 0.1298 | | 3.10 | | 0.437 | | | 0.733 | 411.2 | 0.0754 |
| 13 | 0.3533 | 0.1450 | | 3.09 | | 0.438 | | | 0.748 | 438.7 | 0.0859 |
| 14 | 0.3729 | 0.1547 | | 3.02 | | 0.446 | | | 0.855 | 447.1 | 0.0892 |
| 15 | 0.3925 | 0.1650 | | 2.97 | | 0.451 | | | 0.937 | 458.6 | 0.0938 |
| 16 | 0.4119 | 0.1745 | | 2.92 | | 0.458 | | | 1.024 | 466.4 | 0.0970 |
| 17 | 0.4313 | 0.1847 | | 2.90 | | 0.460 | | | 1.060 | 480.4 | 0.1029 |

NOTE—$a_e$ = effective crack length.
$a_p$ = physical crack length.
$W$ = specimen width.
[a] Unloading slope.
[b] From Fig. 14.
[c] From Fig. 15 using $a_e/w$.

The first five data points represent $K_R$ determined by the E561 elastic-plastic procedure of compliance using effective crack size, $a_e$. When the back edge of the specimen had indicated the 0.2% offset yield strain, the analysis was switched to plastic hinge and crack opening stretch. As shown in Figure 1.6, the analysis uses measurements $V_1$ and $V_2$ to locate the hinge point, and then COS is obtained by interpolation to the position of $a_p$. Unloading slopes continued to indicate physical crack size. The resulting R-Curve is given in Figure 1.7. This computational method had been used over a broad range of materials and specimen sizes and the above pattern of behavior had been generally reproducible. Almost invariably, the transition region between the two computational procedures had been consistent and smooth which promoted some confidence. However, in a special study intended to demonstrate beyond question the specimen size independence of R-Curve by COS, the proof was not very encouraging.(9) Figure 1.8. This was the first evidence of irregularity encountered. Because other elastic-plastic methods were not believed to be available at that time, a second method to prove or reject the apparent geometry effects could not be undertaken.

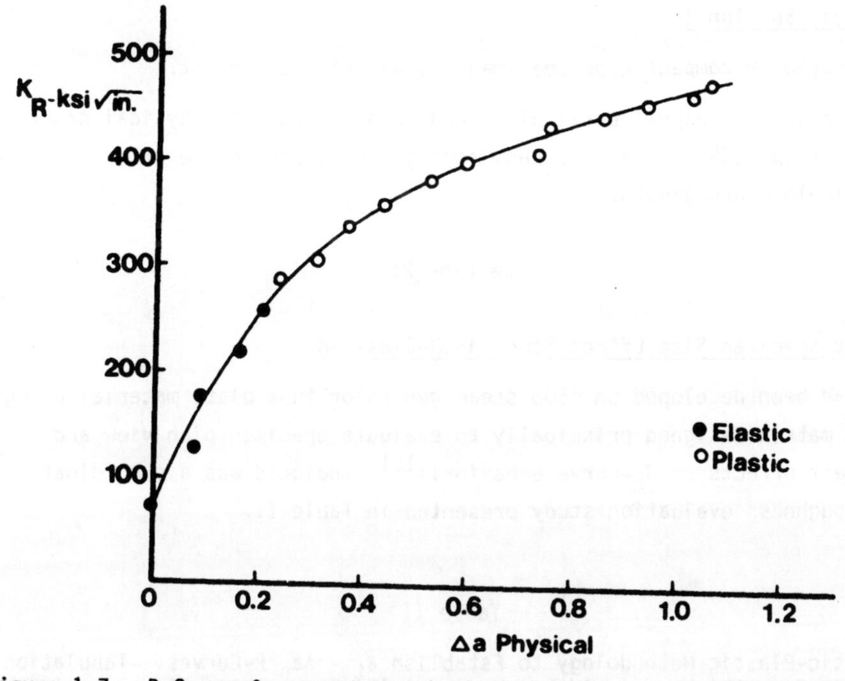


Figure 1.7 - R-Curve for gas transmission line pipe steel; data from Table I.

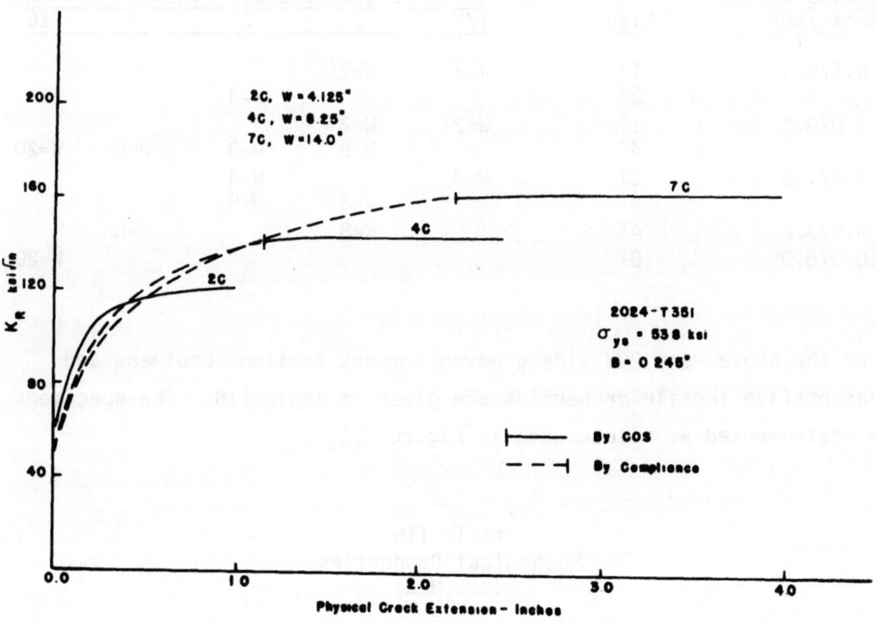


Figure 1.8 - R-Curves developed with three sizes of wedge loaded specimens.

Summary: Section 1

Hinge ratio in compact type specimens is material dependent.

CTOD measured by experimental hinge and inferred for the physical crack tip position (COS) appears to show some geometry dependence after appreciable crack growth.

## Section 2:

### Compact Specimen Size Effect Study A508-Class 2A

Data had been developed on A508 steam generator tube plate material using a test matrix designed principally to evaluate specimen plan view and thickness effects on $J_R$-Curve behavior.[10] Included was a subordinate CTOD toughness evaluation study presented in Table II.

Table II

Elastic-Plastic Methodology to Establish $\delta_t$ - $\Delta a_p$ R-Curves. Tabulation of specimen thickness, widths, W, and initial remaining ligament sizes.

| Thickness Inches Gross/net | Size | Ligament Size - Inches 1/2 | 1 | 2 | 4 | 10 |
|---|---|---|---|---|---|---|
| 0.5/0.4 | 1T | W=2 | W=2 | | | |
| | 2T | | | W=4 | | |
| 1.0/0.8 | 1T | W=2 | W=2 | | | |
| | 4T | | W=8 | W=8 | W=8 | W=20 |
| 2.0/1.6 | 2T | W=4 | | W=4 | | |
| | 4T | | W=8 | W=8 | | |
| 4.0/3.2 | 4T | | W=8 | | W=8 | |
| 10.0/8.0 | 10T | | | | | W=20 |

All of the above were 20% side grooved compact tension specimens and representative tensile properties are given in Table IIB. The specimens were instrumented as illustrated in Figure 2.1.

Table IIB
Mechanical Properties
ksi (MPa)

| Yield Strength | Tensile Strength | % El | Shelf ft-lbs (J) |
|---|---|---|---|
| 55 (379) | 80 (551) | 30% | 95 (130) |

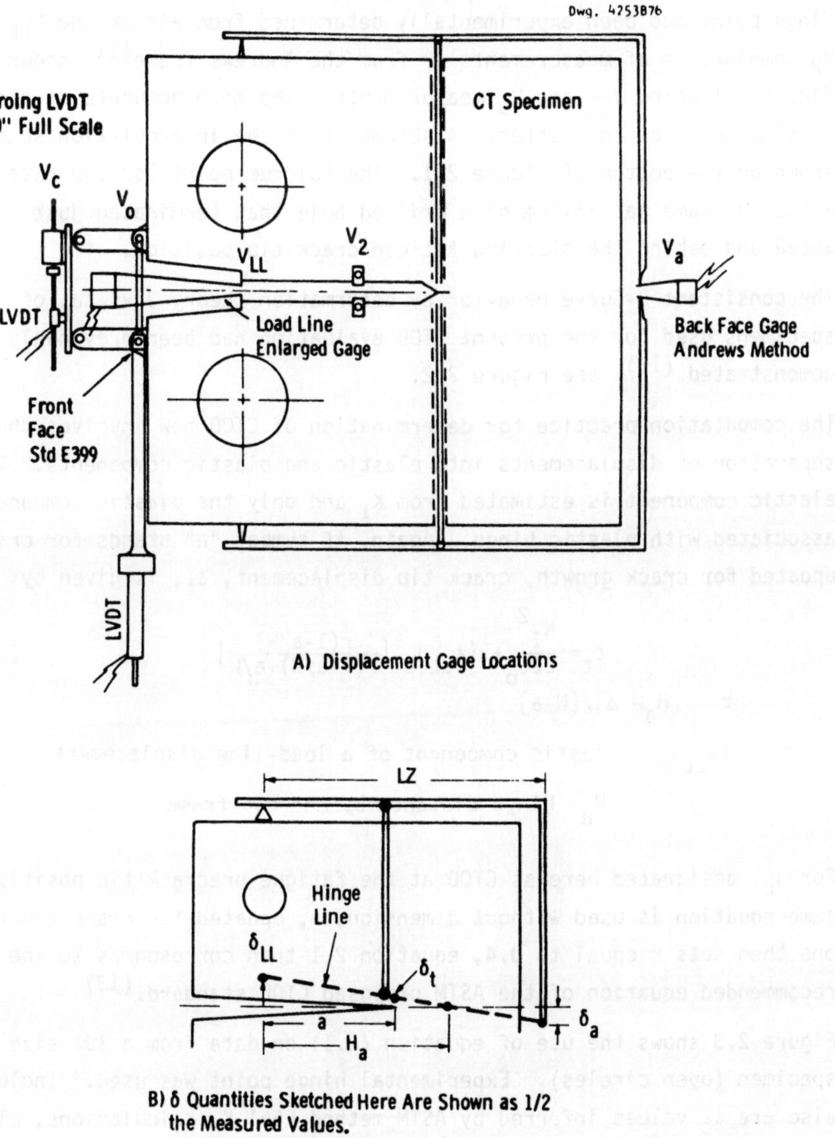


Figure 2.1 - Instrumentation of specimens with five measurement points and schematic representation of plastic hinge displacement.

Hinge point had been experimentally determined from either the $V_{LL}$ and $V_2$ combination of measurements or from the Andrews frame[11] shown in Figure 2.1 using $V_{LL}$ and $V_a$ measurements. The more accurate of the two combinations was the latter. A schematic of the interpolation scheme is shown on the bottom of Figure 2.1. The fulcrum point for the externally attached frame was inside of a drilled hole that terminated just slightly above and behind the starting fatigue crack tip position.

The consistent R-Curve behavior by deformation theory J vs $\Delta a_p$ of specimens used for the present CTOD evaluation had been previously demonstrated.[12] See Figure 2.2.

The computation practice for determination of CTOD now involves the separation of displacements into elastic and plastic components. The elastic component is estimated from $K_I$ and only the plastic component is associated with plastic hinge. Again, if symbol "a" stands for crack size updated for crack growth, crack tip displacement, $\delta_t$, is given by:

$$\delta_t = \frac{K_I^2}{E2\sigma_0} + (\delta_{LL})_p \left[\frac{r(1-a/W)}{r(1-a/W)+a/W}\right] \tag{2.1}$$

$r = (H_a - a)/(W-a)$

$(\delta_{LL})_p$ = plastic component of a load-line displacement.

$H_a$ = hinge distance by Andrews frame

For $\delta_0$, designated here as CTOD at the fatigue precrack tip position, the same equation is used without dimension, a, updated for crack growth. If one then sets r equal to 0.4, equation 2.1 then corresponds to the recommended equation of the ASTM proposed CTOD standard.[13]

Figure 2.3 shows the use of equation (2.1) on data from a 10T size compact specimen (open circles). Experimental hinge point was used. Included also are $\delta_t$ values inferred by ASTM method E561 $K_R$ calculations, closed triangles. As before (Section 1), there is a smooth transition from $K_R$ into hinge indicated toughness by COS. The conversion of $K_R$ into equivalent $\delta_t$ units, however, is reliant on constraint assumptions not used before. This is because of side grooving and an arbitrarily chosen proportionality factor (m=2), nominally for plane strain constraint.

A difference between $\delta_0$ and $\delta_t$ type R-Curves can be expected, the extent of which obviously is a function of crack growth relative to the initial remaining ligament size. In the 10T specimen, growth up to 0.6-inches was

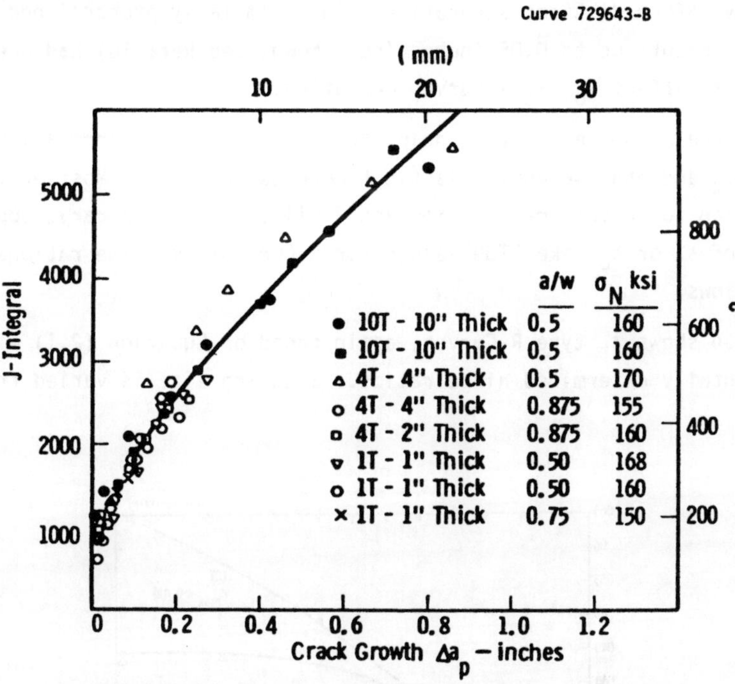


Figure 2.2 - $J_R$-Curve data for specimens with material of equal nominal strength.

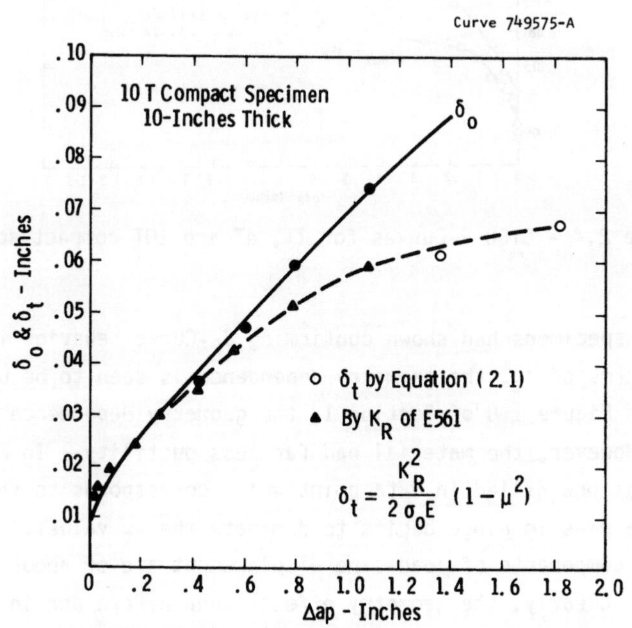


Figure 2.3 -- CTOD at original crack tip, $\delta_o$, and at updated crack position

necessary before distinct separation. In a similarly proportioned 1T specimen, growth up to 0.05 inches (not shown, see Ref. 10) had been required to affect $\delta_o$, $\delta_t$ R-Curve separation.

The evidence presented to this point tends to support a correlation between $\delta_t$ and other elastic-plastic toughness criteria. Most of the information to be presented in Section 3 will, to the contrary, support the use of $\delta_o$ or $\delta_o$ like CTOD values for R-Curve work. The rationale for this follows:

Figure 2.4 shows $\delta_t$ type R-Curves, again based on Equation (2.1) with experimentally determined hinge ratios. Specimen size is varied from 1T to 10T.

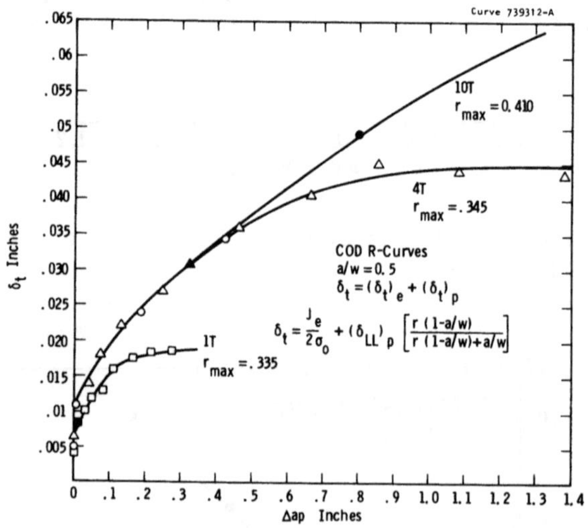


Figure 2.4 - CTOD R-Curves for 1T, 4T and 10T compact specimens.

These same specimens had shown confirming $J_R$-Curve behavior in Figure 2.2. In terms of $\delta_t$, the geometry dependence is seen to be unquestionably strong. In Figure 1.8 of Section 1, the geometry dependence had been quite subtile. However, the material had far less ductility. In Figure 2.4 each data set has one filled in data point which corresponds to the displacement level where plastic hinge begins to dominate the $\delta_t$ values. Here, $(V_{LL})_e$ and $(V_{LL})_p$ components of load-line displacement are of about equal magnitude. Clearly, the geometry effects seen before and in evidence here are experimental hinge related.

Experimentally determined hinge ratios taken from all of the specimens listed in Table II are shown in Figure 2.5.

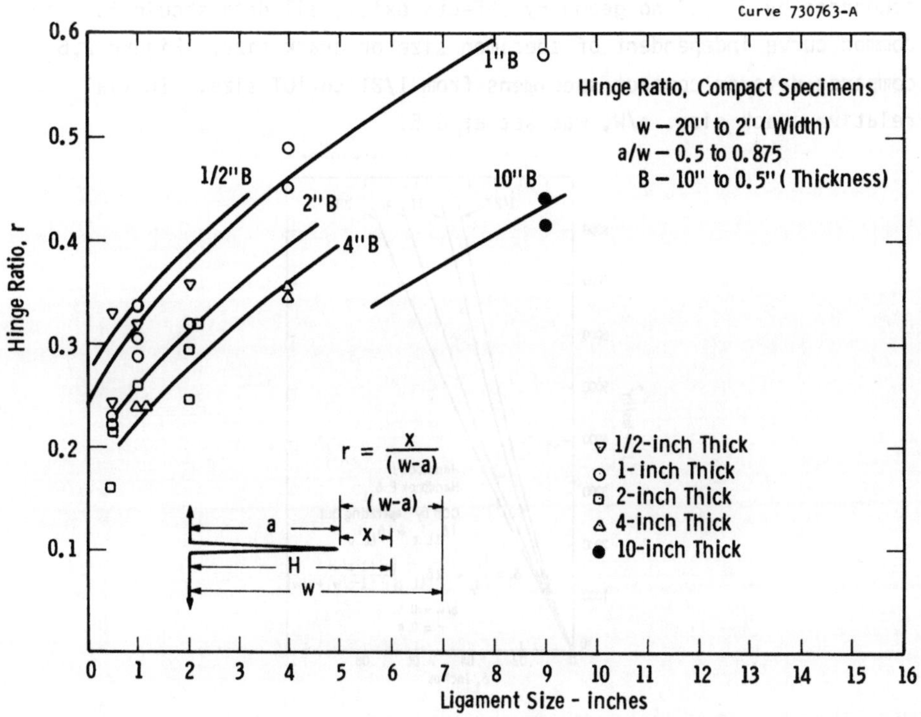


Figure 2.5 - Hinge ratio versus thickness, B, and ligament size, b.

The values are hinge ratios that are stabilized to nearly constant values at fully plastic displacements, and there appears to be a general pattern that develops with ligament size, b, and specimen thickness, B. Hence with increased ligament size and with decreased thickness, upper plateau hinge ratios, r, will tend to increase. For this A508 material, a median hinge ratio of r = 0.4 would appear to be justified as a compromise value over all dimensional variations. This tends to support the ASTM choice of ratio (r = 0.4) to cover all materials and test conditions of concern. The geometry dependence of hinge ratio highlights one of the basic causes for $\delta_t$-$\Delta a_p$ R-Curve geometry dependence. A recently published handbook of elastic-plastic solutions can be used to generate pure data that is not subject to crack growth nor to variabilities of experimentation. With input data of material flow properties, data on J, load, and load line displacement, ($\delta_{LL}$) can be generated. Using this output and assuming a hinge ratio of 0.4 for Equation (2.1), toughness by $\delta_t$ can be compared to toughness by J. If no geometry effects exist, all data should fall on a common curve independent of specimen size or crack size. Figure 2.6 compares data on compact specimens from 1/2T to 10T size. Initial relative crack size, a/W, was set at 0.5.

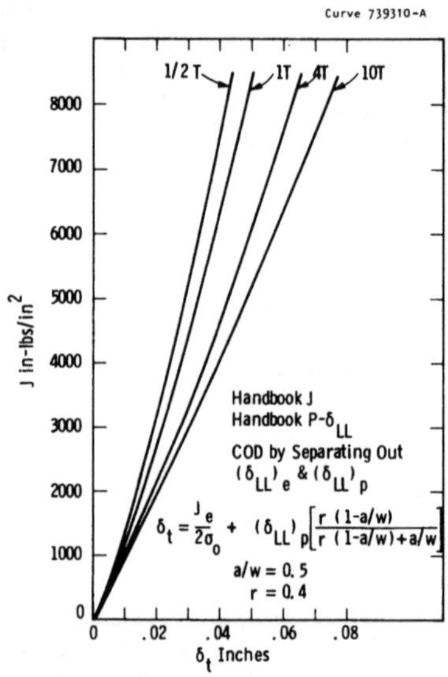


Figure 2.6 - Handbook predicted J and CTOD. Separate J elastic and J plastic contributions.

At low J levels, J and $\delta_t$ compare well over variable size because linear elastic conditions dominate the solutions. The use of r = 0.4 has some strong mitigating influence, but despite this, it appears that size effects override and create the geometry effect seen.

To demonstrate that the foregoing suggestion is plausible, specimens of fixed ligament dimension but having other plan view dimensions should correlate J and $\delta_t$. Figure 2.7 was calculated for 1T and 10T plan view specimens of unit thickness and unit remaining ligament dimension.

The confirmation of the above statement is self evident. An additional convincing demonstration comes from experimental data shown in Figure 2.8.

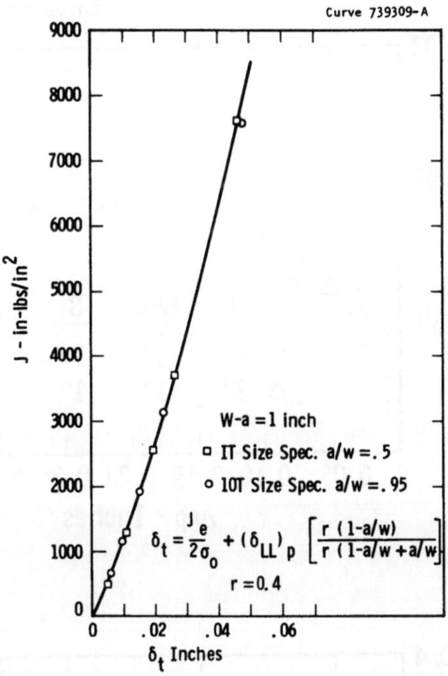


Figure 2.7 - Handbook predicted J and $\delta_t$. 1T and 20T specimens with fixed ligament size.

Although the extent of stable crack growth to the end of the test was not the same in both cases, it is quite apparent that the $\delta_t$ behavior was close; certainly within normal standards of experimental variability.

<u>Summary: Section 2</u>

R-Curves of $\delta_t$ vs $\Delta a_p$ using Equation (2.1) with experimental hinge ratios or with fixed hinge ratio will be geometry dependent.

Problems with geometry dependence develop at a point in the $\delta_t$-R-Curve where the plastic hinge component in Equation (2.1) begins to exert controlling influence on $\delta_t$.

Experimental hinge ratios tend to be dependent on material, ligament size, and specimen thickness.

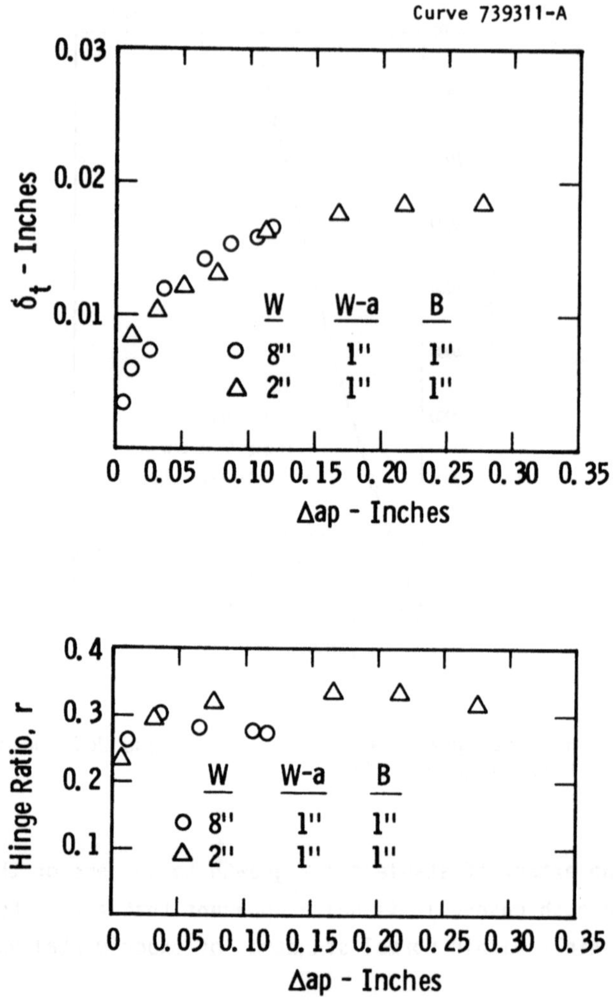


Figure 2.8 - Trend in $\delta_t$ and hinge ratio, r, with crack growth. 4T and 1T plan view compact specimens with 1" × 1" (20% side grooved) ligaments.

## Section 3:

### The Schwalbe Hellman CTOD

The earlier work with compact specimens had been a $J_R$-Curve consistency study, intended to demonstrate that material R-Curves are invarient over geometry changes. This was successful in terms of J and it had been shown to be true for the specific case of compact specimens, thereby representing an important initial step. However, information was needed on a wider range of geometry conditions. Specifically, we needed to know if predominant tension geometries give different $J_R$-Curve behavior than predominant bend. Criteria that are set to assure J controlled growth suggest that there should be no difference, but the few available experimental results of other sources,[14,15] had indicated differently. These previous experimental results were not performed under sufficiently controlled constraint conditions, however, and the genuineness of the comparisons were suspect. Equal constraint is mandatory for fair R-Curve comparisons. Constraint is ligament dimension sensitive and it can be varied consistently within one geometry. However, equal remaining ligament dimensions do not result in equal constraint between compact and center cracked tension geometries. Because the constraint control situation was believed to be an intractable problem; it was set aside in the ongoing work by testing all specimens as thin plate. Hence pure plane stress conditions could be maintained throughout.

The single edge notched geometry (SEN) was chosen to develop variable bend stress to tension stress ratios, $\sigma_B/\sigma_T$. This ratio varies from dominant tension at small relative crack size, a/W, to dominant bend at large relative crack size. Ernst had already studied the SEN geometry[16] and through the use of a Merkle-Corten type of analysis, depicted in Figure 3.1, calculated bend to tension ratios. This rigid rotation model had also been used by Ernst to develop theoretical rotation points and J calibration coefficients.

For experiments, the test specimen design is on the right in Figure 3.2. The specimen on the left is shown for comparison purposes to illustrate the efficiency test material use in the specimen design. Two thirds of each specimen is a reuseable grip. A range of specimen planview sizes and relative crack sizes were tested according to Table III.

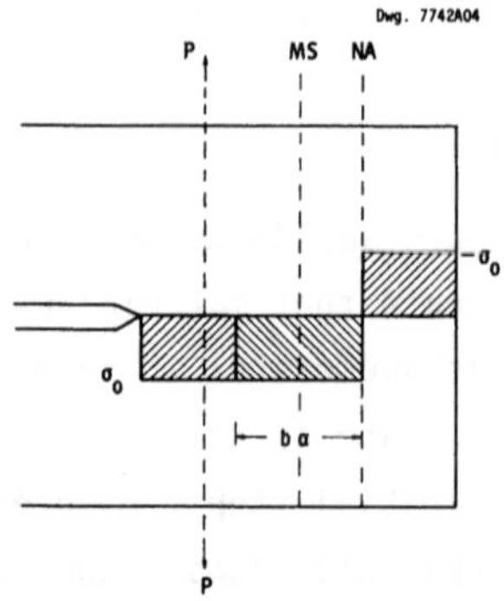


Figure 3.1 - Stress conditions for a ligament subject to arbitrary tension bending loading at plastic limit.

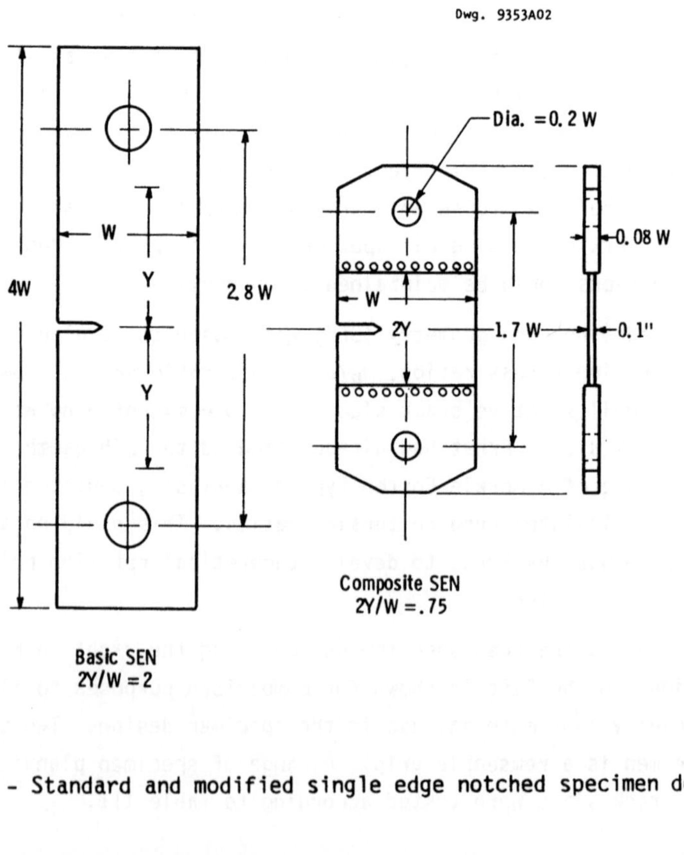


Figure 3.2 - Standard and modified single edge notched specimen designs.

Table III

Test Matrix of SEN Specimens

0.1-inch Thick A533B

| a/W | 0.2 | 0.3 | 0.4 | 0.5 | 0.6 | 0.7 | 0.8 |
|---|---|---|---|---|---|---|---|
| $\sigma_B/\sigma_T$ | 0.75 | 1.3 | 2.0 | 3.0 | 4.5 | 7.0 | 12.0 |
| W=30" | | X | | X | | | |
| W=20" | | X | X | X | X | X | |
| W=10" | | X | | X | | | |
| W=3" | X | X | X | X | X | X | X |

$J_R$-Curves were made on the 10-20-30 inches wide specimens. Modified J was calculated using the following equation:

$$J_M = \frac{K^2}{E} + \frac{1}{B} \int_0^{\delta_p} \frac{\gamma_p}{b} P \, d\delta_p \tag{3.1}$$

where B = thickness

$\gamma_p$ = elastic plastic coefficient

b = remaining ligament

$\delta_p$ = plastic component of load point displacement

Modified J is a new J-like parameter that, unlike conventional deformation theory J, is independent of stable crack growth.

At the time that $J_M$-$\Delta a_p$ R-Curve testing was started, the Schwalbe-Hellmann type CTOD measurement was first proposed.(17) These are linear displacement measurements made at the position of the original crack tip. Unlike the $\delta_t$ values calculated earlier, the measurement is made directly and at the same location, irrespective of the amount of slow stable crack growth. The initial gage span was 0.1-inch, across the fatigue precrack position for all specimens.

Specimen rotation was followed using dial gages affixed to the front and back edges of the SEN grips. See Figure 3.3. The system was crude but effective in developing reasonably accurate hinge characterization.

The material was A533B taken from a reactor nozzle cutout and the properties are given in the following table:

Dwg. 9372A39

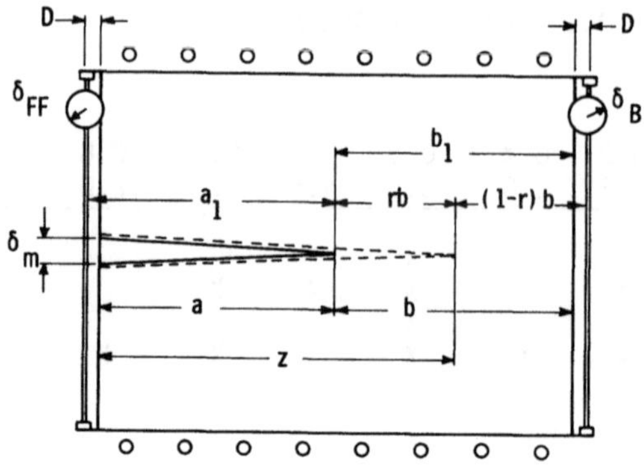


$$\frac{\delta_{FF}}{a_1 + rb} = \frac{\delta_B}{(1-r)\,b_1}$$

$$r = \frac{\left[\frac{\delta_{FF}}{\delta_B}\,b_1 - a_1\right]}{\left[\frac{\delta_{FF}}{\delta_B}\,b_1 + b\right]}$$

$$b = w - a_p$$

$$b_1 = w - a_p + D$$

$$a = a_o + \Delta a_p$$

$$a_1 = a + D$$

$$\delta_o = \left[1 - \frac{a_o}{z}\right]\delta_m$$

$$\delta_t = \left[1 - \frac{a_p}{z}\right]\delta_m$$

Figure 3.3 - Schematic of hinge rotation measurement on SEN specimen.

Table IV

| Yield Str. ksi (MPa) | Tensile Str. ksi (MPa) | CVN ft-lbs (J) | LE mils |
|---|---|---|---|
| 68 ksi (469) | 90 (620) | 92 (126) | 72 |

The $J_M$ R-Curve behavior had been well behaved throughout the test matrix and Figure 3.4 essentially represents this finding.(18) All SEN specimens and compact specimens gave the same $J_M$ R-Curve. The Schwalbe-Hellmann CTOD ($\delta_o$) had been measured on all specimens except for compacts which were tested before the interest had developed. The comparable $\delta_o$-$\Delta a_p$ R-Curve without compact specimen data points is shown in Figure 3.5.

The fact that both types of toughness representation give geometry independent R-Curves suggests a $J_M$-$\delta_o$ correlation.

A slight mistake in specimen preparation helped to provide more correlation confirmation between $J_M$ and $\delta_o$. Antibuckling guides were supported on all specimens by 1/4-inch diameter dowel pins which were inserted through drilled holes in the specimens. These holes were well removed from active elastic-plastic deformation zones in all specimens except for the 10-inches wide SEN specimens with a/W at 0.5. Post test

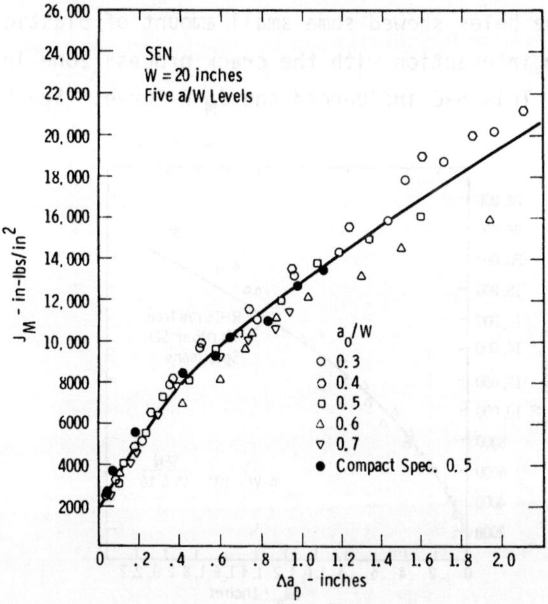


Figure 3.4 - R-Curve data from 20-inches wide SEN specimens.

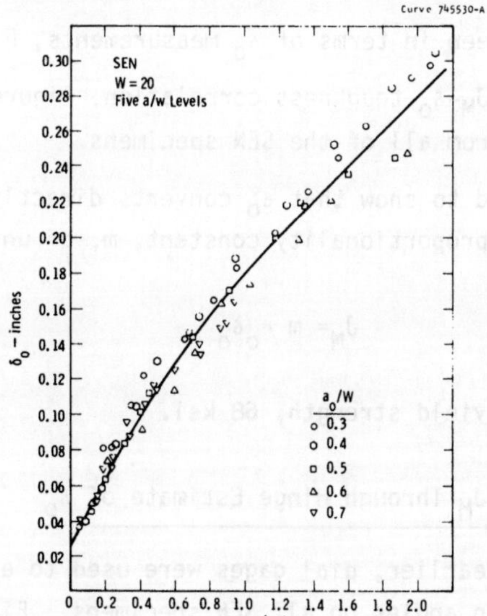


Figure 3.5 - Schwalbe type $\delta_o$ versus crack growth from 20-inches wide SEN specimens.

examination of the holes showed some small amount of plastic deformation had eminated from interaction with the crack process zone in both (duplicate) specimens. This had influenced the $J_M$ R-Curve. See Figure 3.6.

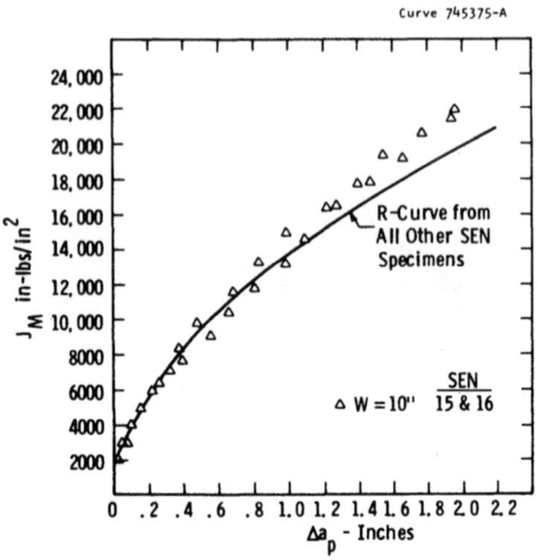


Figure 3.6 - R-Curve data from 10-inches wide SEN specimen a/W = 0.5.

The same problem was seen in terms of $\delta_o$ measurements, Figure 3.7.

To further verify the $J_M$-$\delta_o$ toughness correlation, Figure 3.8 summarizes all $J_M$-$\delta_o$ data pairs from all of the SEN specimens.

This figure can be used to show that $\delta_o$ converts directly to its $J_M$ equivalent by using a proportionality constant, m, of unity.

$$J_M = m\ \sigma_o \delta_o \tag{3.2}$$

where: $\sigma_o$ = material yield strength, 68 ksi.

## 3.1 Determination of $J_M$ Through Hinge Estimate of $\delta_o$

As had been discussed earlier, dial gages were used to experimentally indicate hinge rotation angles on all SEN specimens. Figure 3.3 shows that the measurement locations were slightly removed from the exact specimen edges by spacing, D. The equation used to calculate hinge ratio appears in the same figure. Using information from SEN specimens of all sizes, and compact specimens from the earlier examples, we can demonstrate a consistent pattern of hinge behavior, Figure 3.9.

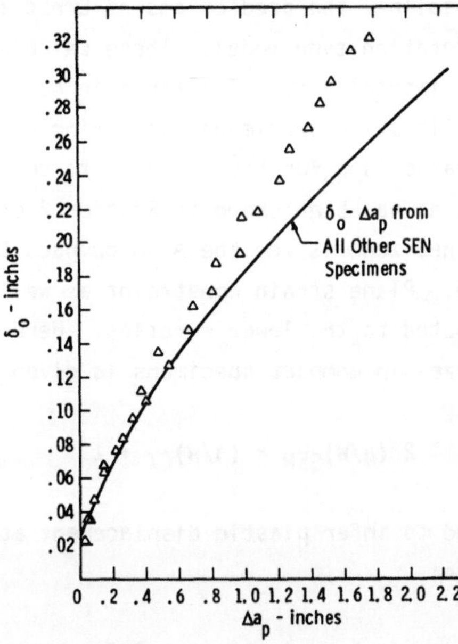


Figure 3.7 - Schwalbe type CTOD ($\delta_o$) versus crack growth from 10-inches wide SEN specimens. $\bar{a}/W = 0.5$.

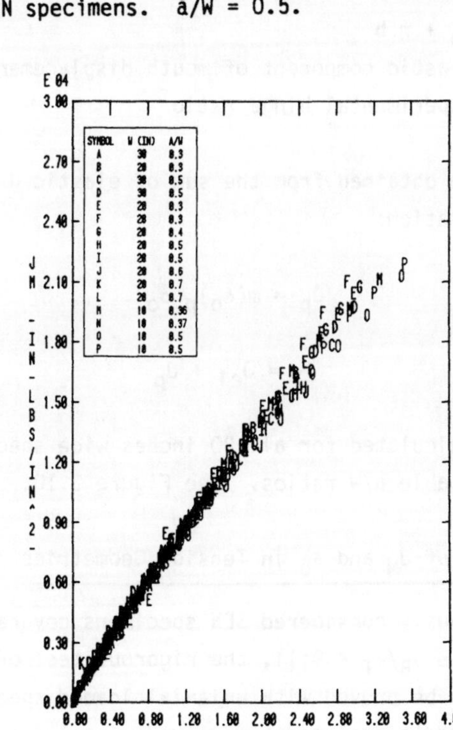


Figure 3.8 - Comparison of modified J and Schwalbe-Hellmann type CTOD.

For the analytical baseline, the predictions of Ernst are shown from the Merkle-Corten rigid rotation type model. These calculations are confirmed by all available experimental data. The three inches wide SEN specimens were blunt notched calibration specimens used principally for J calibration. Hinge ratio is a function of bend to tension ratio as well as remaining ligament as had been shown in Section 2 of this report. The box with a cross-hatched zone is for the A508 compact tension tests in Section 2, Figure 2.5. Plane strain constraint as well as remaining ligament size contributed to the lower r ratios. Here the SEN equivalent for relative crack sizes in compact specimens is given by:

$$2\ (a/W)_{SEN} = (a/W)_{CT} + 1 \tag{3.3}$$

Figure 3.9 can be used to infer plastic displacement at location $a_o$ using the following equation:

$$(\delta_o)_p = (\delta_m)_p\ [Z-a_o]/Z \tag{3.4}$$

where: $Z = a_p + r\ b$

$(\delta_m)_p$ = plastic component of mouth displacement

$r$ = experimental hinge ratio

Total J ($J_T$) can be obtained from the sum of elastic J ($J_{el}$) and plastic J ($J_p$) from hinge rotation:

$$J_p = m(\delta_o)_p\ \sigma_o \tag{3.5}$$

$$J_T = J_{el} + J_p \tag{3.6}$$

Total J ($J_T$) was calculated for all 20 inches wide specimens, covering the full range of available a/W ratios. See Figure 3.10.

### 3.2 Determination of $J_M$ and $\delta_o$ in Tension Geometries

Although the previously considered SEN specimens covered a wide range of $\sigma_B/\sigma_T$ ratios, ($1{:}1 \le \sigma_B/\sigma_T \le 9{:}1$), the rigorous test of the R-Curve uniqueness can only be proved with uniaxial loaded specimens. Accordingly, four additional tests were made involving center cracked tension specimens.[19] The detailed evaluation of these results provided

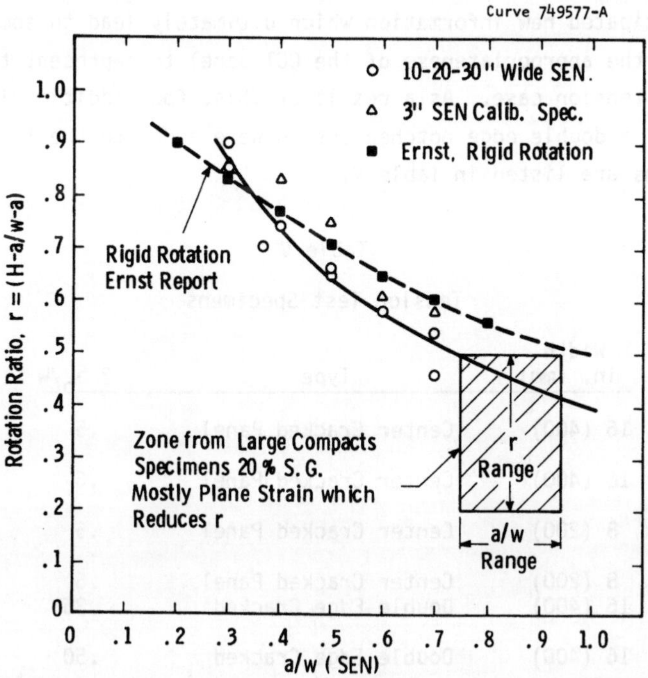


Figure 3.9 - Hinge ratios for SEN and compact specimens.

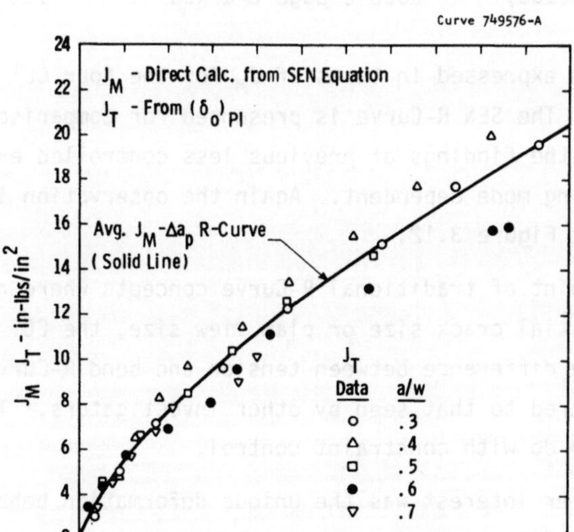


Figure 3.10 - $J_M$-$\Delta a_p$ R-Curve (solid line) versus $J_T$-$\Delta a_p$ data calculated from hinge.

some unanticipated new information which ultimately lead to some skeptical opinions on the appropriateness of the CCT panel to represent the generalized tension case. As a result of this, four additional tension specimens of a double edge notched design were added to the test matrix. The specimens are listed in Table V.

Table V

Tension Test Specimens

| Width in. (mm) | Type | 2 $a_o/W$ |
|---|---|---|
| 16 (400) | Center Cracked Panel | .5 |
| 16 (400) | Center Cracked Panel | .6 |
| 8 (200) | Center Cracked Panel | .5 |
| 8 (200) | Center Cracked Panel | .6 |
| 16 (400) | Double Edge Cracked | .35 |
| 16 (400) | Double Edge Cracked | .50 |
| 8 (200) | Double Edge Cracked | .50 |
| 8 (200) | Double Edge Cracked | .35 |

The R-Curve data expressed in terms of $J_M$ for the four CCT panels is shown in Figure 3.11. The SEN R-Curve is presented for comparison. These data tend to confirm the findings of previous less controlled experiments, that R-Curve is loading mode dependent. Again the observation is supported by $\delta_o$ measurements, Figure 3.12.

From the standpoint of traditional R-Curve concepts where no effects are shown due to initial crack size or plan view size, the CCT data is self consistent. The difference between tension and bend R-Curves was less pronounced compared to that seen by other investigators. This presumably had something to do with constraint control.

Perhaps of greater interest was the unique deformation behavior indicated by supplementary dimension measurements. Width dimension change was measured during the test by dial gages. A strong slip line pattern developed very early in the test, see Figures 3.13 and 3.14.

This pattern of behavior was completely reproduced in all CCT specimens despite relative variations in initial ligament sizes. Although the

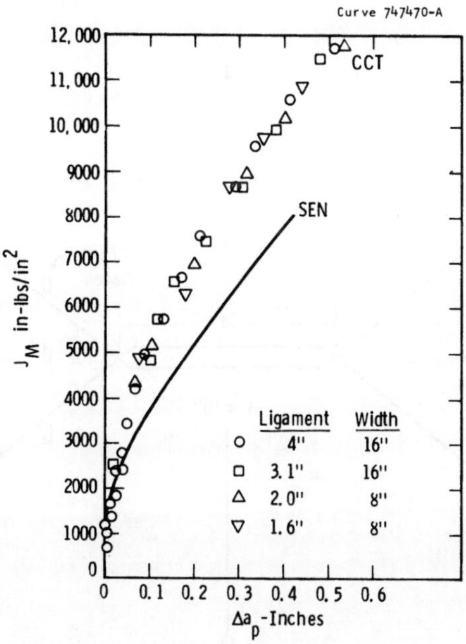


Figure 3.11 - $J_M$ R-Curve for CCT panels compared to an SEN R-Curve.

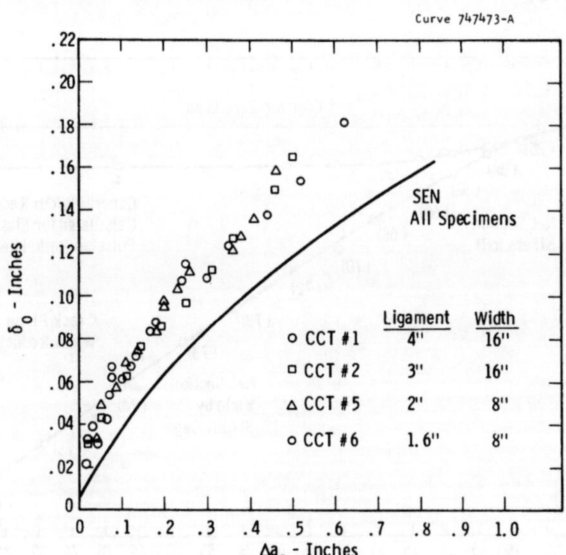


Figure 3.12 - Crack mouth displacement versus crack growth for CCT panels compared to the single edge notched R-Curve.

Dwg. 9363A12

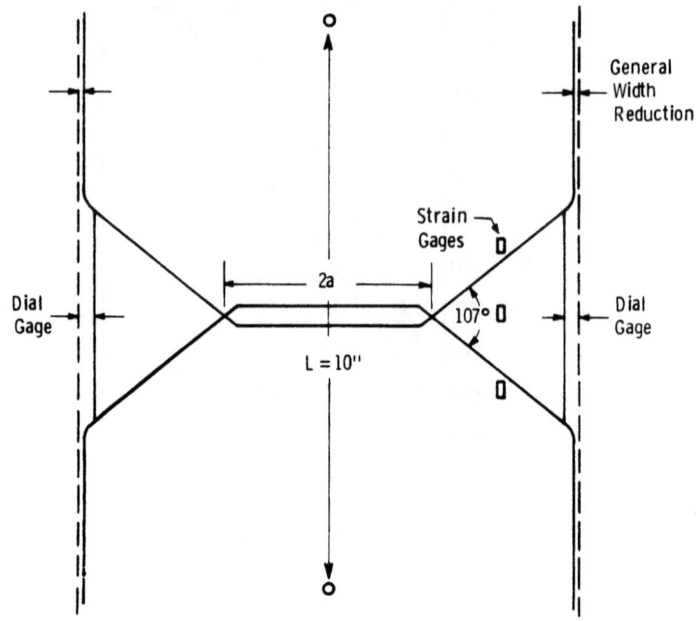


Figure 3.13 - Schematic representation of wedge zone formation in CCT tests.

Curve 747464-A

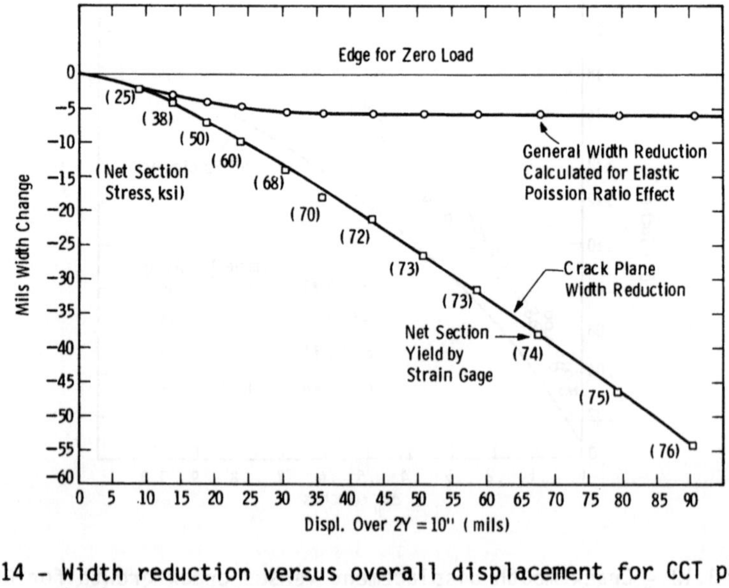


Figure 3.14 - Width reduction versus overall displacement for CCT panel test.

pattern of slip line behavior was expected, the onset so early in the test (well before ligament yield) was a surprise. This, of course, was also well established before onset of stable crack growth and R-Curve data development.

A second choice of tension specimen design was the double edge notched specimen. Here an interaction from opposing crack tips tended to cancel out the unique slip line character that goes with the CCT specimen.

The $J_M$ R-Curve data from the four DEN specimens listed in Table V are shown in Figure 3.15 and $\delta_o$ R-Curve in Figure 3.16. Again data from SEN and compact specimens is included.

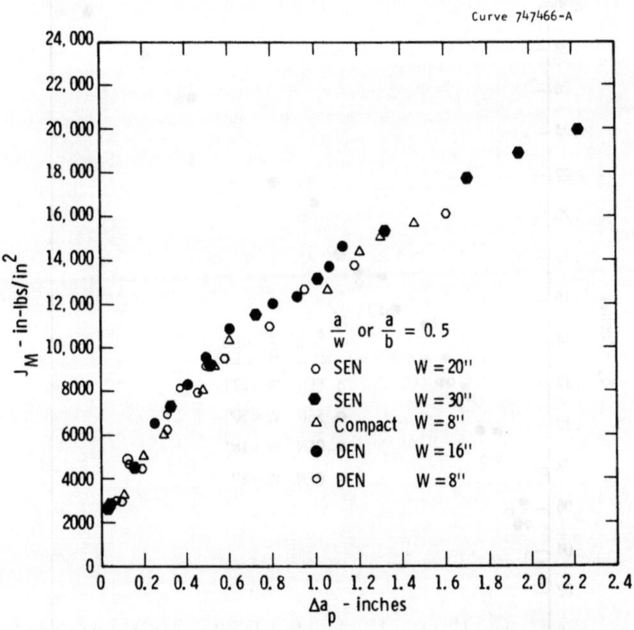


Figure 3.15 - $J_M$ R-Curve data for three specimen types.

These results had confirmed that tension versus bend loading does not of itself create different R-Curve behavior in materials. The $J_M$ R-Curve difference in CCT panels was principally due to the unique (near 45°) slip lines that connect between the crack tip and free edges of the panels.

Experimental confirmation of the plastic flow difference between CCT and DEN is seen in Figures 3.17 and 3.18.

Summary: Section 3

There is a clear and consistent relationship between $J_M$ and $\delta_o$.

Geometry effects when present were indicated independently by $J_M$ and $\delta_o$.

$J_M$ can be reasonably predicted by inference from plastic hinge, where:

$$J_M = J_e + m (\delta_o)_p \sigma_o$$

$(\delta_o)_p$ is obtained by inference.

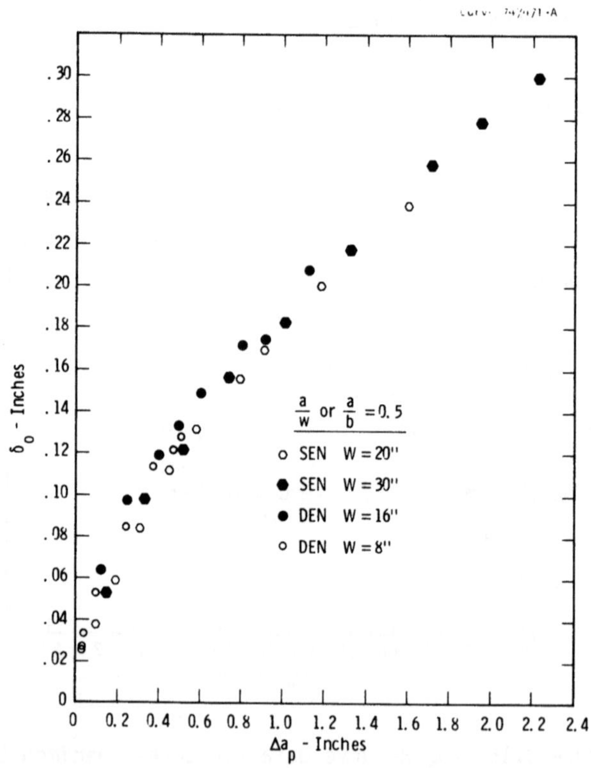


Figure 3.16 - R-Curve by $\delta_o$ versus $\Delta a_p$ for SEN and DEN specimens.

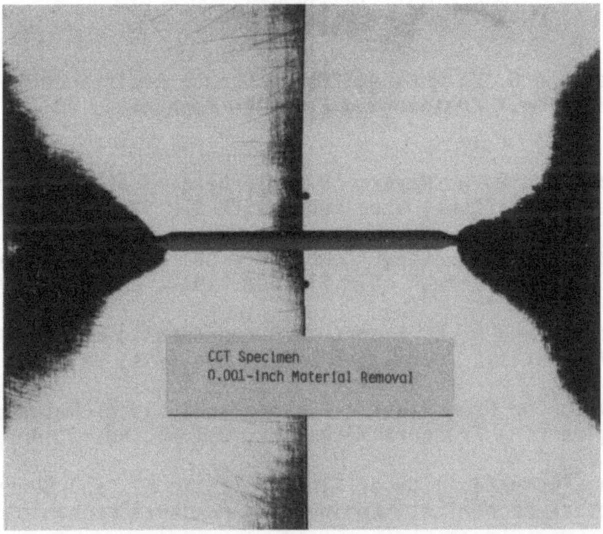


Figure 3.17 -

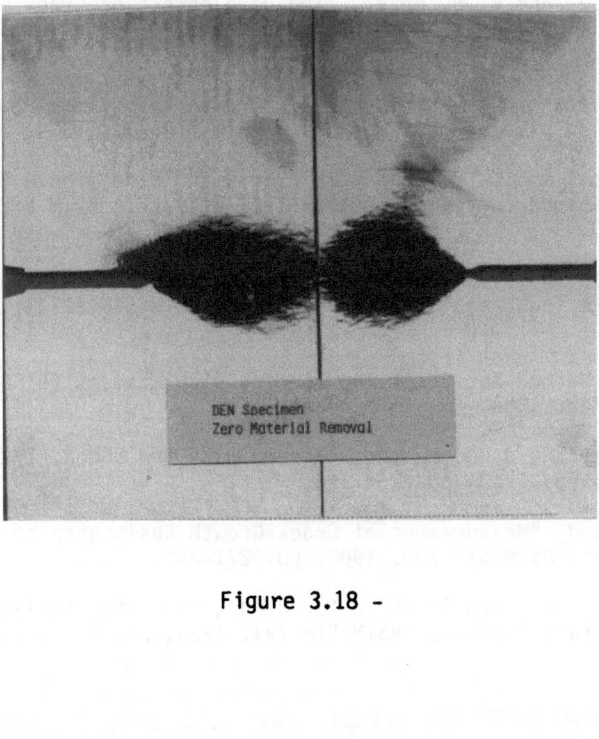


Figure 3.18 -

# REFERENCES


1. R. H. Heyer and D. E. McCabe, "Crack Growth Resistance in Plane Stress Fracture Testing," Engineering Fracture Mechanics, Vol. 4, No. 3, September 1972.

2. D. E. McCabe and R. H. Heyer, "R-Curve Determination Using a Crack-Line-Wedge-Loaded (CLWL) Specimen," ASTM STP 527, 1973, pp. 17-35.

3. D. E. McCabe and G. T. Sha, "Compliance Calibration of Specimens Used in the R-Curve Practice," ASTM STP 632, 1977, pp. 82-96.

4. F. M. Burdekin and D. E. Stone, "COD Approach to Fracture Mechanics in Yielding Materials," Journal of Strain Analysis, Vol. 1, 1966.

5. A. A. Wells, "COD From Elastic Plastic Analysis of Notched Tension Bars," Engineering Fracture Mechanics, Vol. 1, No. 3, April, 1969.

6. G. R. Egan, "Compatability of Linear Elastic $K_{Ic}$ and General Yielding (COD) Fracture Mechanics," Engineering Fracture Mechanics 5, 1973, pp. 167-185.

7. F. M. Burdekin and M. G. Dawes, "Practical Use of Linear Elastic and Yielding Fracture Mechanics to Pressure-Vessel Technology," Institute of Mech. Eng., May 1971.

8. T. C. Harrison, Gas Council Engineering Report, Station report T311, United Kingdom Council of Engineering Research, June, 1970.

9. D. E. McCabe, and R. H. Heyer, "The Development of Crack Growth Resistance Technology," Book: Linear Fracture Mechanics, Envo Publishing, Lehigh Valley, PA, 1976.

10. D. E. McCabe and H. A. Ernst, "Crack Growth Resistance Measurement by Crack Opening Displacement Methods," ASTM STP 833, 1984, pp. 648-665.

11. W. R. Andrews and C. F. Shih, "Thickness and Side Groove Effects on J and δ-Resistance Curves for A533B Steel at 93°C," ASTM STP 668, 1979, pp. 426-450.

12. D. E. McCabe and J. D. Landes, "An Evaluation of the $J_R$-Curve Method for Fracture Toughness Characterization," ASTM STP 803II, 1983, pp. 562-581.

13. Draft ASTM Method for Crack Tip Opening Displacement (CTOD) Testing, Revision, Feb. 1984.

14. J. A. Begley and J. D. Landes, "Serendipity and the J. Integral," Int. J. of Fracture, 12(3) pp. 764-766.

15. S. J. Garwood, "Measurement of Crack Growth Resistance of A533B Wide Plate Tests", ASTM STP 700, 1980, pp. 271-295.

16. H. A. Ernst, "Unified Solution for J Ranging Continuously from Pure Bending to Pure Tension," ASTM STP 791, 1983, pp. 1-499 to 1-519.

17. D. Hellmann and K. H. Schwalbe, "Geometry and Size Effects on J-R and δ-R Curves under Plane Stress Conditions," ASTM STP 833, pp. 577-605.

18. J. D. Landes, D. E. McCabe, H. A. Ernst, "Elastic-Plastic Methodology to Establish R-Curves and Instability Criteria," 7th Semiannual Report, EPRI Contract RP 1238-2, Oct. 1983.

19. J. D. Landes, D. E. McCabe, H. A. Ernst, "Elastic-Plastic Methodology to Establish R-Curves and Instability Criteria," 8th Semiannual Report, EPRI Contract RP 1238-2, July 1984.

# On the Experimental Determination of CTOD Based R-Curves

D. HELLMANN, K.-H. SCHWALBE

GKSS Research Center Geesthacht
D-2054 Geesthacht, Federal Republic of Germany

## Summary

It is demonstrated that the British Standard BS 5762 for measuring the crack tip opening displacement (CTOD) can be modified such that the resulting CTOD is equal to the CTOD measured at the fatigue crack tip. The modifications consist of accounting for crack growth and adjustment of the rotation factor to the material under consideration. The results obtained on six different materials show that the rotation factor depends slightly on the material and is independent of specimen type, crack length, specimen thickness, and amount of crack growth.

## Background

It has recently been shown /1,2/ that a crack growth resistance curve based on the crack tip opening displacement (CTOD) has appreciably greater potential with respect to the amount of valid crack growth than a conventional $J_R$-curve provided both quantities are measured on specimens of identical size and shape. In other words: the CTOD correlates more crack growth unequivocally than the J-integral determined by the usual experimental techniques. Furthermore, it was demonstrated, that a variation of specimen geometry had no effect on the R-curve, i.e. center cracked tension (CCT), single edge notched bend (SENB) and compact (CT) specimens yield the same R-curve. It should be noted that all tests were done on relatively thin specimens which were under plane stress conditions.

In Refs. (1,2) the CTOD was measured on the specimens' side surface at the fatigue crack tip by a special clip gage over a gage span of 5 mm, Fig. 1. The CTOD measured this way will therefore be termed $\delta_5$. It will now be shown that this technique is fully compatible with the procedure described in British Standard BS 5762 /3/ provided that two corrections are applied. It has recently been demonstrated by Ernst /4/ and by Landes, McCabe, and Ernst /5/ that $\delta_5$ is directly related to the modified J-integral, $J_M$. It will therefore be used to calibrate BS 5762 by introducing two modifications: a correction for crack growth and a change of the rotation factor.

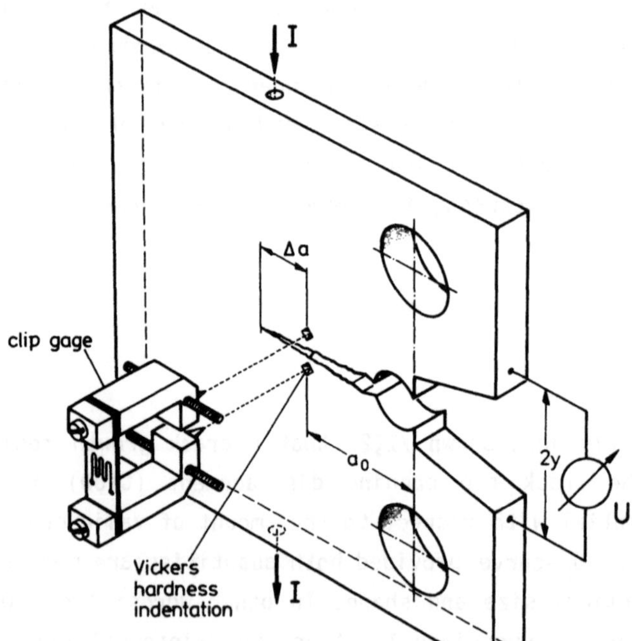


Fig. 1. Experimental setup for measuring the CTOD in terms of $\delta_5$ and the crack length using the electrical potential drop method /1,2/

## Experimental Procedures and Data Evaluation

The details of the experimental procedures have been fully described in Refs. /1,2/. To summarize, R-curve tests were done on the six materials the tensile properties of which are listed in Table 1.

Table 1: Tensile properties

| Material | $\sigma_{0.2}$ MPa | $\sigma_u$ MPa | Work Hardening Exponent, n | $\beta$ |
|---|---|---|---|---|
| 2024-T351 | 317 | 440 | 0.091 | 0.8 |
| 2024-FC | 75 | 217 | 0.294 | 1.5 |
| 35NiCrMo16 | 510 | 726 | 0.161 | 2.5 |
| 20MnMoNi55 | 450 | 586 | 0.155 | 4.0 |
| A572 | 380 | 600 | 0.187 | 2.5 |
| X6NiCr1811 | 240 | 622 | 0.200 | 2.5 |

The constants n and $\beta$ were determined using the Ramberg-Osgood equation $\varepsilon/\varepsilon_Y = \sigma/\sigma_Y + \beta(\sigma/\sigma_Y)^{1/n}$ with $\sigma_Y = \sigma_{0.2}$.

The test parameters varied were specimen type (center cracked tensile, single edge notched bend, and compact specimens, specimen width, W, and pre-crack length, $a_o$. Most of the specimens were 5 mm thick, some tests were done on 10 and 20 mm thick specimens. The electrical potential method was used to measure the crack length and all specimens were instrumented such that crack growth could be correlated with the following quantities:

- J-integral
- crack tip opening displacement (CTOD), $\delta_{BS}$, according to BS 5762, (except center cracked tensile specimens)
- CTOD, $\delta_5$, measured on the specimens' side face at the fatigue crack tip using a special clip gage with 5 mm gage span, Fig. 1.

R-curves based on the J-integral and $\delta_5$ have been shown and discussed at length in Refs. /1,2/.

It is the purpose of the present paper to demonstrate compatibility between $\delta_{BS}$ and $\delta_5$. According to BS 5762 the CTOD is regarded as the sum of a small scale yielding and a fully plastic contribution [1)]:

$$\delta_{BS} = \underbrace{\frac{K^2}{2\sigma_{0.2}E'}}_{\delta_{ssy}} + \underbrace{\frac{v_{pl}}{1+k\,\frac{a_o+z}{W-a_o}}}_{\delta_{fpl}} \qquad (1)$$

with

$$E' = \frac{E}{1-\nu^2} \qquad \text{for plane strain}$$

$$E' = E \qquad \text{for plane stress.}$$

In Eq(1) k characterizes the rotation factor and is set equal to 2.5, $v_{pl}$ is the plastic contribution to the crack mouth opening displacement, v, and z is defined in Fig. 2. When the crack grows the center of rotation moves forward but the location at which the CTOD has to be taken remains at the original fatigue crack tip, Fig. 2a. Therefore, Eq(1) is modified as follows

$$\delta_{BS}^{M} = \frac{K^2}{2\sigma_{0.2}E'} + \frac{v_{pl}}{\dfrac{a+z+\dfrac{W-a}{k}}{a-a_o+\dfrac{W-a}{k}}} \qquad (2)$$

1) In contrast to BS 5762 the pre-crack length was determined as the average of ten measurements along the crack front.

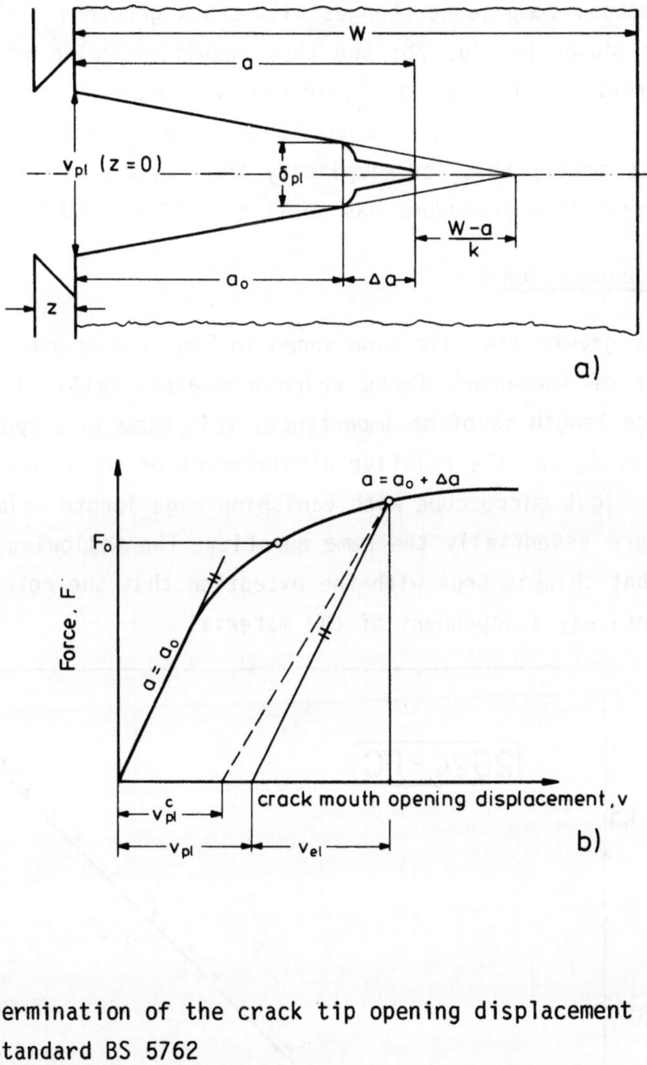


Fig. 2. Determination of the crack tip opening displacement based on British Standard BS 5762

a) geometry of extrapolation procedure

b) determination of the plastic portion, $v_{pl}$, of the crack mouth opening displacement, v.

Since the actual compliance changes with crack growth $v_{pl}$ should be determined as shown in Fig. 2b; the thus resulting value which accounts for crack growth is denoted as $v_{pl}^{c}$ in Fig. 2b. However, example calculations showed that the error introduced by using $v_{pl}$ instead of the correct $v_{pl}^{c}$ was small; thus for simplicity the evaluations were done with $v_{pl}$. This evaluation procedure was applied to CT and SENB specimens.

## Results and Discussion

During crack growth the clip gage shown in Fig. 1 measures the relative displacement of the crack faces which are essentially stress free so that the gage length is of no importance. This view is supported by Fig. 3 which shows $\delta_5$ and the relative displacement of the crack faces, $\delta_{Fl}$ measured by light microscope with vanishing gage length. Similarly, $\delta_{BS}^{M}$ should measure essentially the same quantity. The following results demonstrate that this is true with the exception that the rotation factor, k, is not entirely independent of the material.

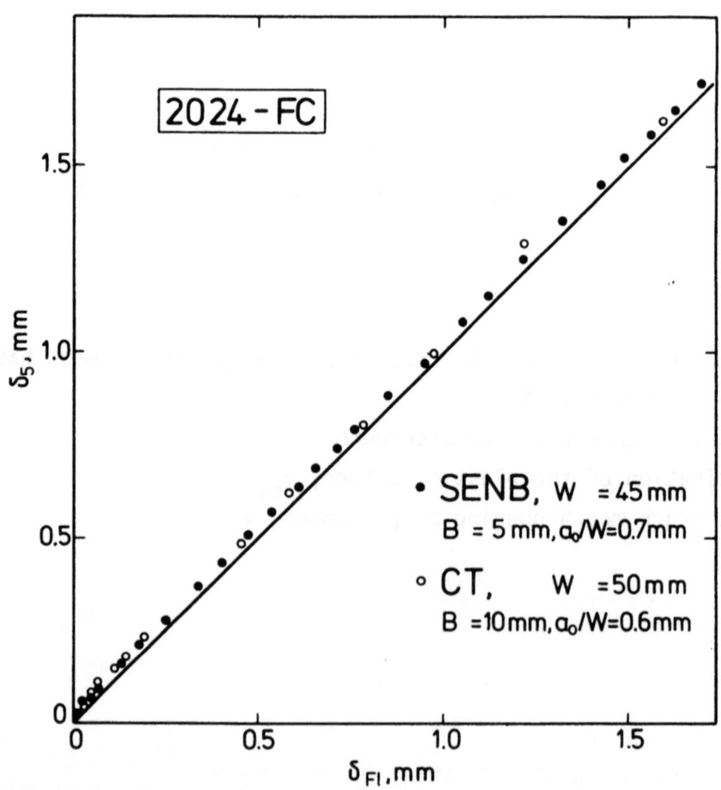


Fig. 3. Correlation between $\delta_5$ and $\delta_{Fl}$ for a growing crack

In Fig. 4 the scatterband of $\delta_{5R}$-curves of the aluminium 2024-FC obtained on all three specimen geometries is plotted; in addition the $\delta^{M}_{BS}$ for one specimen is shown. If the evaluation of $\delta^{M}_{BS}$ is done with k = 2.5 the results lie at the high side of the scatterband. Better representation of the scatterband is obtained for k = 2.9 which was determined as demonstrated below. Furthermore, to show the sensitivity of $\delta^{M}_{BS}$ with respect to k data with k = 4 are also plotted in Fig. 4.

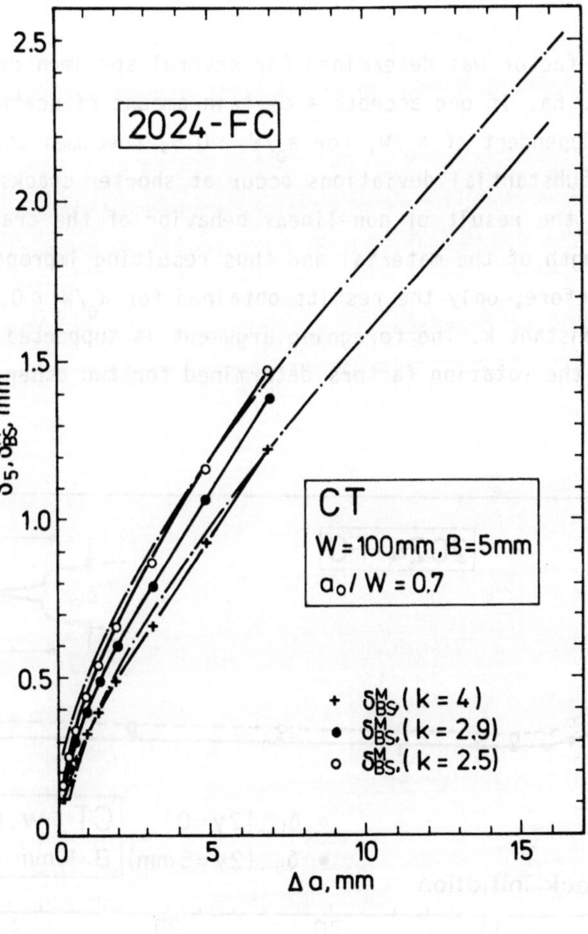


Fig. 4. R-curves based on $\delta_5$ and $\delta_{BS}$ modified, $\delta^{M}_{BS}$, respectively, for aluminium 2024-FC; the scatterband of the $\delta_{5R}$-curve was taken from Refs. /1,2/

Assuming that $\delta_{BS}^{M}$ and $\delta_5$ measure essentially the same quantity the rotation factor can be determined using Eq(2) by setting $\delta_{BS}^{M}$ equal to $\delta_5$. Fig. 5 shows that after an initial transition region k attains a constant value of 2.8 for a CT specimen of the low strength - high hardening aluminium 2024-FC. The average for several specimens amounts to k = 2.9. The graph shows also that prior to crack initiation $\delta_{F1}$ and $\delta_5$ diverge although the difference is less than it may appear from k (see the relatively small influence of k on $\delta_{BS}^{M}$ in Fig. 4). The pre-initiation situation will be investigated further.

The rotation factor was determined for several specimen configurations of 2024-FC, Fig. 6a. If one accepts a certain amount of scatter k can be regarded as independent of $a_0/W$, for $a_0/W \geq 0.5$, specimen thickness and specimen type. Substantial deviations occur at shorter cracks; these are believed to be the result of non-linear behavior of the crack faces due to the low strength of the material and thus resulting improper determination of $\delta_{BS}^{M}$. Therefore, only the results obtained for $a_0/W \geq 0.5$ were taken to evaluate a constant k. The foregoing argument is supported by Figs. 6b and c which show the rotation factors determined for two other materials, the

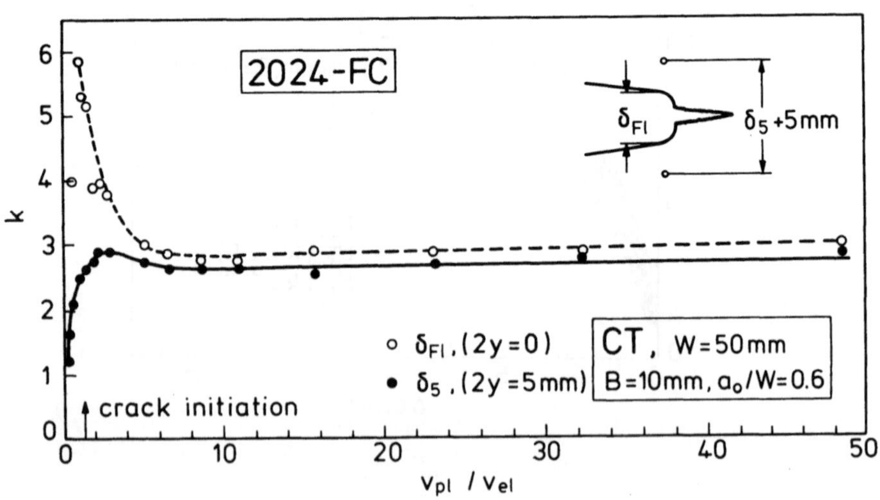


Fig. 5. Rotation factor, k, as a function of the degree of plasticity; k was determined through Eq(2) by setting $\delta_{BS}^{M}$ equal to the experimental values of $\delta_{F1}$ and $\delta_5$, respectively

strength to toughness ratio of which is substantially higher. In all three cases a constant rotation factor is obtained (with the exception mentioned above); k is slightly higher for the high hardening material 2024-FC which means that the center of rotation is somewhat closer to the crack tip.

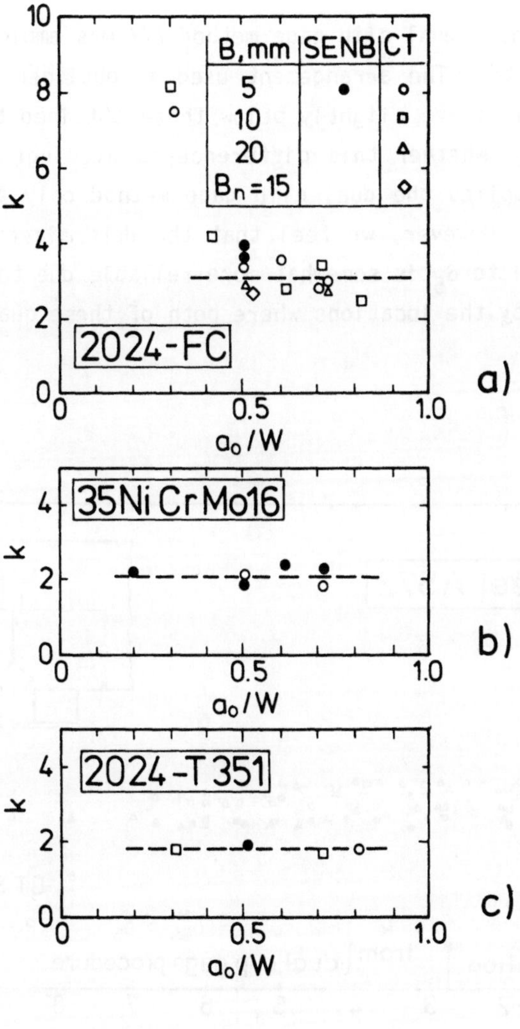


Fig. 6. Rotation factor, k, for determining the modified $\delta_{BS}^{M}$ by Eq(2)
a) aluminium 2024-FC
b) steel 35NiCrMo16
c) aluminium 2024-T351

Further results, obtained on the steel A572 are shown in Fig. 7. This material was tested in the form of CT, three point bend SENB, and four point bend SENB specimens. Again, a constant rotation factor is obtained after a transition region. The data obtained by this test series do not suggest a systematic effect of specimen geometry on k.

For some specimens a dual clip gage method /7/ was employed to determine the rotation factor. The arrangement used is outlined schematically in Fig. 7. The results are slightly below those obtained by comparing $\delta_{BS}^{M}$ with $\delta_5$, Fig. 7. Whether this difference is meaningful, is not clear since we have applied the dual clip gage method only to the few cases shown in Fig. 7. However, we feel that the determination of k through setting $\delta_{BS}^{M}$ equal to $\delta_5$ is somewhat more reliable due to the larger gage length provided by the locations where both of these quantities are measured.

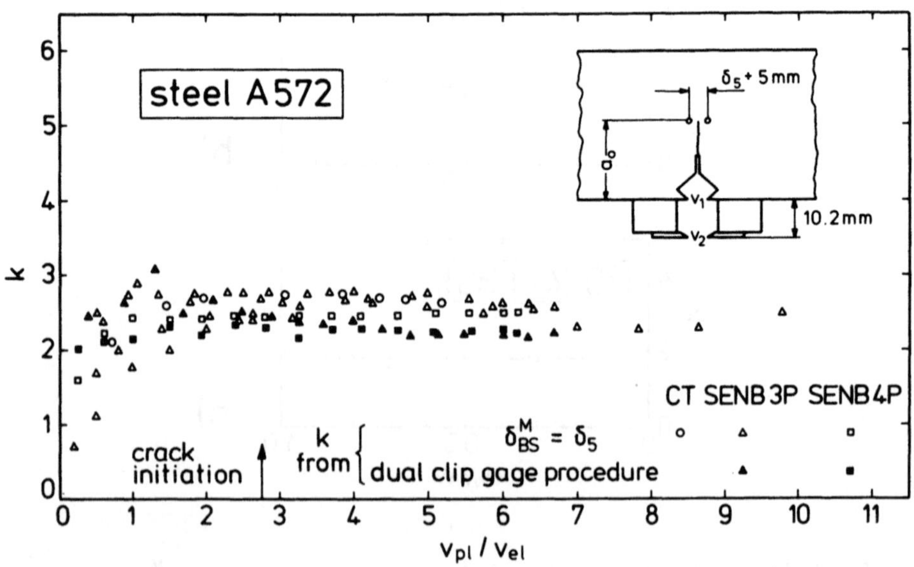


Fig. 7. Rotation factor, k, as a function of the degree of plasticity obtained on A572 steel

Figs. 8 and 9 show R-curves for 2024-T351 and 35NiCrMo16. The parameter $\delta_5$ is consistent with $\delta^M_{BS}$ for the whole range of crack growth. Furthermore, the error in the CTOD is not very large if the standard rotation factor k = 2.5 is used for the evaluation of $\delta^M_{BS}$.

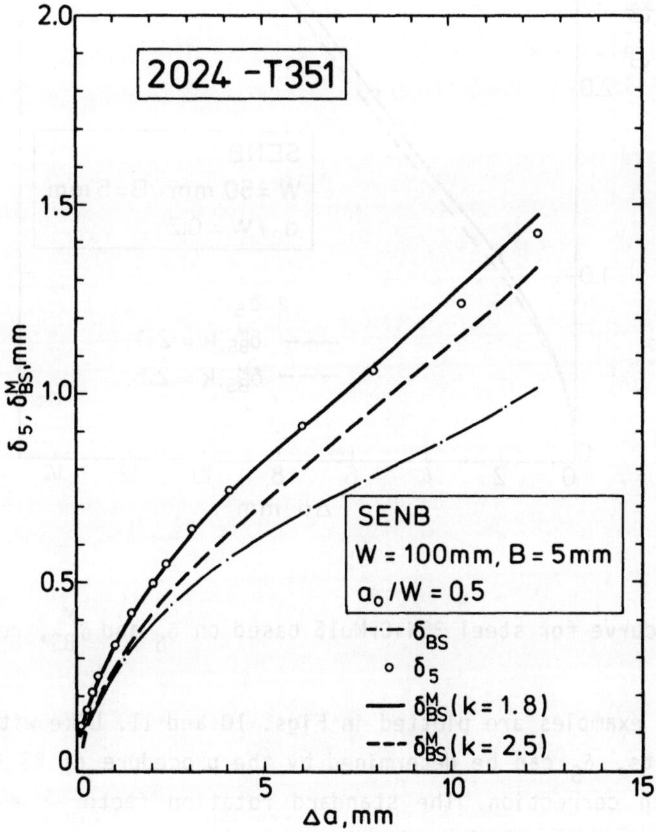


Fig. 8. R-curve for aluminium 2024-T351 based on $\delta_5$ and $\delta^M_{BS}$, respectively

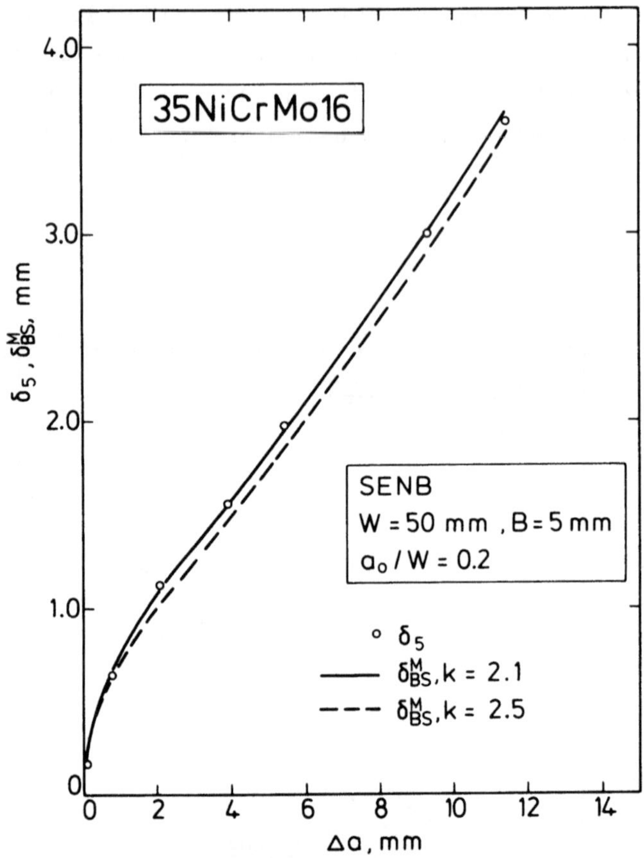


Fig. 9. R-curve for steel 35NiCrMo16 based on $\delta_5$ and $\delta_{BS}^{M}$, respectively

Two further examples are plotted in Figs. 10 and 11. Like with the foregoing results, $\delta_5$ can be determined by the procedure of BS 5762 with a crack growth correction. The standard rotation factor k = 2.5 yields slightly conservative values for low to moderately hardening materials; conservatism increases if the original formula (Eq(1)) without correction for crack growth is used. The results tend to be slightly on the non-conservative side for high hardening materials. Fig. 12 shows that the rotation factor indeed varies consistently with the strain hardening exponent.

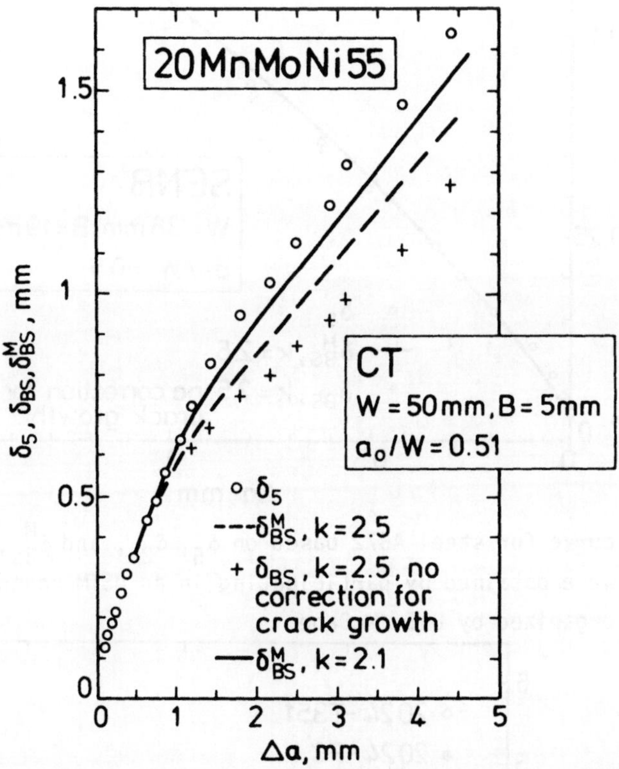


Fig. 10. R-curve for reactor pressure vessel steel 20MnMoNi55 based on $\delta_5$, $\delta_{BS}$, and $\delta_{BS}^{M}$, respectively /6/. The value k = 2.1 was obtained as the average for several specimens

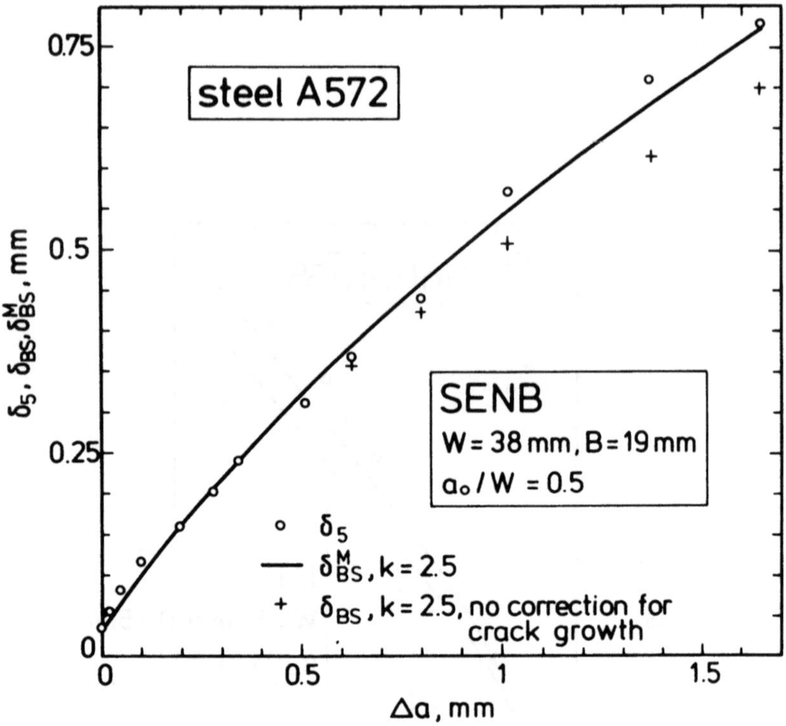


Fig. 11. R-curve for steel A572 based on $\delta_5$, $\delta_{BS}$, and $\delta_{BS}^{M}$, respectively. The results were obtained by participating in an ASTM round robin on the CTOD method organized by D.E. McCabe

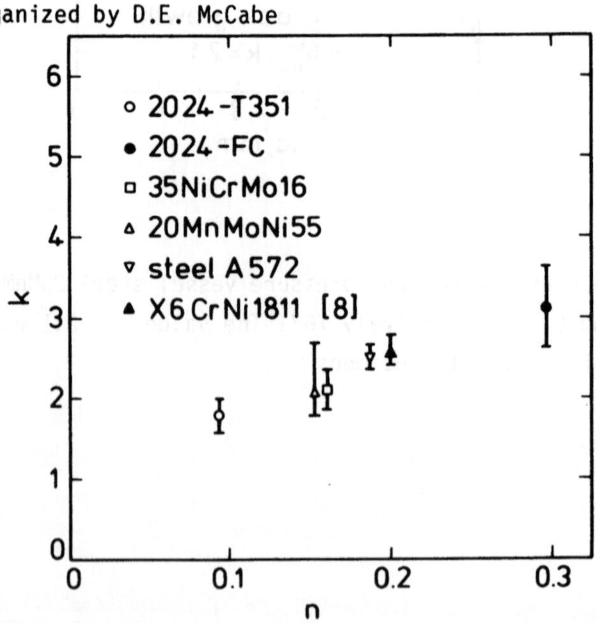


Fig. 12. Influence of strain hardening exponent, n, on the rotation factor k

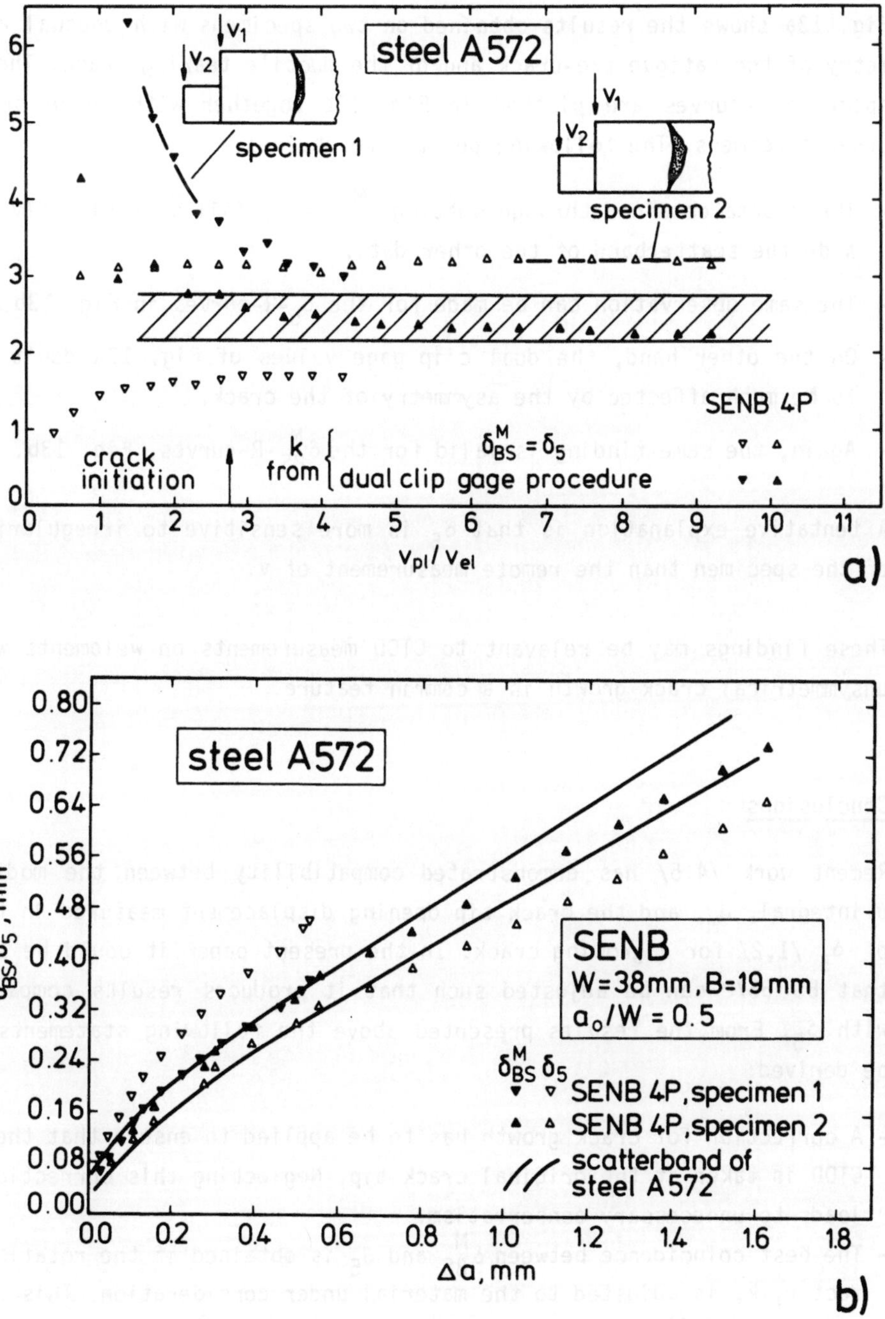


Fig. 13. Additional results obtained on A572 steel relative to
a) rotation factor, k,
b) δ-R-curves

Fig. 13a shows the results obtained on two specimens with unusual asymmetry of the fatigue pre-crack and of the ductile tearing crack. The respective R-curves are plotted in Fig. 13b together with those of the other specimens. The following points are obvious:

- The k data obtained through setting $\delta_{BS}^{M} = \delta_5$ fall significantly outside the scatterband of the other data.
- The same observation can be made for the $\delta_5$-R-curves in Fig. 13b.
- On the other hand, the dual clip gage values of Fig. 13a don't seem to be much affected by the asymmetry of the crack.
- Again, the same finding is valid for the $\delta_{BS}^{M}$-R-curves, Fig. 13b.

A tentative explanation is that $\delta_5$ is more sensitive to irregularities of the specimen than the remote measurement of v.

These findings may be relevant to CTOD measurements on weldments where unsymmetrical crack growth is a common feature.

## Conclusions

Recent work /4,5/ has demonstrated compatibility between the modified J-integral, $J_M$, and the crack tip opening displacement measured in terms of $\delta_5$ /1,2/ for a growing crack. In the present paper it could be shown that BS 5672 can be adjusted such that it produces results comparable with $\delta_5$. From the results presented above the following statements can be derived:

- A correction for crack growth has to be applied to ensure that the CTOD is taken at the original crack tip. Neglecting this correction leads to unnecessary conservatism.
- The best coincidence between $\delta_{BS}^{M}$ and $\delta_5$ is obtained if the rotation factor, k, is adjusted to the material under consideration. This can be accomplished by one calibration test in which $\delta_{BS}^{M}$ and $\delta_5$ are determined simultaneously or by Fig. 12.
- The rotation factor depends slightly on the material under consideration.

- For a given material, the rotation factor is independent of parameters like thickness, crack length, side grooving, specimen type, and amount of crack growth.
- The modified BS 5762 is equally well applicable to CT and SENB specimens.
- If the appropriate k-value cannot be determined in a calibration test or by Fig. 12 the value k = 2.5 of BS 5762 leads to conservative estimates of $\delta_5$ for low to moderately hardening materials.
- The measurement of $\delta_5$ which requires more careful experimentation than simple standard clip gage measurements can obviously be replaced by the modified BS 5762.For other specimen geometries, like center cracked tension or double edge cracked tension specimens, however, there is no substitute and $\delta_5$ has to be measured directly.

List of Symbols

| | |
|---|---|
| $a$ | crack length |
| $a_o$ | length of fatigue pre crack |
| $B$ | specimen thickness |
| $B_n$ | net section thickness of side grooved specimen |
| $E$ | Young's modulus |
| $k$ | rotation factor |
| $K$ | linear elastic stress intensity factor |
| $v$ | crack mouth opening displacement |
| $W$ | specimen width |
| $\delta$ | crack tip opening displacement, general |
| $\delta_{BS}$ | $\delta$ determined using the formalisms of BS 5762 |
| $\delta^M_{BS}$ | $\delta_{BS}$ modified |
| $\delta_{F1}$ | displacement of crack faces near fatigue crack tip |
| $\delta_5$ | $\delta$ measured at fatigue crack tip using a clip gage with 5 mm gage length, see Fig. 1. |
| $\Delta a$ | crack length increment measured by potential drop method |

References

1. Hellmann, D., Schwalbe, K.-H.; Geometry and Size Effects on J-R and $\delta$-R Curves Under Plane Stress Conditions, 15th National Symposium on Fracture Mechanics, College Park, Maryland, 7 - 9 July 1982, published in ASTM STP 833, R.J. Sanford, Ed., American Society for Testing and Materials, Philadelphia, 1984, pp. 577-605.

2. Schwalbe, K.-H., Hellmann, D.; Correlation of Stable Crack Growth with the J-Integral and the Crack Tip Opening Displacement, Effects of Geometry, Size, and Material, GKSS-Research Center Geesthacht GmbH, Report GKSS 84/E/37, Geesthacht 1984.

3. British Standard BS 5762:1979, Methods for Crack Opening Displacement (COD) Testing. The British Standards Institution.

4. Ernst, H.A.,Presentation at ASTM E24.08 Subcommittee meeting, 18-22 April 1983, Louisville, KY.

5. Landes, J., McCabe, D.E., Ernst, H.A.; Elastic-Plastic Methodology to Establish R-Curves and Instability Criteria, Report RP 1238-2, Westinghouse R&D Center, Monroeville, PA, 1983.

6. Heerens, J., Unpublished work.

7. Dawes, M.G.; Technical presentation at the meeting of ASTM Task Group E24.08.05, Dallas, Oct. 23, 1984.

8. Aeberli, K.E.; to be published.

# Relationships Between Various Definitions of the Crack Tip Opening Displacement

K.-H. SCHWALBE, J. HEERENS, D. HELLMANN, and A. CORNEC

GKSS-Forschungszentrum Geesthacht GmbH
Institut für Werkstofftechnologie
Max-Planck-Str.
D-2054 Geesthacht, FRG.

## Abstract

A close examination of the CTOD problem reveals that the crack tip opening displacement is very much a matter of definition. Various definitions are in use either for analytical treatment or for experimental determination. A comparison of these shows that by the widely used experimental methods a quantity is determined which is at variance with the commonly accepted definition as the opening of two points on the crack faces located on two straight lines emanating from the crack tip including an angle of ±45° with the undeformed crack.

## Introduction

The philosophy underlying the CTOD concept [ 1,2 ] is simply that a characteristic displacement occuring at the crack tip - the crack tip opening displacement - controls the conditions at the crack tip, in particular the event of failure, which can be divided into initiation of crack growth plus stable and unstable crack growth. Although plenty of experimental and theoretical work related to the CTOD has been done, there doesn't seem to be any other quantity in fracture mechanics which is as vaguely defined as the crack tip opening displacement.

For the practical application of this concept three problems have to be solved:

- Appropriate definition of CTOD.
  If one looks at the crack tip profile, there does not seem to be a characteristic displacement which must be regarded as the "natural" CTOD. Hence, the CTOD is a matter of definition.
- Reliable experimental techniques are needed in order to determine the critical values of CTOD.
- Relationships between the CTOD and the applied load or strain are needed as transfer mechanisms from the specimen to the cracked structural member under consideration. In other words: we must be able to express the CTOD

as a driving force.

In the following, some definitions of the CTOD and the relevant experimental techniques will be briefly discussed along with analytical models for the prediction of CTOD. The discussion will be restricted to pre-initiation situations, i.e. the so called blunting phase.

## Experimental Techniques

The most widely used technique for the experimental determination of the CTOD is described in the British Standard BS 5762 [ 3 ]. The CTOD is composed of a small scale yielding (see Fig. 3) and fully plastic contribution; the latter is obtained by linear interpolation between the plastic portion of the crack mouth opening displacement, CMOD, and the centre of rotation, Fig. 1. The total CTOD is given by the sum of a small scale yielding and a fully plastic contribution:

$$\delta_{BS} = \frac{K^2(a_o)}{2\ _{0.2}E} + \frac{v_{pl}}{1+k\,\frac{a_o+z}{W-a_o}} \tag{1}$$

$$= \delta_{ssy} + \delta_{pl}$$

where k defines the location of the centre of rotation (k=2.5); for the other symbols see the nomenclature.

Eq (2) can be modified for crack growth in such a way that inspite of the moving centre of rotation the CTOD will always be determined at the position of the original fatigue crack tip [4]:

$$\delta^M_{BS} = \frac{K^2(a)}{2\sigma_{0.2}E} + \frac{v_{pl}}{\dfrac{a+z+\frac{W-a}{k}}{a-a_o+\frac{W-a}{k}}} \tag{2}$$

In addition, $\delta^M_{BS}$ is used with k adjusted to the material's work hardening exponent [4 ]. A technique described in Refs.[5,6 ] measures the displacement, $\delta_5$, on the specimen's side face at the fatigue crack tip over a gage length of 5 mm. Whereas the standardised technique is restricted to bending, the $\delta_5$ technique can be applied to all configurations, for example CCT specimens, through-cracked pipes [7] etc.

Finally, the stretched zone between the fatigue crack and the void calescence regions of a broken specimen can be observed directly; e.g. in a scanning electron microscope. Alternatively, one can take advantage of the infiltration technique by which non-broken specimens can be investigated [8,9]. On the average, the height of the stretched zone, SZH, is equal to one half of the crack tip opening displacement, $\delta_t$, see Fig. 2. Although in principle this is the most appropriate technique it is mostly used in fundamental investigations only (exception: Japanese $J_{Ic}$ standard [10]) since two major difficulties have to be overcome:

- SZH exhibits a large scatter along the crack front.
- It is often difficult to locate the lower and upper boundary of SZH.

## Analytical Models

In the case of elastic-ideally plastic materials the strip yield model and the small scale yielding model both yield a well defined CTOD, Fig. 3., for plane stress. The former one has been solved analytically for the infinitely wide sheet under tension 11 :

$$\delta_D = \frac{8\sigma_Y}{\pi E} a \ln\left[\sec \frac{\pi\sigma}{2\sigma_Y}\right] \tag{3}$$

It should yield realistic values up to $\sigma$ close to $\sigma_Y$. On the other hand, the small scale yielding model [ 1 ] works only up to $\sigma \sim 0.5\ _Y$, however, it applicable to every geometry for which the K solution is known:

$$\delta_{ssy} = \frac{K^2}{E\sigma_Y} \tag{4}$$

By introducing the effective stress intensity factor, $K_{eff}$, which is determined using

$$a_{eff} = a + r_Y \tag{5}$$

$$r_Y = \frac{K^2}{(1+n)\, 2\pi\sigma_Y{}^2}$$

or similar corrections, the small scale yielding model works also close to net section yielding [ 12, 13 ].

Eq(3) can be extended to finitely wide sheets (width) = 2W) [ 14 ]

$$\delta_D = \frac{8\,\sigma_Y}{\pi E}\, a \ln\left\{ \frac{2W}{\pi a} \arcsin\left[ \sin\frac{\pi a}{2W} \sec\frac{\pi\sigma}{2\sigma_Y} \right]\right\} \tag{6}$$

Strain hardening of a material reduces the CTOD. A close examination of the crack tip region using the HRR field solutions reveals rounded crack tip profile, Fig. 4, the curvature of which depends on the material's work hardening exponent, n.

It is obvious, that no "natural" CTOD exists, as it is the case with the models in Fig. 3. The location at which the CTOD has to be taken is a matter of definition. A reasonable definition is shown in Fig. 4 which is compatible with the experimental observation [ 15 ] that the crack bulges forward by roughly one half of the CTOD. Thus, the 45 degree definition takes the CTOD approximately at the position of the original fatigue crack tip. Shih [ 16 ] obtained

$$\delta_{45} = d_n \frac{J}{\sigma_Y} \tag{7}$$

where $d_n$ depends mainly on n and on the state of stress (plane stress or plane strain). An approximate analytical solution [ 17, 18 ] yields

$$d_n = 2\left[\frac{\sigma_Y}{\sqrt{3}}\,\frac{2(1+\nu)}{(1+n)E}\,\frac{(1+n)^2}{n^{n/(1+n)}}\right]^{1+n} \frac{1}{\pi(1+n)}\,\frac{1}{\sigma_Y/E} \tag{8a}$$

for plane stress.
Employing this expression for $d_n$, Eq(7) approaches numerically the Dugdale model for the non-hardening limit.

For plane strain

$$d_n = 2\left[\frac{\sigma_Y}{\sqrt{3}}\,\frac{2(1+\nu)^2}{(1+n)E}\,\frac{(1+n)}{n^{n/(1+n)}}\right]^{1+n} \frac{0.593}{(1-\nu^2)\,\pi(1+n)}\,\frac{1}{\sigma_Y/E} \tag{8b}$$

is obtained [17,18 ].

A further analytical model refers to the experimentally determined $\delta_5$. With $F_Y$ being the load at incipient net section yielding and $\delta_Y$ being the value of $\delta_5$ at $F = F_Y$, it is stated that in the net section yielding regime, ie. at $F > F_Y$ for simple geometries $\delta_5$ can be determined by [12]

$$\delta_5 = \delta_Y \left[ \frac{F}{F_Y} \right]^{1/n} \quad (9)$$

## Comparison of Various Experimental and Analytical Results

The main emphasis of this section is placed on a comparison of the experimental "remote" techniques ($\delta_{BS}$, $\delta_5$) with the real situation at the very crack tip as given by SZH and SZW measurements or by theory ($\delta_{45}$, $\delta_D$, $\delta_{ssy}$).

It has been shown [15] that SZH values determined on fracture surfaces can be very well predicted by $\delta_{45}$, see also Ref.[9]. A necessary prerequisite is a proper power law fit to the material's stress strain curve [19]. It can thus be concluded that the HRR field is a good representation of the near-tip behaviour of a crack in an real material.

On the other hand, the data compiled in Figs. 5 through 7 demonstrate that the remote techniques do not correspond with the true crack tip events:

$\delta_{BS}$ and $\delta_5$ tend to be larger than $\delta_{45}$ (or 2 SZH).

The matter of compatibility between $\delta_{BS}$ and $\delta_5$ has already been discussed in Ref.[4]. Both techniques measure essentially the same quantity if the influence of the material on the rotation factor is taken into consideration (in Fig. 5b the $\delta_{BS}$ line could be matched with the $\delta_5$ data if a modified rotation factor would be employed, for details see Ref.[4]. Although for a given material there is a unique relationship between $\delta_5$ and J independent of the specimens in-plane geometry [6,13], Fig. 7, $\delta_5$ and $\delta_{BS}$ are in general not a measure of the CTOD as defined by $\delta_{45}$ or SZH.

A simple geometrical consideration demonstrates that the straight line extrapolating the crack mouth opening displacement to the crack tip hits the vertical line through the crack tip at a location which is different from that of $\delta_{45}$, Fig. 8.

These findings are relevant to the experimental determination of initiation values of the CTOD since the proposed CTOD Test Method of ASTM [20] utilises a blunting line

$$\Delta a = 0.5\ \mathrm{CTOD} = 0.5\delta_{BS} \tag{10}$$

which in view of the above described findings cannot be adequate. This is supported by the data in Fig. 9 which show that $\Delta a$ = SZW is always less than 0.5 $\delta_{BS}$. It is therefore suggested that the GKSS blunting line proposed for the $J_{Ic}$ determination [15] be transferred to the CTOD method. This blunting line is based on the experimental observation that

$$\mathrm{SZW} = \Delta a \approx 0.4\ \delta_{45} \tag{11}$$

With Eq(7) one obtains

$$\Delta a = 0.4\ d_n \frac{J}{\sigma_Y} \tag{12}$$

Usually J is not determined in a CTOD test, but it can be replaced by the plasticity corrected strain energy release rate, $G_{eff} = K^2_{eff}/E'$, provided that the applied load , F, is not bigger than the yield load, $F_Y$, which is determined using the limit load solution for the specimen geometry under consideration evaluated with the materials yield strength, $\sigma_Y$. Thus

$$\Delta a = 0.4\ d_n \frac{G_{eff}}{\sigma_Y}, \quad F \leq F_Y \tag{13}$$

may be an appropriate expression for the CTOD blunting line, Fig. 9. For the determination of $K_{eff}$ see Ref.[13].

Finally, it is interesting to note that the non-hardening solutions (Eqs(3,4,6)) match with measurements on hardening materials: up to $F = F_Y$ the experimental $\delta_5$ values are in very good agreement with Eq(4) [12,13]. An example is shown in Fig. 10 where experimental data obtained on different specimen geometries form a common scatterband which is very well represented by Eq(4). It is of particular interest that the relationship between $\delta_5$ and K remains the same (or between $\delta_5$ and J for $F > F_Y$, see Fig. 7,) also after the crack had started growing, i.e. the $\delta_5$ at the original fatigue crack tip is a measure of the events at the actual crack tip even after crack growth. This means that a quasi-stationary treatment of a growing crack with respect to $\delta_5$ (which is essentially equal to the separation of the crack flanks) is appropriate provided that the actual crack length is used for calculation.
Fig. 11 shows a further example: finite element calculations from Ref.[21]

for an austenitic steel are compared with Eq(4). Again, the non-hardening solution is an appropriate measure of $\delta_5$ even for a highhardening material in the regime $F \leq F_Y$.

A qualitative explanation of this effect is simply that for contained yielding the influence of strain hardening vanishes if one leaves the very crack tip; at locations remote from the crack tip (e.g. crack mouth, load points) plasticity corrected linear elastic solutions are appropriate, whereby the effect of hardening is on the plastic zone size only, and this effect is minute compared to the hardening effect on $\delta_{45}$. Thus, the strains sampled over the gage length of 5mm are affected by hardening only over a short distance compared to the whole gage length, and the sum of the strains results in a displacement only minutely affected by hardening. This explanation can only hold within a certain range of load and gage length; a more detailed discussion will be given in Ref.[ 22].

Beyond $F_Y$, hardening plays a role in the sense of Eq(9): the CTOD increases with the normalised load, $(F/F_Y)$, raised to the (1/n)th power. This simple model could be verified on a number of materials and specimen geometries [ 12,13 ]. An example is shown in Fig. 12.

Conclusions

Various experimental and analytical definitions of the CTOD have been identified. These are:

- $\delta_{BS}$, the CTOD determined according to BS 5762,
- $\delta_5$, measured on the specimen's side surface at the position of the fatigue pre-crack tip with a gage length of 5mm,
- twice the stretch zone height, 2 SZH.

Whereas by $\delta_{BS}$ and $\delta_5$ measure essentially the same quantity, 2 SZH can be substantially smaller than both of these.

- $\delta_D$, the CTOD calculated on the basis of the Dugdale model, non-hardening materials,
- $\delta_{ssy}$, the small scale yielding solution for non-hardening materials,
- $\delta_{45}$, the solution for strain hardening materials.

A comparison of experimental and analytical results shows that below the yield load $\delta_{BS}$ and $\delta_5$ measure quantities which don't exhibit strain hardening effects, i.e. they can be approximated by $\delta_{ssy}$ and $\delta_D$, respectively. On the other hand, the stretch zone can be well described by $\delta_{45}$. This means that the standard methods don't determine the real tip displacement, $\delta_{45}$. Nevertheless, they can be related to the applied load in both contained yield and fully plastic conditions and thus serve for structural assessments.

## List of Symbols

| | |
|---|---|
| $a$ | crack length |
| $a_o$ | starting crack length |
| $E$ | Young's modulus |
| $K$ | linear elastic stress intensity factor |
| $K_{eff}$ | K with plasticity correction |
| $n$ | strain hardening exponent |
| $W$ | specimen width (half width of CCT specimen) |
| $\delta$ | crack tip opening displacement, general, various definitions of $\delta$ are explained in text |
| $\sigma$ | applied nominal stress |
| $\sigma_{0.2}$ | 0.2% proof stress |
| $\sigma_Y$ | yield strength, general |

References

[ 1] Wells, A.A.: Application of fracture mechanics at and beyond general yield, British Welding Journal, 1963, pp. 563-570.

[ 2] Wells, A.A.: Crack opening displacements from elastic-plastic analyses of externally notched tension bars, Engineering Fracture Mechanics, 1969, Vol. 1, pp. 399-410.

[ 3] BS 5762:1979. Methods for crack opening displacement (COD) testing. The British Standards Institution.

[ 4] Hellmann,D., and K.H. Schwalbe: On the experimental determination of CTOD-based R-curves, this workshop.

[ 5] Hellmann, D.,and K.-H. Schwalbe: Geometry and size effects on $J_R$ and $\delta_R$-curves under plane stress conditions, in: ASTM STP 833, 1984, pp. 577-605.

[ 6] Schwalbe, K.-H., and D. Hellmann: Correlation of stable crack growth with the J-integral and the crack-tip opening displacement, effects of geometry, size, and material, GKSS 84/E/37.

[ 7] Aeberli, K.E., H.D. Schulze, K.-H. Schwalbe, P. Morawietz, H. Fuhlrott, and J. Heerens: Fracture mechanics tests on axially cracked pipe sections subjected to internal pressure, this workshop.

[ 8] Robinson, J.N., and A.S. Tetelman: Measurement of $K_{Ic}$ on small specimens using critical crack tip opening displacement, in: ASTM STP 559, 1974, pp. 139-158.

[ 9] Knauf, G., and H. Riedel: A comparative study of different methods to measure the crack opening displacement, in: D. François, ed., Advances in Fracture Research, vol. 5, 1981, pp. 2547-2553, Pergamon Press.

[10] Kobayashi, H., H. Nakamura, and J. Nakazawa: Comparison of $J_{Ic}$ test methods recommended in ASTM and JSME, to be published in ASTM STP 856.

[11] Goodier, J.N., and F.A. Field, in: Fracture of Solids, Wiley, 1963.

[12] Schwalbe, K.-H.: A simple engineering treatment of center cracked tension panels in the regime of non-linear fracture mechanics under plane stress conditions, GKSS 84/E/38, 1984.

[13] Schwalbe, K.-H.: Extension of the Engineering Treatment Model (ETM) to bending configurations under plane stress, to be published.

[14] Schwalbe, K.-H.: A modification of the COD concept and its tentative application to the residual strength of center cracked panels, in ASTM STP 700, 1980, pp. 500-512, GKSS 81/E/46.

[15] Schwalbe, K.-H., and J. Heerens: Proposed modifications of ASTM E813-81 standard test method for $J_{Ic}$, in: Proceedings of the 10th MPA Seminar, Stuttgart 10-13 October 1984: Safety and Reliability of Pressure Components with Special Emphasis on Fracture Exclusion.

[16] Shih, C.F.: Relationships between the J-integral and crack opening displacement for stationary and extending cracks, General Electric Report No. 79CRD075, 1979.

[17] Schwalbe, K.-H.: Approximate calculation of fatigue crack growth, Int. J. of Fracture, vol. 9, No. 4, 1973, pp. 381-395.

[18] Rice, J.R.: Private communication.

[19] Cornec, A.: A new theoretical blunting line determination, in preparation.

[20] "Draft ASTM Test Method for Crack Tip Opening Displacement (CTOD) Testing," Working Document of Committee E24 on Fracture Testing, American Society for Testing and Materials, Philadelphia, Pa., February 1984.

[21] Amstutz, H., and T. Seeger: Bruchmechanische Analyse eines austenitischen Stahles (X6CrNi1811) an zwei Standardrißscheiben mit ebenem Spannungszustand, F1-2/1984, Technical University of Darmstadt, 1984.

[22] Cornec, A.: Some theoretical considerations on the crack tip opening displacement $\delta_5$ for a stationary crack in thin structures, in preparation.

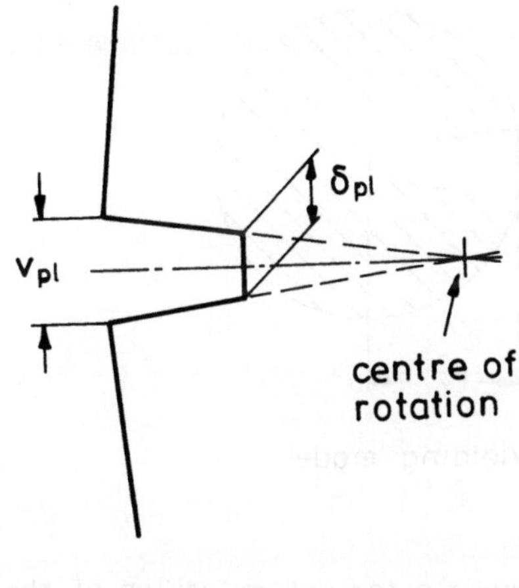


Fig. 1: Schematic showing determination of fully plastic portion of $\delta_{BS}$ from $v_{pl}$.

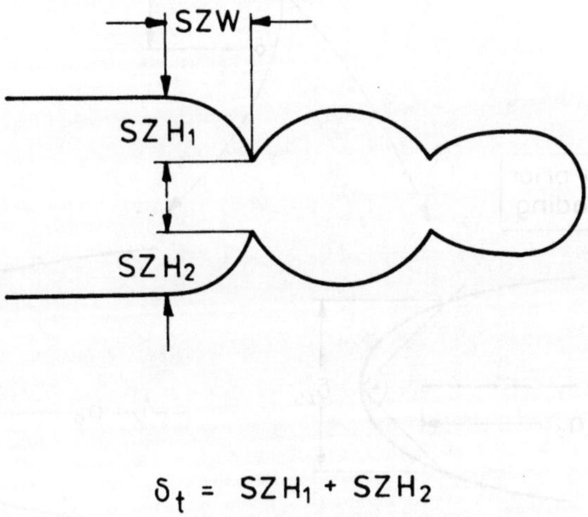


Fig. 2: Crack tip opening displacement defined as height of strech zone.

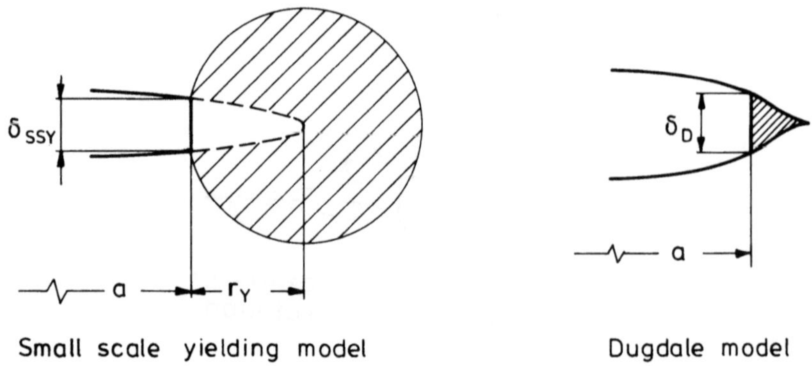


Fig. 3: Analytical models for determination of the CTOD for non-hardening materials under plane stress conditions.

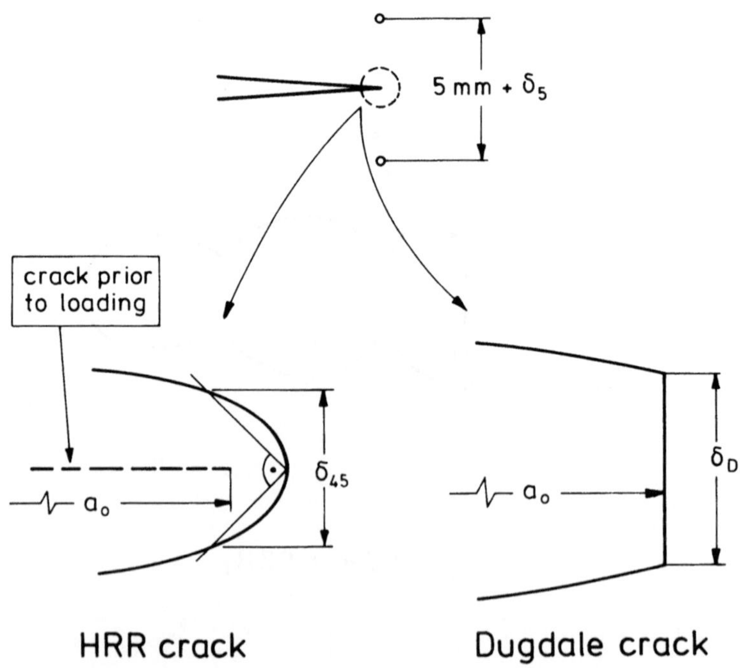


Fig. 4: Comparison of $\delta_{45}$, $\delta_5$, and $\delta_D$.

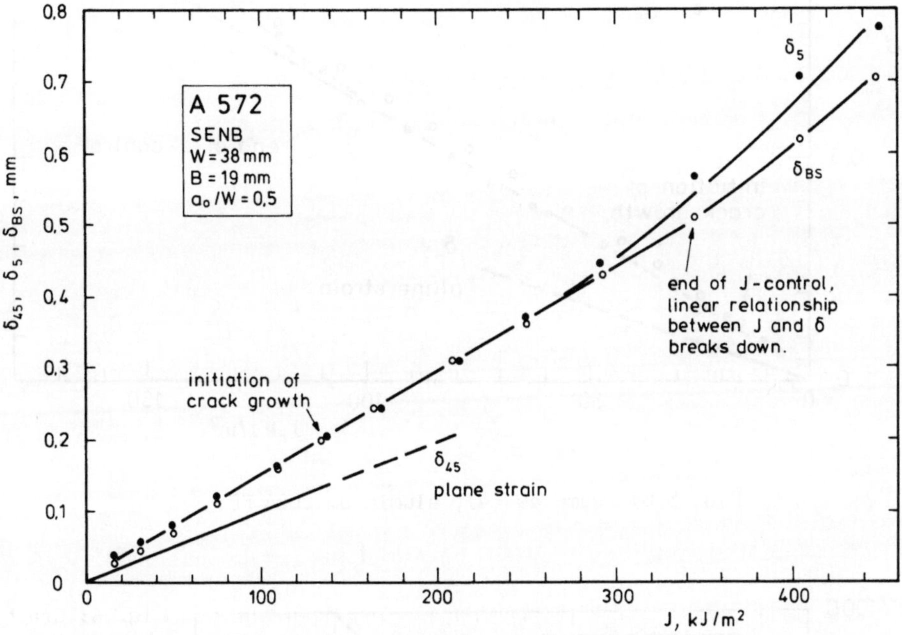


Fig. 5 a) CTOD in terms of $\delta_{45}$, $\delta_5$, and $\delta_{BS}$ for the steel A572 as a function of the J-integral. The statement "end of J-control" refers to the findings of Ref. [ 5 ] where J control was assumed to breack down at the splitting points of J-R-curves obtained on specimens with various ligament sizes. This coincides with the end of linearity between J and $\delta_5$.

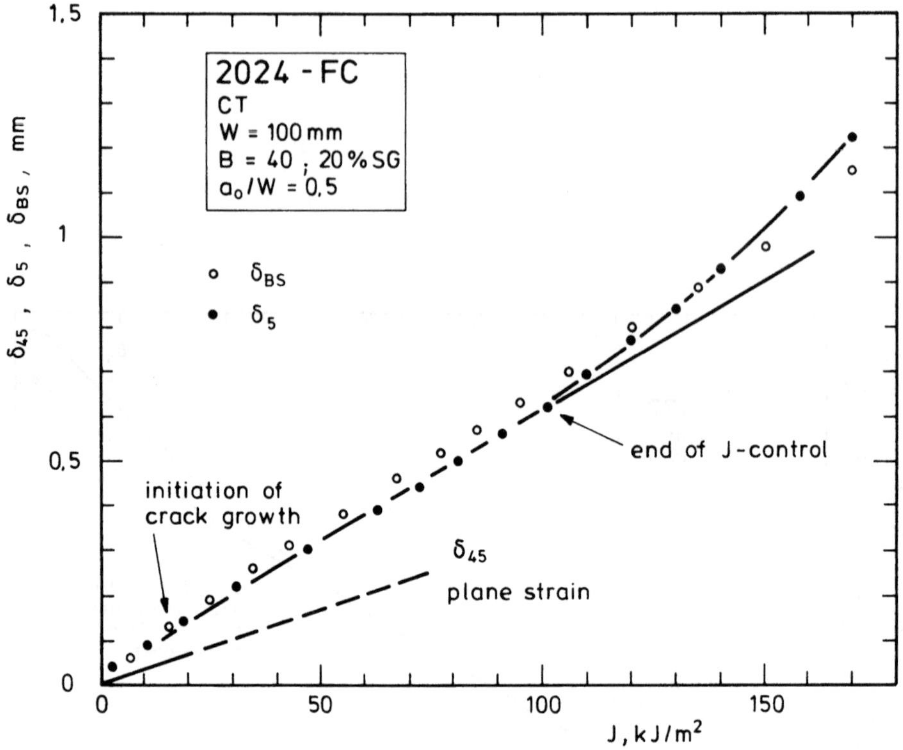


Fig. 5 b) same as (a), aluminium 2024-FC.

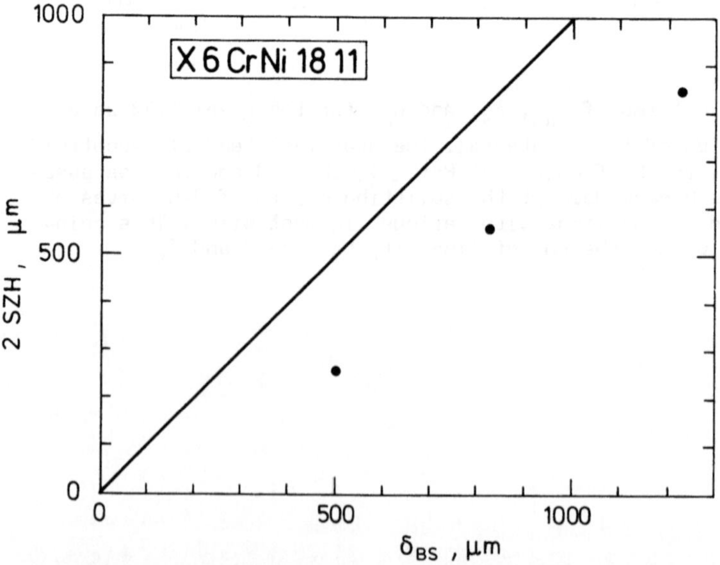


Fig. 6: Graph showing that CTOD determined as stretch zone height is at variance with standardised experimental method determining $\delta_{BS}$, austenitic steel X6CrNi1811.

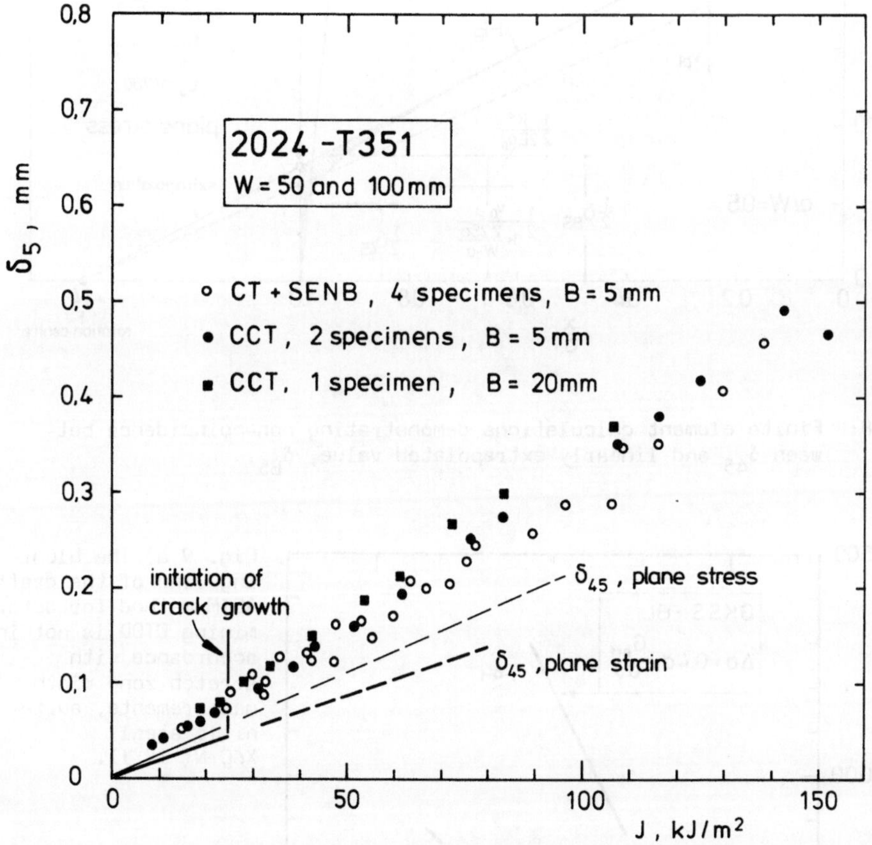


Fig. 7: Graph showing that $\delta_5$ is at variance with the analytical $\delta_{45}$ and showing unique relationship between $\delta_5$ and J independent of specimen geometry, aluminium alloy 2024-T351.

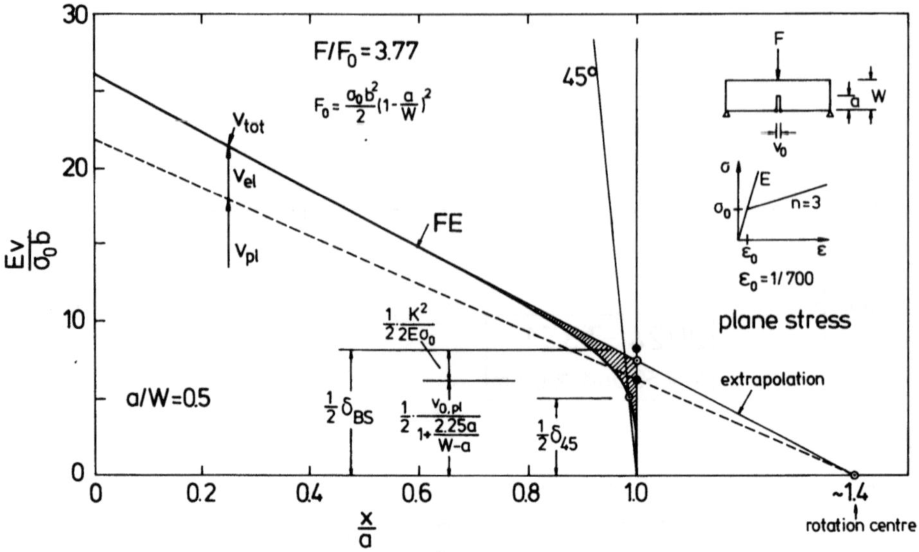


Fig. 8: Finite element calculations demonstrating non-coincidence between $\delta_{45}$ and linearly extrapolated value, $\delta_{BS}$.

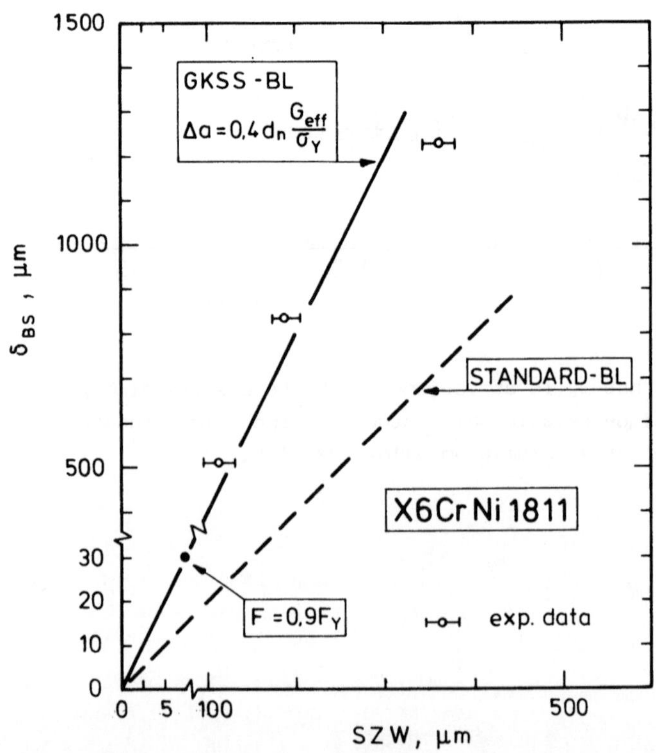


Fig. 9 a) The blunting line of the draft ASTM method for determining CTOD is not in accordance with stretch zone width measurements, austenitic steel X6CrNi 18 11.

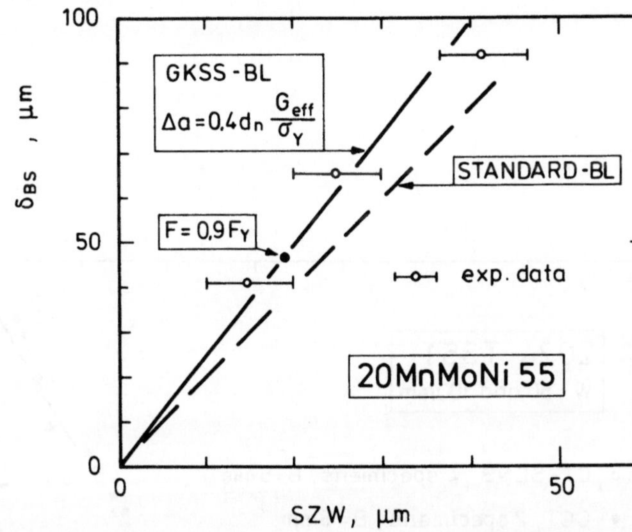


Fig. 9 b) same as (a), pressure vessel steel 20MnMoNi55,

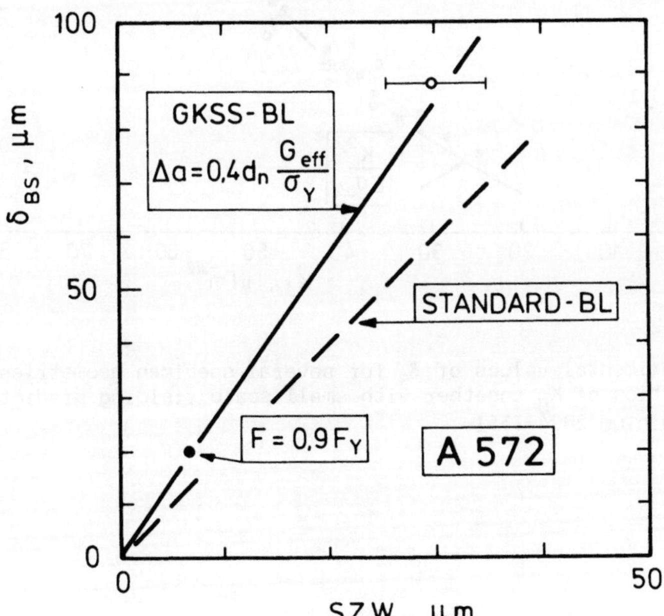


Fig. 9 c) same as (a), steel A572.

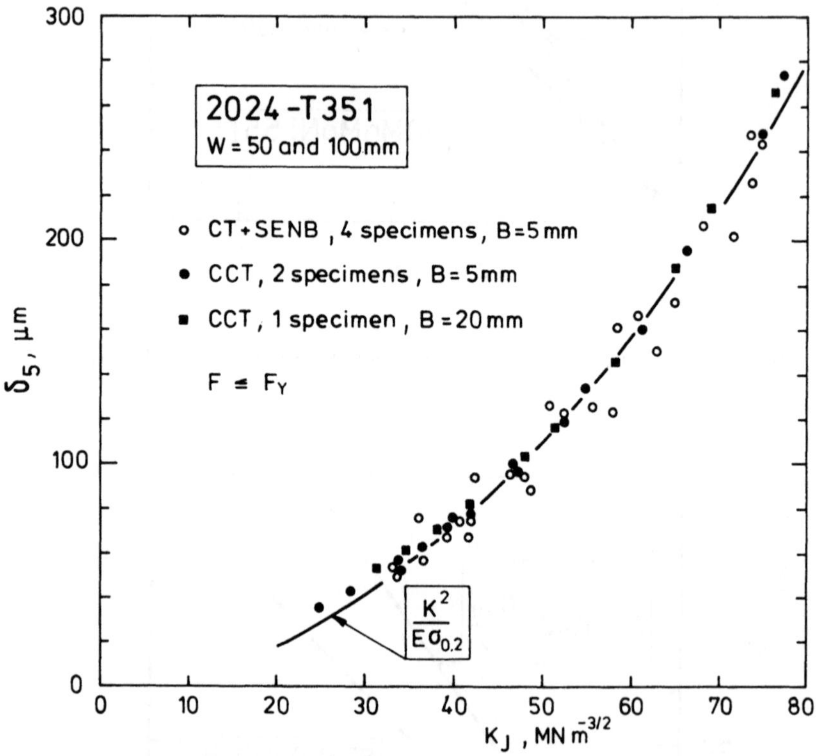


Fig. 10: Experimental values of $\delta_5$ for several specimen geometries as a function of $K_J$ together with small scale yielding prediction, aluminium 2024-T351.

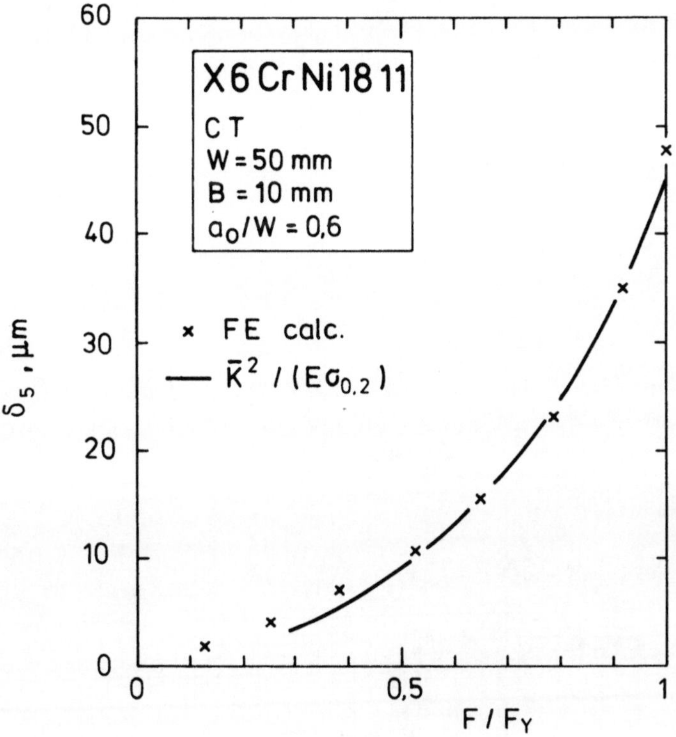


Fig. 11: Finite element values of $\delta_5$ for a CT specimen [21] together with small scale yielding prediction, austenitic steel X6CrNi 18 11.

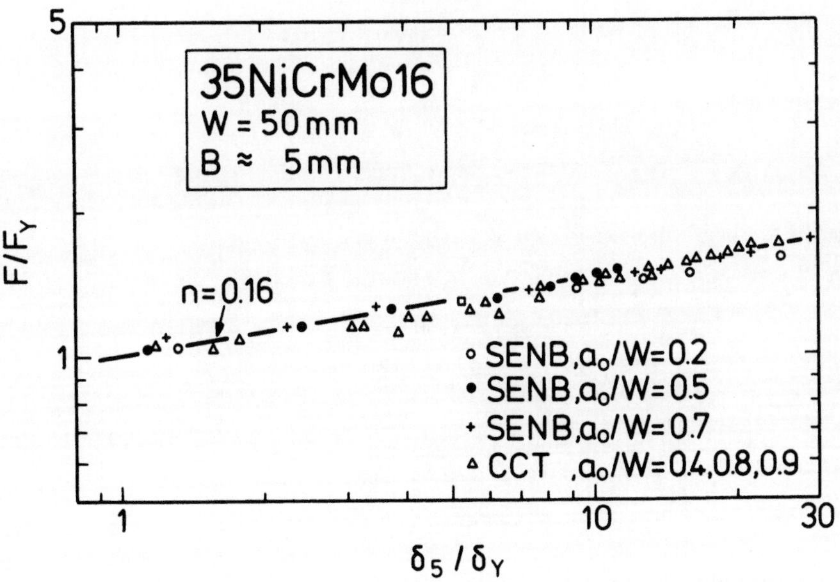


Fig. 12: Experimental values of $\delta_5$ as a function of the applied load, both normalised by their respective values at incipient net section yielding, steel 35NiCrMo16.

# CTOD and Crack Growth Including J-CTOD Relationship

# Evaluation of the Crack-Tip-Opening-Displacement ($V_R$) Resistance Curve Method

J. C. NEWMAN, JR.

NASA Langley Research Center
Hampton, Virginia 23665-5225

## Summary

Stable crack growth data were used to calculate $V_R$ curves for eleven material and thickness combinations. The materials were either aluminum alloys or steels. The specimens were either compact or middle-crack-tension specimens ranging in thickness from 0.5 to 25 mm. For some of the material and thickness combinations, both types of specimens were analyzed.

The $V_R$ curves were independent of crack length, specimen width, and specimen type but were dependent upon material and specimen thickness. The $V_R$ curves were also used to predict stable crack growth and maximum failure loads. Comparisons with experimental data showed that the $V_R$ curve concept accurately predicts stable crack growth and maximum failure loads on cracked metallic materials.

## INTRODUCTION

In damage-tolerant and structural-integrity analyses, the residual strength of a flawed component must be evaluated. The residual strength is the maximum load carrying capacity of the flawed structure. For brittle materials, linear-elastic fracture mechanics (LEFM) concepts, such as $K_{Ic}$ (plane-strain fracture toughness), are used. But for materials that exhibit large amounts of plasticity at the crack tip and stable crack growth prior to failure, LEFM concepts are not applicable. For these materials, methods which account for plasticity and stable crack growth should be used.

Dugdale's model [1] leads to a very simple analysis that simulates the effects of plasticity on plastic-zone size and crack-tip-opening displacement (CTOD) for thin materials. The model does not, however, accurately model yielding under plane-strain conditions. But the model is adopted herein because of its mathematical simplicity. The Dugdale model concept has been used with middle-crack tension (formally center-crack) specimens in several fracture analyses [2-4] for cracked metallic materials. These fracture analyses, which have had varying degrees of success, have lacked the versatility of the resistance-curve concepts in accounting for stable crack growth.

Resistance curve methods are used to characterize the resistance to fracture during slow-stable crack extension in metallic materials. Unlike brittle fracture, which is characterized by a single value of fracture toughness ($K_{IC}$), the resistance curve provides a toughness record as a crack is driven stably into the plastic zone caused by increasing applied loads. The resistance curve should be unique for the material and thickness of interest. It would be dependent upon temperature, environmental conditions, and strain rate. It should be independent, however, of crack length, specimen width, and specimen type for through-the-thickness cracks. The resistance curve can then be used to predict stable crack growth and instability (or maximum) load for cracked structural components.

The $V_R$ resistance curve method [5] is quite similar to the $K_R$ or $J_R$ resistance curve methods except that the "crack drive" is written in terms of crack-tip-opening displacement for a "stationary" crack instead of K or the J integral. Relationships between $V_R$, crack length, specimen width, and tensile properties have been derived from the Dugdale model [5]. The Dugdale model solutions are easily obtained from superposition of two elastic crack problems. Consequently, the $V_R$ curve method can be applied to any crack configuration for which these two elastic solutions have been obtained.

In the present paper, stable crack growth data (load against physical crack extension) were used to calculate the $V_R$ curves for eleven material and thickness combinations. The materials were either aluminum alloys or steels. The specimens were either compact or middle-crack-tension specimens ranging in thickness from 0.5 to 25 mm, see Fig 1 and Table 1. For some material and

Table 1. $V_R$ curve constants for materials, thicknesses, and specimen types analyzed.

| Material | Ref | B, mm | Specimen type[a] | $V_i$, m | C | n |
|---|---|---|---|---|---|---|
| 2024-T3 | 13 | 1.5 | MT | 7.21E-05 | 3.44E-03 | 0.61 |
| 2024-T3 | 12 | 2 | MT | 1.14E-06 | 6.08E-04 | 0.54 |
| 2024-T351 | 11 | 5 | CT, MT | 8.71E-06 | 4.48E-03 | 0.53 |
| 2024-T351 | 10 | 12.7 | CT, MT | 1.89E-05 | 2.21E-03 | 0.56 |
| 2024-T81 | 13 | 1.5 | MT | 0 | 3.89E-04 | 0.34 |
| 7075-T6 | 13 | 1.6 | MT | 3.52E-05 | 9.43E-05 | 0.23 |
| 7075-T6 | 12 | 2 | MT | 9.80E-07 | 4.15E-04 | 0.51 |
| 7075-T651 | 15 | 12.7 | CT, MT | 0 | 2.47E-04 | 0.45 |
| AM350CRT | 13 | 0.5 | MT | 6.47E-05 | 2.98E-03 | 0.54 |
| AM355CRT | 13 | 0.5 | MT | 0 | 1.69E-03 | 0.47 |
| HY-130 | 14 | 25.4 | CT | 0 | 0.028 | 0.71 |

[a] Compact tension (CT) and middle-crack tension (MT) specimen.

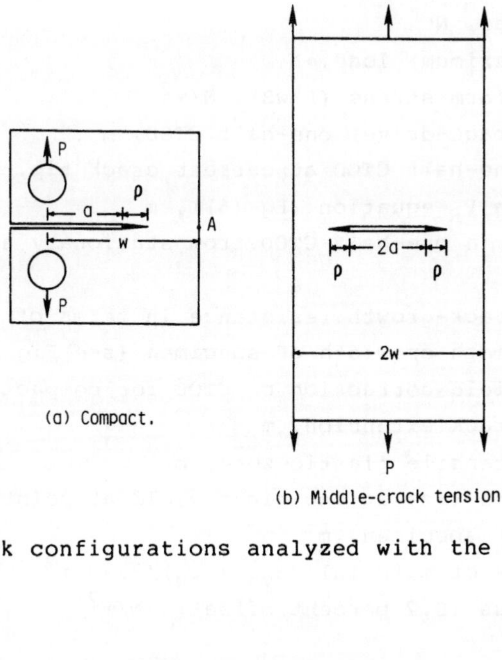


Fig.1. Crack configurations analyzed with the Dugdale model.

thickness combinations, both types of specimens were analyzed. For brevity, the $V_R$ curves for only some of the eleven material and thickness combinations analyzed are shown.

The analysis procedures used to predict stable crack growth and instability under load-control (or maximum load under displacement-control) of any through-the-thickness crack configuration made of the same material and thickness, and tested under the same environmental conditions, is presented. Comparisons are made between experimental and predicted load-against-physical crack extension.

NOMENCLATURE

| | |
|---|---|
| $a$ | crack length, m |
| $a_o$ | initial crack length, m |
| $B$ | specimen thickness, m |
| $C$ | constant in $V_R$ equation (Eq (5)) |
| $E$ | modulus of elasticity, $N/m^2$ |
| $J_R$ | material crack-growth resistance in terms of J, N/m |
| $K$ | stress-intensity factor, $N/m^{-3/2}$ |
| $K_R$ | material crack-growth resistance in terms of K, $N/m^{-3/2}$ |
| $n$ | constant in $V_R$ equation (Eq (5)) |
| $P$ | applied load, N |
| $P_f$ | failure (maximum) load, N |
| $S$ | remote uniform stress (P/wB), $N/m^2$ |
| $V_A$ | applied (crack-drive) one-half CTOD, m |
| $V_c$ | critical one-half CTOD at current crack tip, m |
| $V_i$ | constant in $V_R$ equation (Eq (5)), m |
| $V_{pw}$ | difference in one-half CTOD from stationary and growing crack, m |
| $V_R$ | material crack-growth resistance in terms of CTOD, m |
| $w$ | one-half width or width of specimen (see Fig 1), m |
| $Y$ | backface yield correction on CTOD for compact specimen |
| $\Delta a$ | physical crack extension, m |
| $\rho$ | length of tensile plastic zone, m |
| $\rho_o$ | plastic-zone size at incipient yield at point A (see Fig 1) for compact specimen, m |
| $\sigma_o$ | flow stress of material $(\sigma_{ys} + \sigma_u)/2$, $N/m^2$ |
| $\sigma_{ys}$ | yield stress (0.2 percent offset), $N/m^2$ |

$\sigma_u$ ultimate tensile strength, $N/m^2$
$\phi$ unit load function for CTOD due to applied loading
$\psi$ unit stress function for CTOD due to flow stress

## THE $V_R$-CURVE METHOD

Figure 2 illustrates the definition of $V_R$. A cracked plate with an initial crack length, $a_o$, is subjected to an applied load P which causes the crack to stably tear into the active plastic zone and leaves behind the residual plastic deformation zone. The critical crack-tip-opening displacement ($V_c$) has been shown to be constant during stable crack growth both experimentally [6] and analytically [7,8]. The amount of stable crack growth has been shown, analytically, to be controlled by the residual plastic deformation zone [8]. The "plastic wake" profile is defined as the difference in the crack surface displacements for a stationary crack and for a growing crack (shaded region). $V_{pw}$ is the plastic-wake displacement. The displacement $V_R$ is defined as the sum of $V_c + V_{pw}$. In the $V_R$ hypothesis, it is assumed that $V_R$ and $V_{pw}$ are uniquely related to the amount of crack extension, $\Delta a$. Thus, the $V_R$ curve is used to characterize the influence of the plastic wake on crack extension.

In Fig 2, note that $V_R$ is equal to the crack-tip-opening displacement for a "stationary" crack. For simplicity, the Dugdale model is used to calculate $V_R$, as shown in Fig 3.

Dugdel model solutions are easily obtained from superposition of two elastic crack problems - the remote loading P and the crack surface stress $\sigma_o$. Consequently, the $V_R$ curve method can be applied to any crack configuration for which these two elastic solutions have been obtained.

The compact and middle-crack tension specimens, being the most widely tested, were considered herein. Relationships between $V_R$, crack length, specimen width, and tensile properties have been derived from the Dugdale model. The following is a brief description of the procedures used to calculate $V_R$. Further details are given in Ref 5.

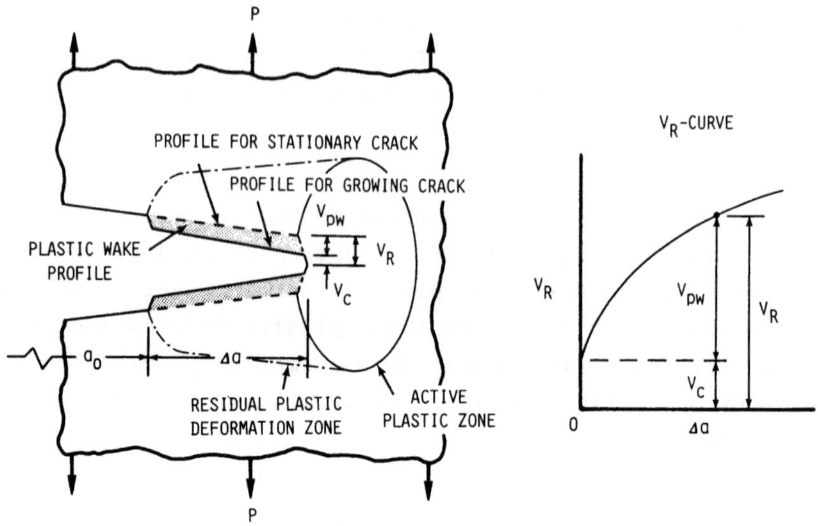


Fig.2. The $V_R$ resistance curve concept.

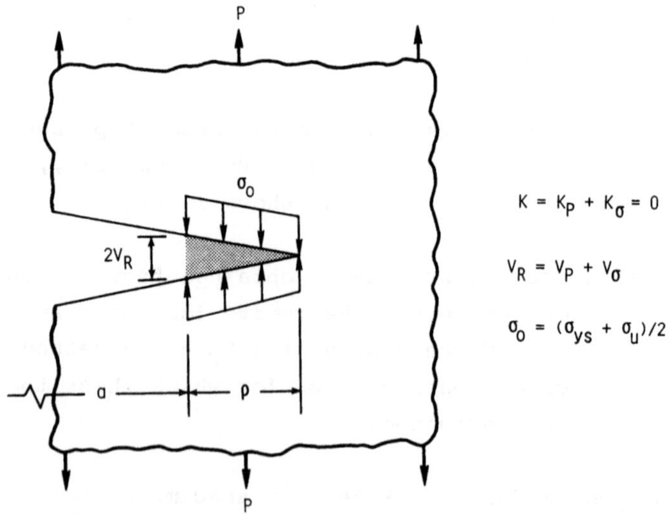


Fig.3. Crack-growth resistance, $V_R$, calculated from the Dugdale model.

Compact specimen.- From load-crack-extension ($P,\Delta a$) data calculate the plastic-zone size [9] as

$$\rho = \frac{\pi}{8}\left(\frac{K}{\sigma_o}\right)^2 F_o \qquad (1)$$

where K is the elastic stress-intensity factor at the current load and crack length, $\sigma_o$ is the average between yield stress and ultimate tensile strength, and the function $F_o$ is defined in Ref 5 or 9. Next, calculate $V_R$ as

$$V_R = P\phi(a,w,B,\rho) - \sigma_o\psi(a,w,B,\rho) \qquad (2)$$

for each P-$\Delta a$ data point. The functions $\phi$ and $\psi$ are defined in Ref 5.

In the compact specimen, the material at point A in Fig 1(a) is in compression. At a certain load, this material will yield in compression. The influence of backface yielding on $V_R$ has been accounted for by using an approximate equation developed from the results of a finite-element-strip-yield analysis [5]. The backface yield correction (Y), a multiplying factor in $\phi$ and $\psi$, is given by

$$Y = 1 + 0.25\left(\frac{\rho}{\rho_o} - 1\right)^4\left(\frac{w}{a}\right) \quad \text{for } \rho > \rho_o \qquad (3)$$

where

$$\frac{\rho_o}{w-a} = 0.48 - 0.84\,\frac{a}{w} + 0.56\left(\frac{a}{w}\right)^2 \qquad (4)$$

When $\rho$ is less than $\rho_o$, the material at point A is elastic and Y is equal to unity. The $\rho_o$ value is the crack-tip-plastic-zone size at incipient yielding at point A.

For convenience, fit the equation

$$V_R = V_i + C(\Delta a)^n \quad (5)$$

to the $V_R$ against $\Delta a$ data using a least-squares procedure described in Ref 5. (Although $V_i$ could be denoted as $V_c$, the determination of the displacement at initiation is difficult and the fitted constant $V_i$ may not necessarily be equal to $V_c$.)

Middle-crack tension specimen.- Using the load-crack-extension ($P,\Delta a$) data, calculate the plastic-zone size [5] as

$$\rho = a\left\{ \frac{2w}{\pi a} \sin^{-1}\left[\sin\left[\left(\frac{\pi a}{2w}\right) \sec\left(\frac{\pi S f}{2\sigma_o}\right)\right] - 1\right\}\right. \quad (6)$$

where $f = 1 + 0.22\left(\frac{a}{w}\right)^2$. Next, calculate $V_R$ as

$$V_R = P\phi(a,w,B,\rho) - \sigma_o\psi(a,w,B,\rho) \quad (7)$$

for each $P$-$\Delta a$ data point. The functions $\phi$ and $\psi$ are also defined in Ref 5. Again, fit the $V_R$ equation (Eq (5)) to the $V_R$ against $\Delta a$ data for the middle-crack tension specimens.

## ANALYSIS OF TEST DATA

Fracture data from the literature [10-15] for a number of aluminum alloys and steels were analyzed to determine the $V_R$ curve. For convenience, Eq (5) was fitted to the $V_R$ data using a best-fit procedure described in Ref 5. The fracture constants ($V_i$, C and n), thickness, and specimen type(s) for the materials analyzed are shown in Table 1. For brevity, the results from only some of the eleven material and thickness combinations analyzed are shown herein.

### Aluminum Alloys

McCabe and Ernst [10] conducted displacement-control fracture tests on a compact and middle-crack tension specimen made of 2024-T351 aluminum alloy. They determined the applied load against physical crack extension for each specimen using either the unloading compliance method or visual observation.

For each applied load and crack extension value (P,$\Delta$a) measured on the compact specimen, the crack-growth resistance, $V_R$, was calculated from Eq (2). The values of $V_R$ and corresponding $\Delta$a are plotted in Fig 4(a) for both compliance and visual measurements. Solid symbols denote crack extension values beyond maximum load. Up to maximum load, the visual and compliance values agreed well. Beyond maximum load, however, the results were in considerable disagreement. The reason for this behavior is not understood. Thus, at present, the determination of the $V_R$ data must be restricted to crack extension values up to maximum load. (Recently, some progress has been made in the treatment of $V_R$ data beyond maximum load, see Ref 16.) For convenience, Eq (5) was fitted to the $V_R$ data (open symbols) and the corresponding constants ($V_i$, C and n) are given on the figure.

Figure 4(b) shows a comparison between $V_R$ data calculated on the compact and on a middle-crack tension (MT) specimen. For the MT specimen, the $V_R$ data was calculated from Eq (7). Again, up to maximum load, the results are in very good agreement and indicate that the $V_R$ curves are independent of specimen type.

Hellman and Schwalbe [11] conducted fracture tests on different size compact and middle-crack tension specimens made of a thinner 2024-T351 aluminum alloy than that shown in Fig 4. Figure 5 shows the $V_R$ data calculated from their test results. The solid symbols show results that exceed various limits on the application of the $V_R$ equations (see Ref 5). The open symbols show that the $V_R$ curve is independent of specimen size and specimen type for this material. Again, Eq (5) was fitted to the valid (open symbol) data.

Schwalbe and Setz [12] conducted fracture tests on a wide range of different size middle-crack tension specimens made of a thin 7075-T6 aluminum alloy material. A wide range of initial crack sizes was also tested. The specimens were restrained against

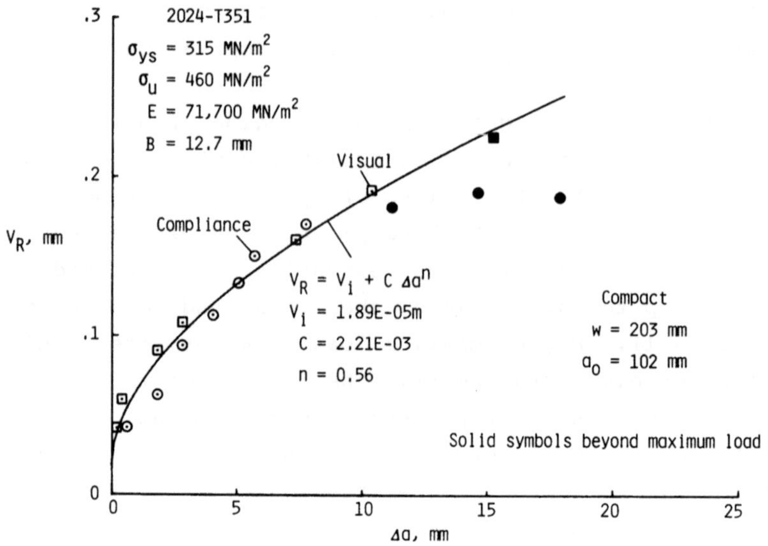


Fig.4(a). Crack-growth resistance curve for 2024-T351 aluminum alloy compact specimen [10].

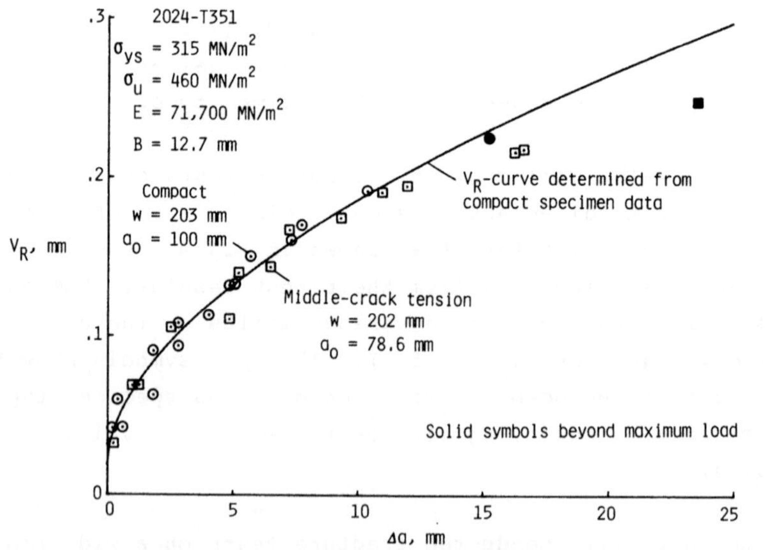


Fig.4(b). Crack growth resistance curve for 2024-T351 aluminum alloy compact and middle-crack specimens [10].

buckling. The $V_R$ data calculated from Eq (7) and shown in Fig 6 is nearly independent of crack length and specimen size. Two tests (A and B) have been identified as being the most discrepant. The significance of these particular test specimens will be discussed in the section on "Prediction of Stable Crack Growth and Instability". Again, the solid curve shows the $V_R$ equation fitted to these results.

## Steels

Forman [13] conducted fracture tests on middle-crack tension specimens made of very thin AM350CRT and AM355CRT stainless steel (see Table 1). The initial crack-length-to-width ($a_o/w$) ratio was 0.25 and the specimen width ranged from 76 to 305 mm. Again, these specimens were restrained against buckling. Only the results for AM355CRT are shown herein.

Figure 7 shows the $V_R$ data calculated from Eq (7) using the P-Δa data on AM355CRT. Again, the $V_R$ curve is independent of specimen size for this material. The $V_R$ curve, fitted to these data, is shown as the solid curve.

McCabe and Landes [14] conducted a fracture test on a large thick compact specimen made of HY-130 steel. The specimen was 25.4 mm thick and was 610 mm wide. Eq (2) was, again, used to calculate $V_R$ (see Fig 8). The $V_R$ equation fitted to these data will be used to predict stable crack growth and instability on other specimen sizes in the next section.

## PREDICTION OF STABLE CRACK GROWTH AND INSTABILITY

In the following section, the $V_R$ curves determined for some of the aluminum alloys and steels will be used to predict stable crack growth and instability. Comparisons will be made between experimental and calculated (or predicted) load-against-crack-extension (P-Δa) behavior. The "calculated" results mean that the test results were also used to obtain the $V_R$ curve, whereas "predicted" results mean that the test results were not used in determining the $V_R$ curve.

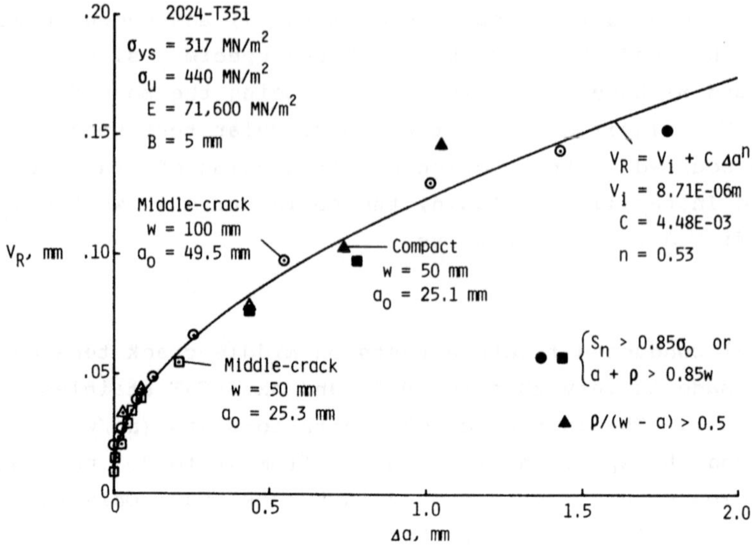


Fig.5. Crack-growth resistance curve for thin 2024-T351 aluminum alloy compact and middle-crack specimens [11].

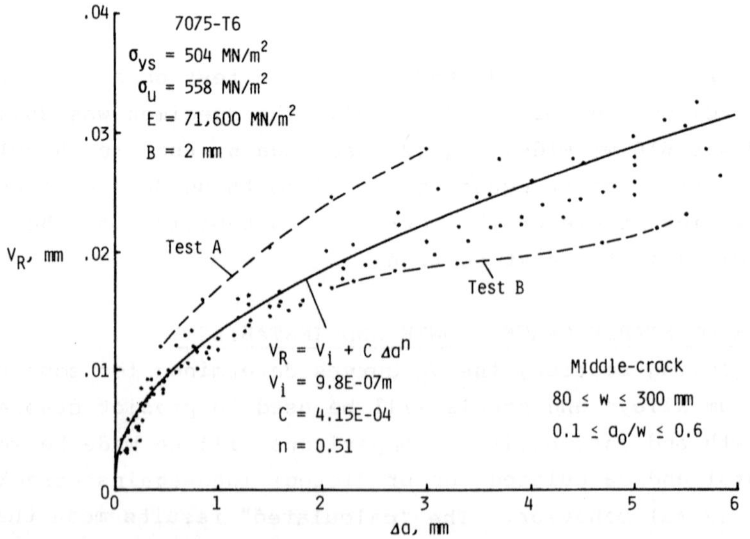


Fig.6. Crack-growth resistance curves for 7075-T6 aluminum alloy middle-crack sheet specimens [12].

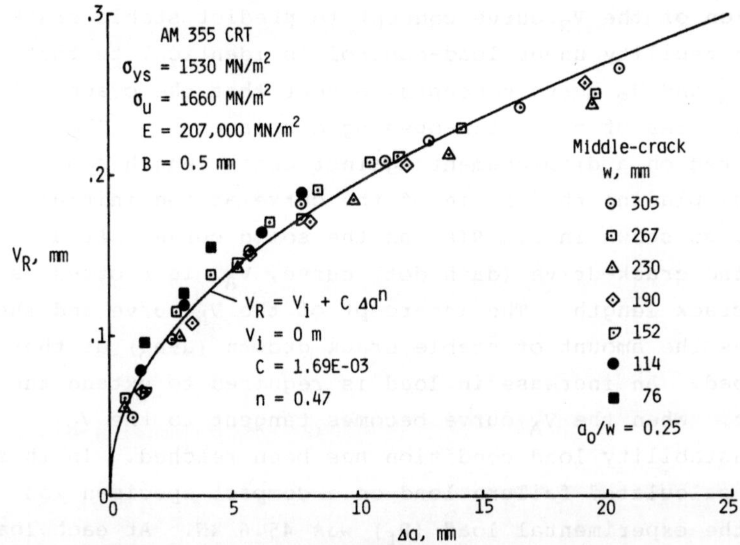


Fig.7. Crack-growth resistance curve for AM 355 CRT middle-crack sheet specimens [13].

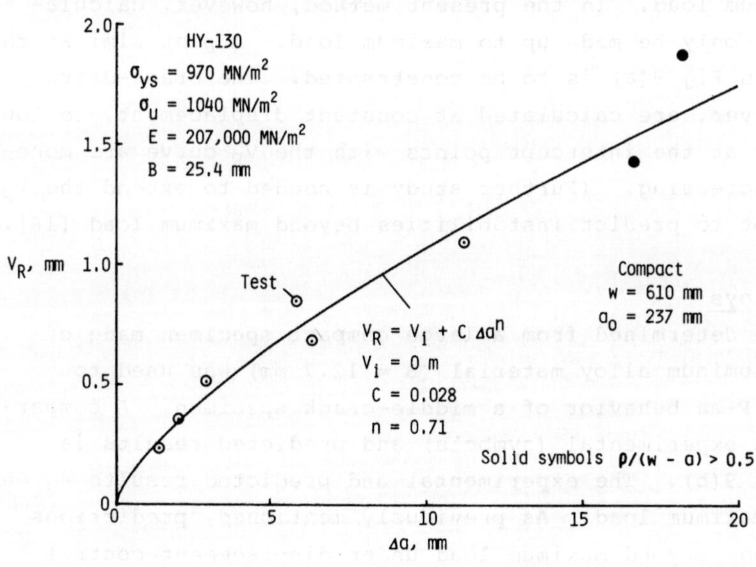


Fig.8. Crack-growth resistance curve for HY-130 steel compact compact specimen [14].

The application of the $V_R$ curve concept to predict stable crack growth and instability under load-control is identical to that used in the $K_R$ and $J_R$ curve concepts, except that the crack drive is written in terms of crack-tip-opening displacement. The $V_R$ curve is plotted on a displacement against crack length coordinate system by placing the origin of the curve at the initial crack length, as shown in Fig 9(a) as the solid curve. At a fixed load, the crack-drive (dash-dot) curve, $V_A$, is plotted as a function of crack length. The intercept of the $V_A$ curve and the $V_R$ curve gives the amount of stable crack growth ($a-a_o$) at that particular load. An increase in load is required to extend the crack further. When the $V_A$ curve becomes tangent to the $V_R$ curve, the instability load condition has been reached. In this example, the calculated failure load on a compact specimen was 45.3 kN and the experimental load ($P_f$) was 45.6 kN. At each load step, the various limits on the method or equations must be checked (see Ref 5).

For displacement-control conditions, crack instability may occur beyond maximum load. In the present method, however, calculations should only be made up to maximum load. A plot similar to that shown in Fig 9(a) is to be constructed. The crack-drive curves, however, are calculated at constant displacement, so long as the loads at the intercept points with the $V_R$ curve are monotonically increasing. (Further study is needed to extend the $V_R$ curve concept to predict instabilities beyond maximum load [16].)

Aluminum Alloys

The $V_R$ curve determined from a large compact specimen made of 2024-T351 aluminum alloy material (B = 12.7 mm) was used to predict the P-$\Delta a$ behavior of a middle-crack specimen. A comparison between experimental (symbols) and predicted results is shown in Fig 9(b). The experimental and predicted results agreed well up to maximum load. As previously mentioned, predictions cannot be made beyond maximum load under displacement-control conditions [16].

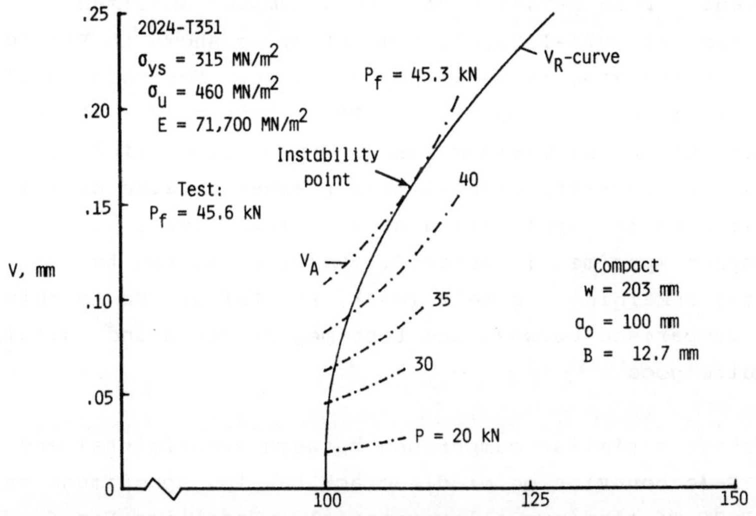


Fig.9(a). Application of $V_R$-curve concept to predict stable crack growth and failure load on a 2024-T351 aluminum alloy compact specimen [10].

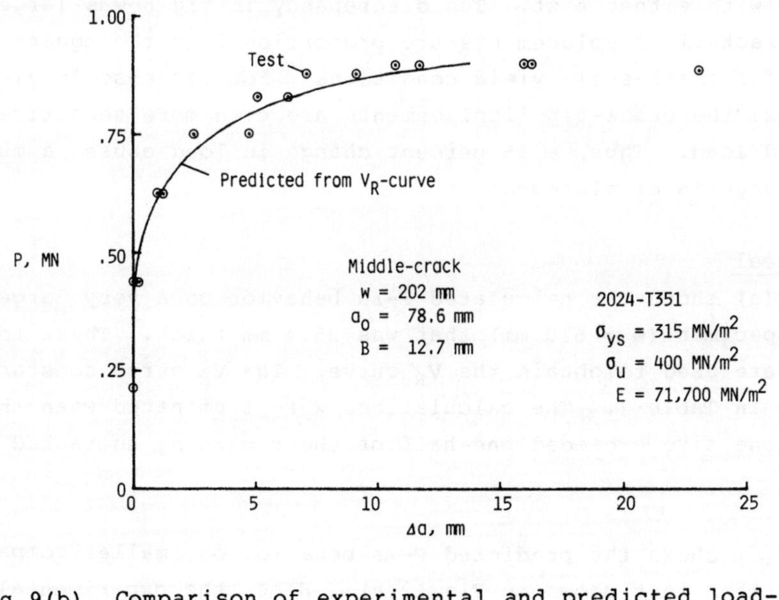


Fig.9(b). Comparison of experimental and predicted load-against-crack-extension on 2024-T351 aluminum alloy middle-crack specimen [10].

The experimental P-$\Delta a$ behavior of a thin compact specimen (B = 5 mm) made of 2024-T351 aluminum alloy is shown in Fig 10. The calculated behavior is shown as the curve. For this specimen, backface yielding at point A in Fig 1 begins (Y = 1) at about 75 percent of the maximum load. The backface yield correction, Y, on the crack-tip displacement reaches a value of 1.13 when the limit on the application of the displacement equations for the compact specimen is reached. Here, $\rho$ exceeds the limit of 0.5 of the remaining ligament (w-a), see Ref 5. Up to this point, the comparison between the test measurements and calculations is quite good.

Figure 11 shows a similar comparison between experimental and calculated P-$\Delta a$ behavior on middle-crack tension speciimens made of a thin 7075-T6 aluminum alloy material. Recalling Fig 6, there were two tests (A and B) which showed a large discrepancy from the mean $V_R$ curve. The calculated load using the mean $V_R$ curve is within ±15 percent of the test results. Thus, for a given amount of crack extension the calculated load is in fair agreement with either test. The discrepancy in Fig 6 was larger because crack-tip displacements are proportional to the square of the load for small-scale yield conditions. For large-scale yield conditions, the crack-tip displacements are even more sensitive to applied load. Thus, a 15 percent change in load causes a much larger change in displacement.

<u>HY-130 Steel</u>

Figure 12(a) shows the calculated P-$\Delta a$ behavior on a very large compact specimen (w = 610 mm) that was 25.4 mm thick. These test results were used to obtain the $V_R$ curve. The $V_R$ curve constants are given in Table 1. The calculations were terminated when the plastic-zone size exceeded one-half of the remaining uncracked ligament.

Figure 12(b) shows the predicted P-$\Delta a$ behavior on smaller compact specimens than that shown in Fig 12(a). Here, the experimental results show some erratic behavior but the predicted behavior is in general agreement with the experimental trends. Again, the predictions were terminated when $\rho$ exceeded 0.5 (w-a).

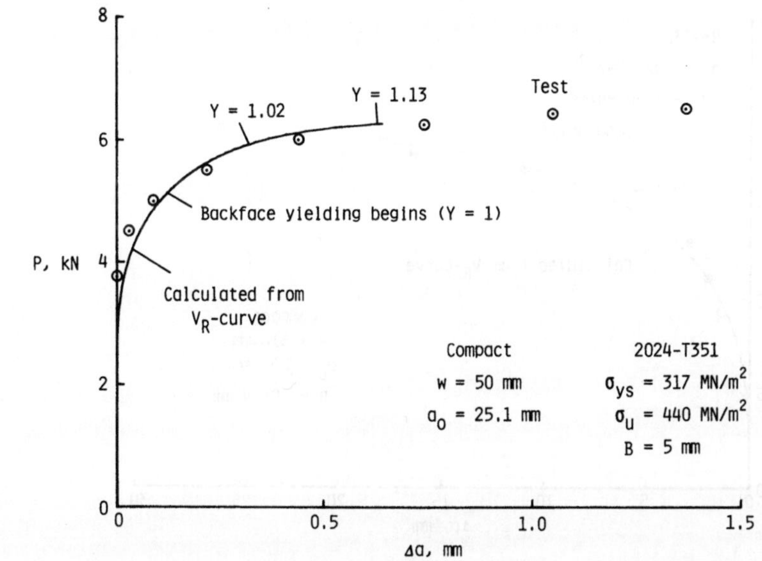


Fig.10. Comparison of experimental and calculated load-against-crack-extension on thin 2024-T351 aluminum alloy compact specimen [10].

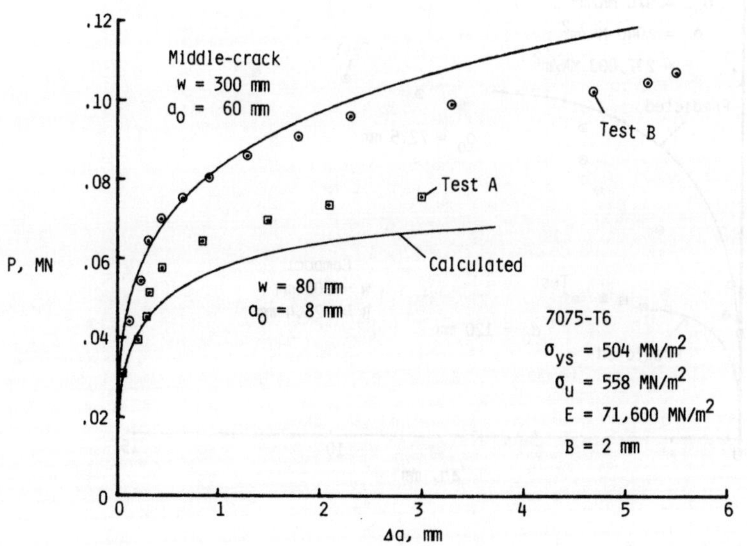


Fig.11. Comparison of experimental [12] and calculated load-against-crack extension for Test A and B shown in Figure 6.

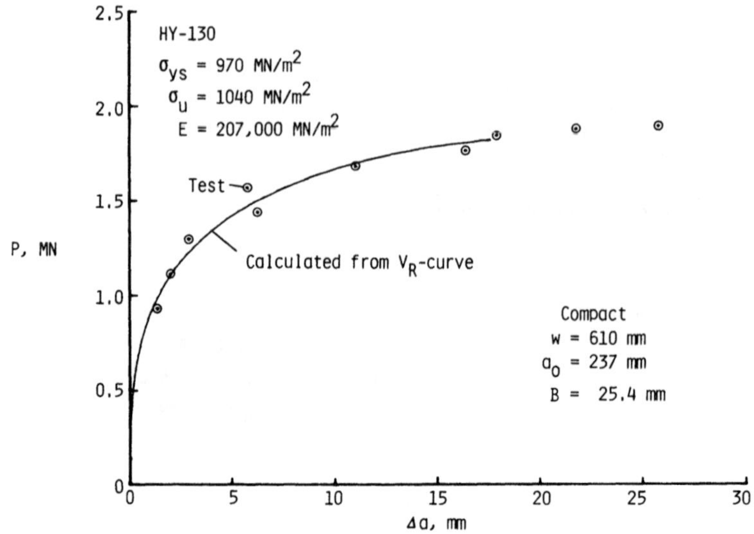


Fig.12(a). Comparison of experimental and calculated load-against-crack-extension on HY-130 steel compact specimen [14].

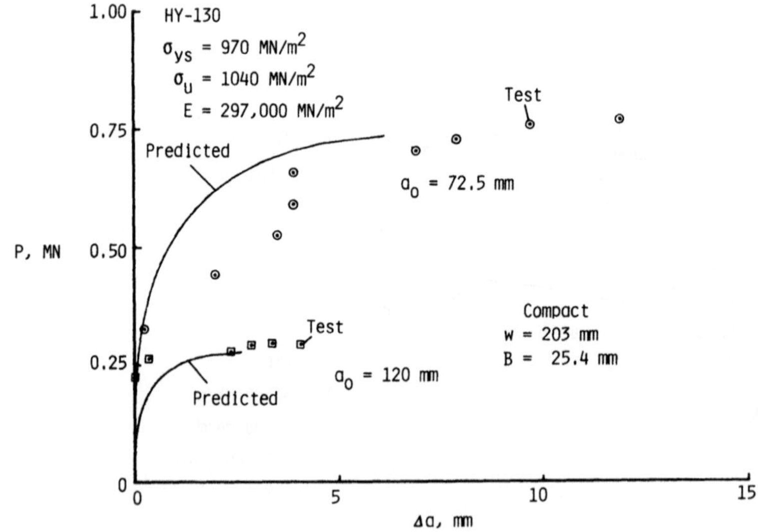


Fig.12(b). Comparison of experimental and predicted load-against-crack-extension on small HY-130 steel compact specimens for two different initial crack lengths [14].

## CONCLUSIONS

A resistance curve method based on crack-tip-opening displacement for a "stationary" crack, the $V_R$ curve method, was applied to eleven different materials and thickness combinations. The $V_R$ resistance curve concept is based on elastic stress-intensity factors and elastic crack-surface displacements (or the Dugdale model). The materials were either aluminum alloys or steels. And the specimens were either compact or middle-crack tension specimens ranging in thickness from 0.5 to 25 mm. For some material and thickness combinations, both types of specimens were analyzed. For each specimen type, a wide range in specimen widths and crack lengths was analyzed. The results presented in this paper support the following conclusions.

(1) The $V_R$ resistance curve is independent of crack length, specimen width, and specimen type up to maximum load.

(2) The $V_R$ resistance curve may be determined from load against crack extension data.

(3) The $V_R$ resistance curve concept can accurately predict stable crack growth and instability (or maximum load) of cracked metallic materials.

## References


1. Dugdale, D.S.: Yielding of steel sheets containing slits. Journal the Mechanics and Physics of Solids 8 (1960) 100-104.

2. Newman, J.C., Jr.: Fracture of cracked plates under plane stress. Engineering Fracture Mechanics 1, 1 (1968) 137-154.

3. Heald, P.T.; Spinks, G.M.; Worthington, P.J.: Post-yield fracture mechanics. Material Science and Engineering 10 (1972) 129-138.

4. Larsson, L.H.: Use of EPFM in design. Advances in Elasto-Plastic Fracture Mechanics, Applied Science Publishers, Ltd., Essex, England, 1979, 261-278.

5. Newman, J.C., Jr.: Prediction of stable crack growth and instability using the $V_R$ curve method. Elastic-Plastic Fracture Mechanics Technology, ASTM STP 896, J.C. Newman, Jr. and F.J. Loss (eds.) American Society for Testing and Materials (1985) 139-166.

6. Luxmoore, A.; Light, M.F.; Evans, W.T.: A comparison of energy release rates, the J-integral and crack tip displacements. Int. J. of Fracture 13 (1977) 257-259.

7. de Koning, A.U.: A contribution to the analysis of slow stable crack growth. The Netherlands, National Aerospace Laboratory Report NLR MP 75035 U, 1975.

8. Newman, J.C., Jr.: Finite-element analysis of initiation, stable crack growth, and instability using a crack-tip-opening displacement criterion. NASA TM-84564, National Aeronautics and Space Administration (1982).

9. Newman, J.C., Jr.; Mall, S.: Plastic zone size and CTOD equations for the compact specimen. International Journal of Fracture 24 (1984) R59-R63.

10. McCabe, D.E.; Ernst, H.A.: A perspective on R-curves and instability theory. Fracture Mechanics: Fourteenth Symposium - Vol. I: Theory and Analysis, ASTM STP 791, J.C. Lewis and G. Sines (eds.) American Society for Testing and Materials, (1983) I561-I584.

11. Hellman, D.; Schwalbe, K.H.: Geometry and size effects on $J_R$ and $\delta_R$ curves under plane stress conditions. Fracture Mechanics: Fifteenth Symposium, ASTM STP 833, R.J. Sanford (ed.) American Society for Testing and Materials (1984) 577-605.

12. Schwalbe, K.H.; Setz, W.: R-curve evaluation for center cracked panels. Fracture and Fatique, J.C. Radon (ed.) Pergamon Press, Proceedings Third Colloquium on Fracture, London, Sept. 8-10, 1980, 277-285.

13. Forman, R.G.: Experimental program to determine effect of crack buckling and specimen dimensions on fracture toughness of thin sheet materials. AFFDL-TR-65-146, January 1966.

14. McCabe, D.E.; Landes, J.D.: Elastic-plastic R-curves. Journal of Engineering Materials and Technology 100 (1978) 258-266.

15. Newman, J.C., Jr.: An evaluation of fracture analysis methods. Elastic-Plastic Fracture Mechanics Technology, ASTM STP 896, J.C. Newman, Jr. and F.J. Loss (eds.) American Society for Testing and Materials (1985) 5-96.

16. Newman, J.C., Jr.; McNeill, S.R.; Sutton, M.A.: Application of the $V_R$ resistance curve to fracture of various crack configurations. Presented at the Eighteenth National Symposium on Fracture Mechanics, Boulder, Colorado, June 25-27, 1985.

# CTOD and J Relationships for a Growing Crack

W. R. Andrews

Turbine Technology Laboratory, General Electric Company,
Building 55, Room 163, Schenectady, NY 12345.

ABSTRACT

Tests were performed on three steels, A533-B, A508 Class 2, and ASTM A240 Thype 304 stainless. These materials exhibit a wide range of properties; their strain-hardening exponents n are within the range of 8 to 12 while their flow stresses sigma-0 are within the range of 45 to 90 ksi.

A technique for measuring crack growth termed the displacement based key-curve method was used.

The development of this method used a model of a growing crack that incorporated a "virtual" non-growing crack profile at the current crack tip. This assumed crack-growth model, having the mathematical form of the Rice, Drugan and Sham model for a growing crack, also permitted the derivation of an expression for the crack-tip opening displacement required to evaluate the proposed relationship between J and delta-t.

Comparison of R curves developed using the displacement based key-curve crack-extension model and multiple specimen tests correlated closely validating both the key-curve method and the assumed constancy of an empirical parameter 1/rho that is interpreted as a component of the crack-opening angle. The displacement data and the J-integral values obtained from the tests were shown to be correlated by the proposed relationship (Eq. 1) over a wide range of large-scale plasticity. For smaller J values in the range of small scale plasticity, the relationship varied from that proposed.

INTRODUCTION AND BACKGROUND

The relationship between crack-tip opening displacement CTOD and the J integral J for a stationary crack tip is relatively well established and is summarized by Eq. 1 [1]:

$$\delta_t = d_n \frac{J}{\sigma_o} \tag{1}$$

where $d_n$ is a constant and delta-t is the CTOD.

The relationship for an advancing crack is less clearly defined. Recent works by Rice, Drugan and Sham [2] and Sham [3] have established the theoretical formulation for the deformation field around an advancing crack. Ernst [4] has introduced a J-like parameter $J_M$ that is consistent with the requirements of a growing crack. The resulting relationship for a growing crack is given by Eq. 2:

$$\frac{\delta_c}{r_m} = \frac{d_n}{\sigma_o} \cdot \frac{dJ}{da} + \frac{\beta' \sigma_o}{E} \ln\left(\frac{eR}{r_m}\right) \tag{2}$$

More recently, the author [5] has suggested a definition for CTOD at the tip of a growing crack that seems to be consistent with Equations 1 and 2 and with the concept of $J_M$.

The objective of this article is to present materials data demonstrating the relationship between J and CTOD for a growing crack.

MEASUREMENT OF CTOD

The method of measurement of CTOD and the method for determining crack growth used in this study was presented elsewhere [5]. In that presentation, the author demonstrated that the model used has the same form as does the Rice-Drugan-Sham model [2] shown here as Eq. 2.

RELATIONSHIP OF CTOD TO J

The following sections will develop the necessary relationships to relate measured displacements in a compact specimen to the J integral. This task will be broken into two parts. The first will deal with a stationary crack. The relationship developed is essential since it is inherent in the virtual non-growing crack assumption applied to a growing crack. Calculated values of delta-t are needed for correlation with J. The second part will deal with a growing crack.

Displacement Relationships for a Stationary Crack

A fracture test specimen may be viewed as two symmetrical solid blocks separated at the crack plane. The blocks rotate relative to each other about an axis located within the ligament and on the crack plane. This simple concept, assuming all displacements are rotational about the axis, defines the crack-tip opening displacement, Delta-T. Delta-T is not generally the same value as the fracture criterion delta-t, defined in Fig. 1 because of the large strains near the crack tip. (The relationship between delta-t and delta-T is shown in Fig. 2.)

To calculate delta-t, having measured delta-T, depends upon assuming the displacements of the crack faces remote from the tip are simply rotations and blend smoothly into the nonlinear displacements near the tip. This concept leads to two conditions involving the crack profile (Fig. 2) that must be satisfied:

1. Continuity of the crack faces at distances r from the tip.

2. Matching slopes between the linear far-field crack faces and the nonlinear, near-tip crack faces at their conceptual juncture located at r*.

These two conditions apply to the near-tip opening profile of the crack tip defined by Shih [1]:

$$\delta_{(r)} = \frac{2\alpha\sigma_0}{E}\left[\frac{E}{\alpha\sigma_0^2}\cdot\frac{J}{I_n}\right]^{\frac{n}{n+1}} * r^{\frac{1}{n+1}} * \delta' \qquad (3)$$

where:

alpha = a coefficient of the Ramberg-Osgood stress-strain

relationship.

delta-prime and $I_n$ = integration constants.

The two conditions also apply to the far-field displacements and are described by:

$$\delta(r) = V_L\left[1 - \frac{a_o - r}{X_o}\right] \tag{4}$$

at a point where the near and far fields join at the distance r* from the crack-tip. The crack opening given by Eq. 4 defines a center of rotation, or hinge point, at a distance ($X_o$ minus $a_o$) behind the crack tip. That hinge point may be determined by test.

The simultaneous solution of Eq. 3 and Eq. 4 shows that in general:

$$r^* = \frac{X_o - a_o}{n} \tag{5}$$

To derive the relationship between delta-t and the measurable delta-T, we note that:

$$\delta_T = \delta(r^*) = r^* \left.\frac{d\delta(r)}{da}\right|_{r^*} \tag{6}$$

Substituting delta-t defined by Fig. 1

$$\delta_t = \frac{2\alpha\sigma_o}{E}\left[\frac{E}{\alpha\sigma_o^2}\cdot\frac{J}{I_n}\right]^{\frac{n}{n+1}}\left(\frac{\delta_t}{2}\right)^{\frac{1}{n+1}} * \delta' \tag{7}$$

taking the ratio of Eq. 7 to Eq. 6:

$$\frac{\delta_t}{\delta_T} = \left(\frac{\delta_t}{2r^*}\right)^{\frac{1}{n+1}} * \frac{n+1}{n} \tag{8}$$

Rearranging and substituting in Eq. 5 yields:

$$\delta_t = \delta_T^{\frac{n+1}{n}} \left[\frac{n}{2(X_o-a_o)}\right]^{\frac{1}{n}} \left(\frac{n+1}{n}\right)^{\frac{n+1}{n}} \tag{9}$$

Displacement Relationships for a Growing Crack

In this section, the definition of delta-t for a stationary crack Eq. 1 is generaliized to that for a growing crack.

Shih [1] and Rice et. al. [2] have suggested that the crack-opening angle measured near the current crack tip is a viable criterion for continued crack growth. This angle is represented by an opening delta-c at a distance $r_m$ from the current tip (Fig. 3). The opening angle is approximated by Eq. 2. Dean and Hutchenson [8] suggested that for small scale yielding:

$$R = C\left(\frac{K_I}{\sigma_o}\right)^2 = \frac{CEJ}{\sigma_o^2} \tag{10}$$

where C is a constant which establishes the extent of the zone of intense plasticity.

They found that the values of C were approximated by the constant value of 0.28 for a variety of finite element determinations using materials of differing n values. The results of another finite element study by Sham [3] indicated the values range from 0.113 to 0.133 for the crack-growth histories studied.

Additional information about a growing crack can be gathered by examining the general relation for a crack opening profile [1]:

$$\delta = C_m J^{\frac{n}{n+1}} r^{\frac{1}{n+1}} \tag{11}$$

and assigning:

$$r_m = \frac{\delta_t}{2} \tag{12}$$

It has been shown [19] that differentiation of Eq. 11 and substitution of Eq. 12 results in:

$$\frac{d\delta_t}{da} = \frac{d_n}{\sigma_o} \cdot \frac{dJ}{da} \tag{13}$$

Substitution for J in Eq. 2 and using Eq. 11 results in the relationship:

$$\frac{2\delta_o}{\delta_t} = \frac{d\delta_T}{da} + \frac{\beta'\sigma_o}{E} \ln\left(\frac{2eCE}{d_n\sigma_o}\right) \tag{14}$$

Relating Eq. 14 to Fig. 3 gives a graphic visualization of the proposed definitions for a growing crack. The crack-tip opening displacement is defined for the growing crack using a virtual profile indicated by the dashed lines. Remotely measured opening displacements at the load line $V_L$ and the critual near-tip profile define the parameter delta-t the crack-tip opening displacement for a growing crack. The virtual profile is the same as that of a static crack of the same length and opening $V_L$. The local angle delta-c/$r_m$ (see Eq. 2) is more precisely defined by Eq. 13 as 2*delta-c/delta-t and differs from d(delta-T)/da by a nearly constant value, the second right-side term. The second right-hand term depends strongly on material properties and weakly on geometry.

## MATERIALS

The materials used were A533-B steel, A508 Class-2 steel, and ASTM A-240 Type 304 stainless steel. The compositions are shown in Table 1. The tensile properties are shown in Table 2. The flow stresses sigma-0, and the strain hardening exponents, n, obtained from analysis of the tensile data are also listed in Table 2.

The A508 Class 2 steel was obtained from the Westinghouse Research Laboratory and came from the center-thickness portions of the TAMPA tube-sheet forging used for EPRI Project RP1238-2 [7].

The 304 stainless steel was purchased through a local distributor.

J-INTEGRAL CALCULATION

The J-integral values used here were calculated from the load and the load-line displacements recorded during each test. The increamental formulation of ASTM E813-81, "Standard Test Method for $J_{Ic}$, a Measure of Fracture Toughness" was used. This formulation approximates the deformation theory value corrected for a growing crack:

$$J_{i+1} = \left[J_i + \left(\frac{f(\frac{a}{W})}{c}\right)_i \cdot \frac{A_{i,i+1}}{B_N}\right] \cdot \left[1 - \left(\frac{Y}{c}\right)_i \left\{(a_p)_{i+1} - (a_p)_i\right\}\right] \quad (15)$$

where:

$Y = 1 + 0.76(c_i/W)$

A = Area under the load, load-line displacement record.

$B_N$ = Specimen net thickness.

c = Uncracked ligament, W-a.

The subscripts i and i+1 indicate functions evaluated in the current (i+1) calculational step or the preceding (i) step. The term $A_{i,i+1}$, refers to the area bounded by the actual test record trace and lines of constant displacement, $(V_L)_i$, and $(V_L)_{i+1}$.

The values of $J_M$ have not been calculated. This procedure will be accomplished in the future.

EXPERIMENTAL RESULTS

Figure 4 presents the data for A533-B steel at room temperature using 2T compact specimens of varied crack length. Of the three results, specimen U21 and U5 form one R curve, while specimen U21 forms a second. The R curves formed by specimens U1B1 and U5 were further verified by heat-tint results obtained using additional specimens. The cause of the separate R curve resulting for specimen U21 is uncertain however, the correspondence between R curves for specimens U1B1 and U5, as well as all the remaining heat-tint results, suggest it is erroneous and should be discounted.

Figure 5 shows the R-curve data obtained for the A508, Class 2, steel. The open points are displacement based key-curve crack-extension values while the filled-in data are heat-tint crack-length data from the present investigation. The curve shown represents the data reported by Landes, et.al.[7].

Figure 6 shows the R curve obtained for the 304 stainless steel using 2T specimens. The open circles and the small dots are key-curve crack-length results, while the large, filled-in circles are heat-tint results.

## DISCUSSION

The objective of these tests was to examine the proposed relationship between J and delta-t. The data obtained will be presented to show the level of agreement with Eq. 1. assuming the model of Fig. 3. This will be done in two steps. First it will be shown that the displacement based key-curve method of crack growth calculation is valid for constant 1/rho. Second, the crack-opening displacement data delta-t and the J-integral obtained independently of one another from the current tests will be cast in a form which permits assessment of the validity of Eq. 1

### Evaluation of the Crack Growth Model

To determine the validity of the displacement based key-curve crack growth model, single-specimen R curves were determined [5] and compared to multispecimen R curves. These comparisons are presented in Figures 4, 5, and 6.

Comparison of the data presented in Figures 4 through 6 shows good agreement between the methods. Moreover, the author has obtained equally good agreement in other unpublished work performed by the General Electric Company for materials ranging from 2024T351 aluminum to CrMoV steel alloy. In light of the demonstrated agreement, it can be concluded that the displacement based key-curve model for crack extension [5] is valid for constant 1/rho.

Evaluation of the J-delta-t Relationship

Eq 1 provides the proposed relationship between J and delta-t. To assess the validity of this expression, it is necessary to recast the current experimental results in the form:

$$d_n = \frac{\delta_t \sigma_0}{J} \tag{16}$$

As suggested by Shih [1], $d_n$ should be constant and exhibit values ranging from 0.4 to 0.8 for common structural steels. Calculated J and delta-t values were obtained and the results were plotted as functions of $J_I$/(sigma-0 * c), where c is the uncracked ligament, in Fig. 7.

From the results shown in Fig. 7, it is evident that $d_n$ is essentially constant as proposed by Shih [1] for 0.01<=J/(sigma-0*c)<=0.04. In fact, $d_n$ may be considered to be constant up to J/(sigma-0*c) equals 0.1 based on the results in Fig. 7 however, $d_n$ was observed to vary for smaller values of J. This deviation may be due to inherent difficulties in measuring small displacements, but the trend is consistent among the specimens and the materials tested. The deviation is toward higher values of delta-t and $d_n$ for lower values of J. Further investigation is needed to resolve the relationship for small-scale yielding.

Two complimentary theories exist which relate J and delta-t. These are represented by Equations 1 and 2. The theory employed in the current work which results in Eq 16 is based on deformation theory which excludes crack growth however, as noted above, the relationship was shown to be almost constant over a wide range of crack growth. The use of a "virtual" crack-tip to define the crack-tip-opening displacement at the current crack length provided the means necessary to extend the applicability of Eq. 16 to a growing crack. The complimentary theory for a growing crack was provided by Rice, et. al. [2] (Eq 2). The author [5] has suggested that the displacement based key-curve relationship has the same form as Eq 2 so consequently Eq 17 must follow:

$$\frac{1}{\rho} = \frac{\beta' \sigma_o}{E} \ln\left(\frac{2eCE}{d_n \sigma_o}\right) \tag{17}$$

Examination of the data obtained in the current work (Fig 8) shows material dependence different from that indicated in Eq 17. In the Figure, the right-hand term of Eq 17 is shown to increase with sigma-O/E while 1/rho decreases with sigma-O/E. Possible explanations exist. First, the definition for delta-t in this study differs from that of Rice, Drugan and Sham [2]. Second, this study deals with large-scale plasticity whereas Rice et.al [2] were concerned with small-scale plasticity.

These results show that the $J_I$ related term d-delta-t/da of Eq 14 becomes truly dominant only for the stainless steel. In the case of the stainless steel, the crack opening angle is very large for the early stages of crack growth. The crack opening angles for the other materials were much smaller and the dominance of the d-delta-T/da term fades for the A508, Class 2, steel. The degree of J dominance may be deduced from Fig 8. For most ductile steels, the parameter d-delta-T/da is much larger than 1/rho. Since d-delta-T is a measure of $J_I$ and 1/p is not, J dominance is implied.

For less ductile steels, d-delta-T/da becomes smaller relative to 1/rho, implying a reduced dominance of $J_I$. This observation is in accord with Shih's analysis [1] in which the right-hand term of Eq 17 represented the J-independent term. As a consequence of these observations, it appears that further experimental work is needed to assess the relative dominance of J in continuing fracture. The author is encouraged by the in introduction of the parameter $J_M$ [4] since it has properties that are consistent with Eq 2 while the displacement model [5] developed independently has the form of Eq 2.

## CONCLUSIONS

1. The relationship between J and delta-t was experimentally determined for growing cracks over a range from small-scale plasticity to full scale plasticity for three materials having different strain-hardening and strength characteristics. J and

delta-t were observed to be related by a constant coefficient $d_n$ for a wide range of the large-scale plasticity region, thus confirming the relationship proposed by Shih [1]. For small-scale yielding however, the coefficient varies.

2. An alternative definition was proposed for the crack-tip opening displacement delta-t for a growing crack as Eq 9. This definition can be characterized as a virtual non-growing crack profile at the current crack tip (Fig 3).

3. These results suggest that the $J_I$ model is the dominating describer of the growing crack-tip field for very ductile steels but $J_I$ becomes less important for less ductile steels.

4. The correspondence in form between the stably growing crack model of Rice, Drugan and Sham (Eq 2) and the displacement based key-curve model lends support to the concept of $J_M$ [4] suggested by Ernst.

*Acknowlegements*

The author is grateful to those who have supported this work with their encouragement and critical review, especially R. Jones and D. Norris of EPRI, D. F. Mowbray and A. F. Anderson of GE Co.., and C. F. Shih of Brown University. The financial support came primarily from Electric Power Research Institute under Contract RP1237-8.

REFERENCES

1.) Shih, C. F., Journal of Mechanics and Physics of Solids, Vol. 29, No.4, 1981, pp.305-326.

2.) Rice, J. R., Drugan, W. J., and Sham, T-L. in Fracture Mechanics, Proceedings of the Twelfth National Symposium on Fracture Mechanics, STP 700, American Society for Testing and Materials, Philadelphia, 1980, pp. 189-221.

3.) Sham, T-L., "A Finite Element Study of the Asymptotic Near-Tip Fields for Mode I Plane-Strain Cracks Growing Stably in Elastic-Ideally Plastic Solids," in Elastic-Plastic Fracture: Second Symposium, Volume I-Inelasitc Crack Analysis, ASTM STP 803, C. F. Shih and J. P. Gudas, Eds., ASTM, 1983, ppI-52-I-79.

4.) Ernst, H. A., "Material Resistance and Instability Beyond J-Controlled Crack Growth," Elastic-Plastic Fracture: Second Symposium, Volume I - Inelastic Crack Analysis, ASTM STP 803, C. F. Shih and J. P. Gudas, eds., American Society for Testing and Materials, pp I-191-I-213.

5.) Andrews, Wayne R., "Determining Crack Extension Using a Displacement Based Key-Curve Method." Elastic-Plastic Fracture Test Methods: The User's Experience, ASTM STP 856, E. T. Wessel and F. J. Loss, Eds, American Society for Testing and Materials, 1984, pp 308-321.

6.) Andrews, W. R. and Shih, C. F., in Elastic-Plastic Fracture, ASTM STP 668, eds. J. D. Landes, J. A. Begley, and G. A. Clarke,

American Society for Testing and Materials, 1979, pp426-450.

7.) Landes, J. D., McCabe, D.E., Ernst, H. A., Wessel, E. T. and Pryle, W. H., "Elastic-Plastic Methodology to Establish R-Curves and Instability Criteria," Sixth Semiannual Report - Contract RP1238-2, January 1, 1982 to June 30, 1982, Electric Power Research Institute, August 4, 1982.

8.) Dean, R. H., and Hutchenson, J. W., "Quasi-Static Steady Crack Growth in Small Scale Yielding," in Fracture Mechanics, ASTM STP 700, American Society for Testing and Materials, 1980, pp 383-405.

9.) Kumar, V., German, M.D., Wilkining, W.W., Andrews, W.R., deLorenzi, H.G., and Mowbray, D.F. "Advances in Elastic-Plastic Fracture Analysis" EPRI NP 3607, Project 1237-1, Final Report, August, 1984.

## ADDENDUM - Clarification

The author, upon submitting the manuscript, has realized that the interpretation used may be misleading. A second interpretation is more satisfying and probably more correct.

Specifically, the interpretation of the basic DBKC formula in light of the differential form of the Rice-Drugan-Sham model [2] leads to the conclusions illustrated in Fig. 8 which are misleading. The more satisfying interpretation is by means of the integrated form of the RDS model given by Eq. 18 [2]:

$$= \beta \frac{\sigma_o}{E} \dot{a} \ln\left(\frac{\bar{R}}{r}\right) + \frac{\alpha \dot{J}}{\sigma_o} + \mu \dot{a} \tag{18}$$

Using this model, it is straight foreward to derive the DBKC model using the simplest of algebra. Making use of Eq 1 and Eq 18 leads to the DBKC model expressed in the terminology use by Rice et. al.

$$\dot{a} = \frac{\dot{\delta} - \alpha \dot{J}/\sigma_o}{\beta(\sigma_o/E)\ln(\bar{R}/r) + \mu} \tag{19}$$

The interpretation of the opening angles must be different than presented in the body of the paper to maintain consistancy with Eq 18.

Taking note of Eq 19, it is apparent that by this interpretation, the opening delta-dot is the opening at the original crack tip. The second term in the numerator is equivilaent to delta-T calculated at the original crack tip under the assumption of no crack growth. The adjustible parameter 1/rho is relatable to the first term on the right side of Eq 19. Finally, the parameter $\mu$ is the rate of change of delta-T with respect to crack growth measured at the original crack tip. With regard to $\mu$, it is accurate for small crack extensions only in as much that it serves to project the calculation of delta-T to the current crack tip.

The points made in the preceding paragraph are summarized by Equations 20-23.

$$\dot{\delta} = \delta_o \tag{20}$$

$$\frac{\alpha \dot{J}}{\sigma_o} \cong \frac{d_\infty J}{\sigma_o} = \delta_T \tag{21}$$

$$\frac{1}{\rho} \cong \beta \frac{\sigma_o}{E} \ln\left(\frac{\bar{R}}{r}\right) \tag{22}$$

$$\mu = \frac{d\,\delta_T}{d a} \tag{23}$$

Table 1

MATERIAL COMPOSITION
WEIGHT PERCENT

| C | Mn | P | S | Cu | Si | Ni | Cr | Mo | Al | Co | N | Sn | V |
|---|---|---|---|---|---|---|---|---|---|---|---|---|---|
| | | | | | | A533-B | | | | | | | |
| .20 | 1.22 | 0.01 | .016 | | .15 | .65 | | .54 | | | | | |
| | | | | | | A508-2 | | | | | | | |
| .20 | .73 | 0.10 | .008 | .004 | .03 | .73 | .33 | .60 | .006 | | | .001 | .3 |
| | | | | Type 304 Stainless Steel | | | | | | | | | |
| .03 | 1.62 | .034 | .015 | .31 | .43 | 10.04 | 18.3 | .31 | | .15 | .07 | | |

Table 2

MATERIALS
STRESS STRAIN PROPERTIES

| | Ramberg-Osgood Parameters(1,2) | | | | |
|---|---|---|---|---|---|
| Specification | $\sigma_o$ ksi | $\varepsilon_0$ | n | E ksi | $\alpha$ |
| A533-B | 60 | 0.002 | 9.71 | 30,000 | 1.12 |
| A508 Cl. 2 Steel | 87 | 0.0029 | 11.41 | 30,000 | 1.04 |
| A204/304 Stainless | 24.5 | 0.00079 | 8.04 | 31,000 | 1.03 |

(1) $\varepsilon/\varepsilon_0 = \sigma/\sigma_o + \alpha(\sigma/\sigma_o)^n$

$\varepsilon$ = Engineering Strain

$\sigma$ = Engineering Stress

(2) Data taken at strains less than 30%.

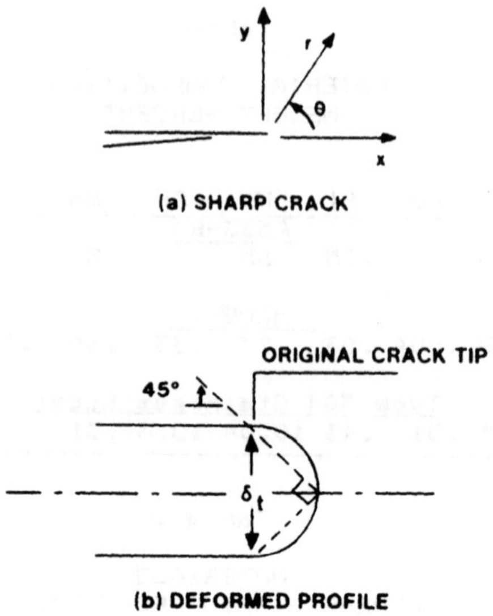


Figure 1. Sharp crack and deformed crack profiles illustrating the 45-deg construction for defining crack-tip opening displacement.

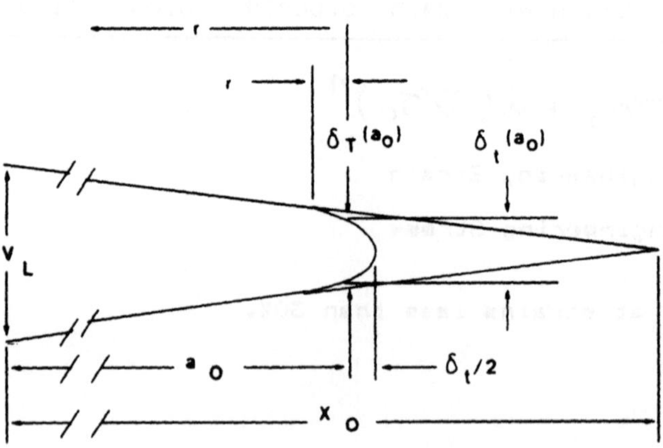


Figure 2. Definition of terms relating to a crack tip.

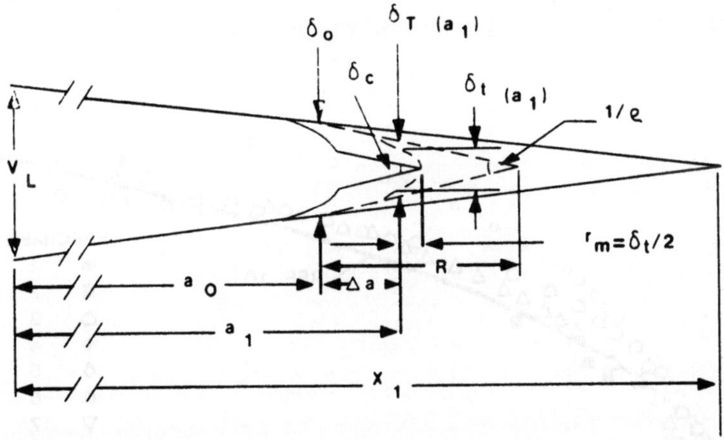


Figure 3. Schematic profile of a growing crack.

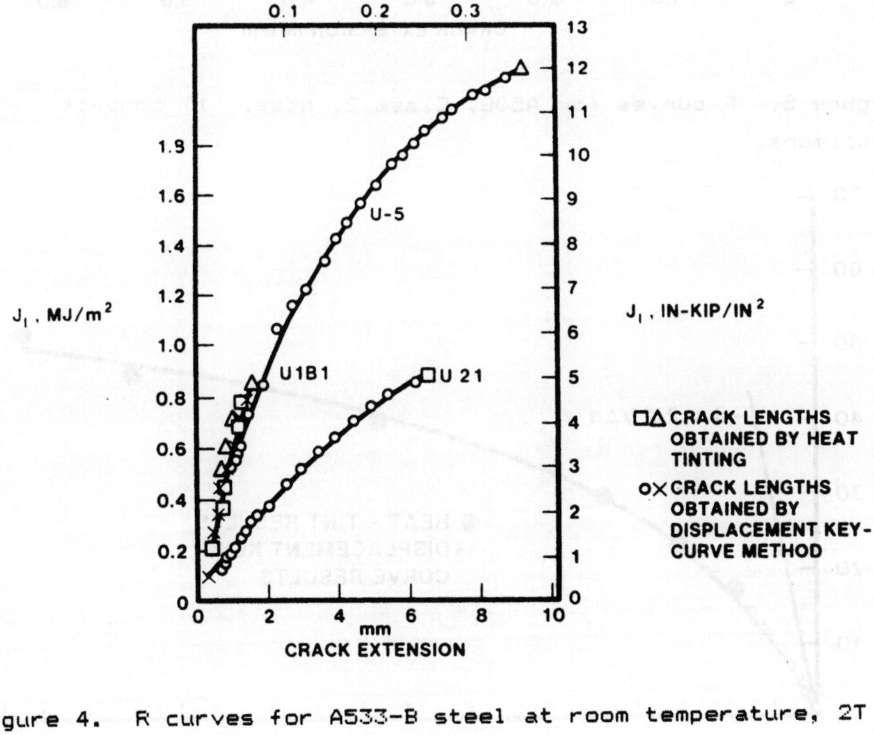


Figure 4. R curves for A533-B steel at room temperature, 2T compact specimens.

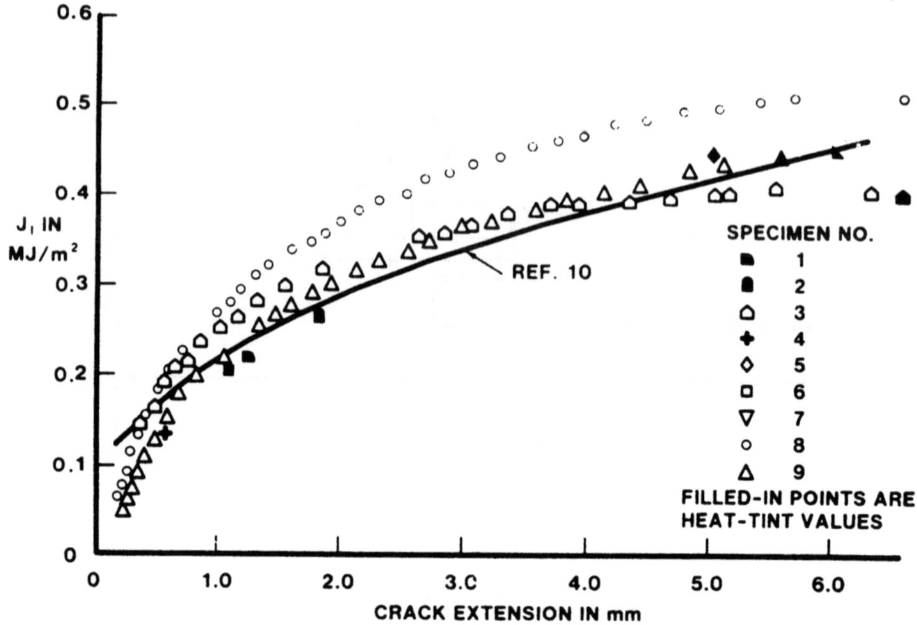


Figure 5. R curves for A508, Class 2, steel, 1T compact specimens.

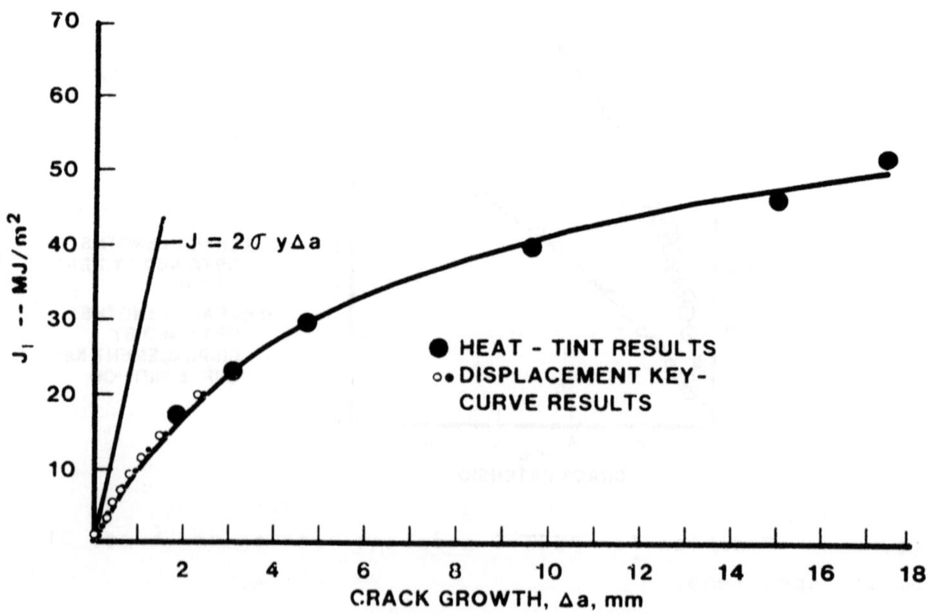


Figure 6. R curves for Type 304 stainless steel at room temperature, 2T compact specimens.

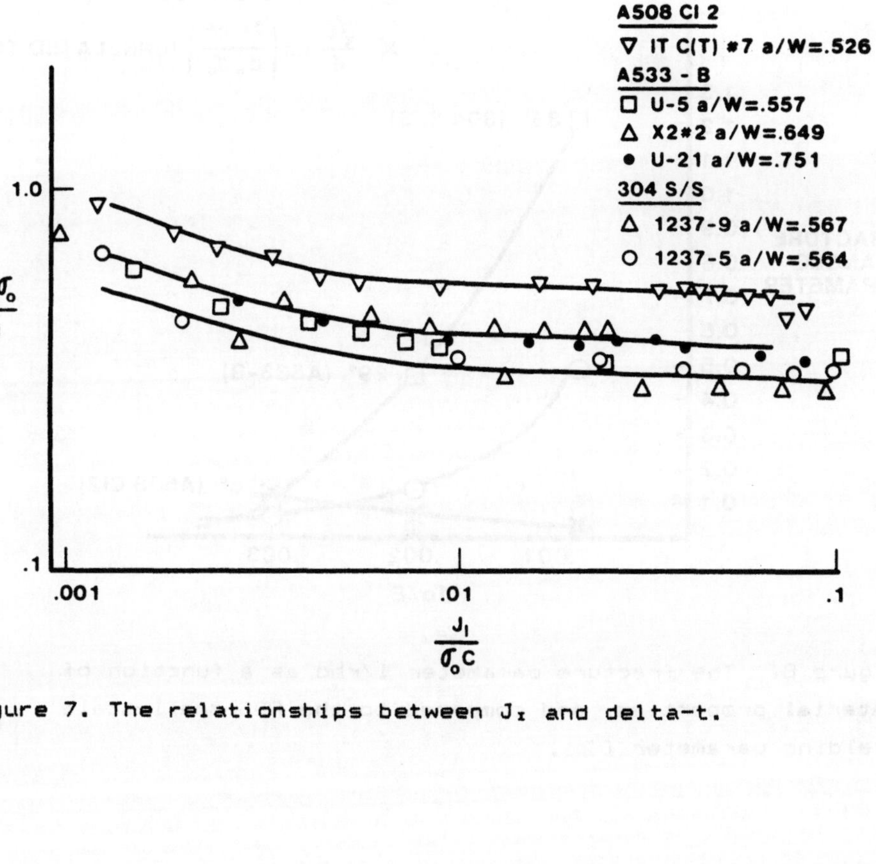


Figure 7. The relationships between $J_I$ and delta-t.

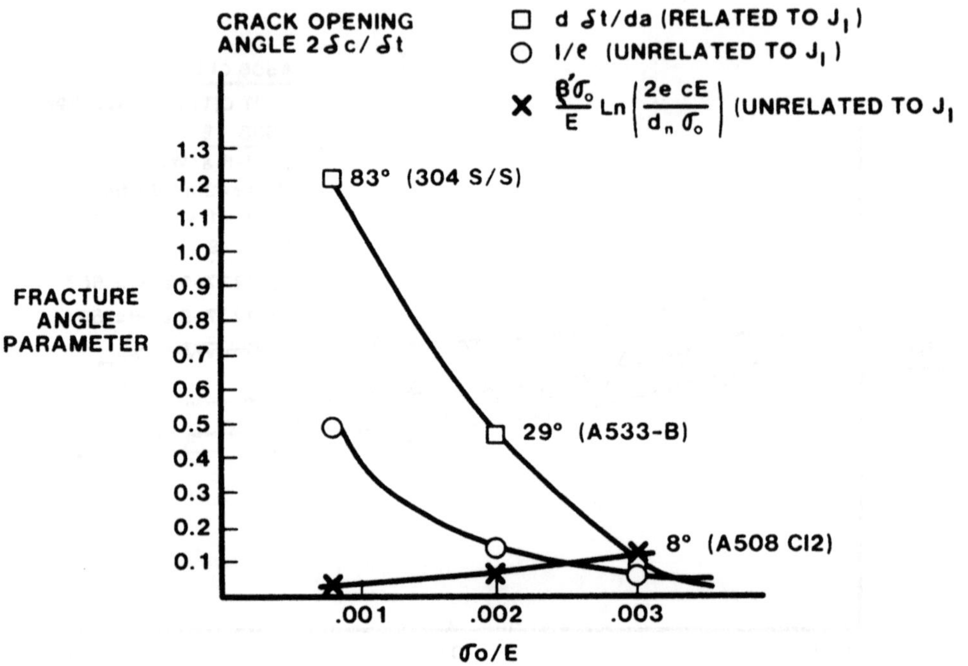


Figure 8. The fracture parameter 1/rho as a function of material properties, and compared to the RDS small-scale yielding parameter [2].

# Relations Between the Crack Tip Opening Displacement, $\delta_5$, and the Modified J, $J_M$

H. A. ERNST
TECHINT SA R & D Department
Cordoba 320 - Buenos Aires - Argentina

Abstract

In this paper different candidate parameters to describe the process of crack growth and plastic deformation are discussed: in particular the crack opening displacement, $\delta_5$, and the modified J, $J_M$. It is shown that a one to one correlation is expected between them, independently of the specimen characteristics.
Experimental data show the adequacy of the approach.

Key words mechanics, materials, fracture, elastic plastic, resistance curve, Modified J, Delta five, crack growth.

In the last several years significant effort has been devoted to study the process of crack initiation and growth in a real elastic plastic material.

The onset of crack extension has been, often, identified with the attainment of a critical strain in a small region ahead of the crack. This is consistent with the attainment of a critical value of a field parameter such as J or the crack opening displacement COD, ie. $J_{Ic}$ or $COD_c$.

On the other hand, the highly irreversible process of crack extension accompanied by extensive plastic deformation is still subject of intense study and discussion. It is not clear yet which is the best parameter capable of describing the irreversible process of both crack growth and plastic deformation, and what local fracture criterion should be used.

The use of a field parameter such as J is objected, since it is based on deformation theory of plasticity, ie. no unloading allowed, and it is obvious that after some crack extension portions of the body will see a decrease in the stresses.

Nevertheless, it has been shown [1] that for small amounts of crack growth J still can be used if a certain condition is met i.e. $\omega = (b/J) \cdot (dJ/da) >> 1$.
That is, if the amount of crack growth is very small (about 6 % of the ligament b) and the $\omega$ condition is satisfied the J-resistance curve is essentially geometry independent. On the other hand, for larger amounts of crack extension and very different specimen sizes, i.e. size ratios of 20 to 1, the J-R and the COD-R curve show their instrinsic geometry dependence [2-3].
To overcome these difficulties several candidate parameters have been proposed. Among others the COD taken as the relative vertical distance between two points originally at 5 mm, fixed in the body, at the original crack tip position called $\delta_5$ [4-5], and a modified version of the J-integral, $J_M$ [6], defined as

$$J_M = J - \int_{a_o}^{a} \left.\frac{\partial J_{pl}}{\partial a}\right|_{\delta_{pl}} da \qquad (1)$$

or equivalently

$$J_M = G + \int_{o}^{\delta_{pl}} \left.\frac{\partial P}{\partial b}\right|_{\delta_{pl}} d\delta_{pl} \qquad (2)$$

where the integrals in the above equations are calculated <u>following the crack growth.</u>

With

| | | |
|---|---|---|
| $J$ | : | deformation theory J based on current values of a and $\delta$ |
| $\delta$ | : | load point displacement |
| $\delta_{pl}$ | : | plastic part of the load point displacement |
| $a_o$, $a$ | : | initial and current values of the crack length |
| $G$ | : | Griffth's energy release rate or the linear elastic part of J |
| $P$ | : | load per unit thickness |
| $b$ | : | remaining ligament. |

Both parameters $J_M$ and $\delta_5$, have shown their adequacy for significantly extending the geometry independent range of the resistance curve [4-6]. It is then important to explore the possible connection between them.

## Crack Opening Displacement Profiles for Growing Cracks

In a recent work [7], it has been found that the rate of increase of the crack opening displacement $\delta_t$ of a point at a relative distance r from the moving crack tip is

$$\dot{\delta}_t = \frac{\alpha}{\sigma_o}\dot{J}_x + \beta\frac{\sigma_o}{E}\dot{a}\ln\frac{R}{r} \tag{3}$$

where

| | | |
|---|---|---|
| $\alpha$, $\beta$ | : | are material parameters approximately constants |
| $\sigma_o$ | : | yield strength |
| E | : | Young modulus |
| $\dot{J}_x$ | : | rate of increase of a J like parameter |
| $\dot{a}$ | : | rate of increase of crack length |
| R | : | length parameter approximately equal to the radius of the plastic zone size. |
| r | : | relative distance from the point of interest to the moving crack tip. |

This study was performed assuming an elastic rigid plastic material under plane strain small scale yielding conditions. The values of $\alpha$ and $\beta$ obtained under those conditions were 0.65 and 5.08 respectively. It is argued that, for other conditions, a similar form applies with possible variations of the value of $\alpha$ and $\beta$.

Equation (3) was integrated [7] to give

$$\delta_t(r,a,J_x) = \frac{\alpha}{\sigma_o} r \frac{dJ_x}{da} + \beta\frac{\sigma_o}{E} r \ln\frac{eR}{r} \tag{4}$$

which gives $\delta_t$ as a function of r for each stage of the growth process. It is importand to note, at this point that the J like parameter $J_x$ suggested, but not clearly defined in [7], has been identified with $J_M$ in [6]. In fact the property that $J_x$ is required to have in [6], was the basis for the definition of $J_M$.
Alternatively, the same Equation can be integrated following a different scheme, in order to give $\delta_t$ <u>at the original crack tip location</u> i.e. $r = \Delta a$ (amount of crack extension) [8]

$$\delta_t\ (r = \Delta a,\ a,\ J_M) \equiv \delta_5\ (\Delta a) = \tag{5}$$

$$= \frac{\alpha}{\sigma_o} J_M + \int_{a_o}^{a} \beta \frac{\sigma_o}{E} \ln \frac{eR}{a-a_o}\, da$$

For situation where

$$\frac{\alpha}{\sigma_o} J_M >> \int_{a_o}^{a} \beta \frac{\sigma_o}{E} \ln \frac{eR}{a-a_o}\, da \tag{6}$$

then

$$\delta_5\ (\Delta a) \approx \frac{\alpha}{\sigma_o} J_M \tag{7}$$

It is to be noted that the model developed in [7] (Equation 3) is only valid for $r \to 0$. So it is not expected to be valid for large r. Nevertheless, the $\delta_5$ is taken at a distance $r = \Delta a$ that can be considered "large", in the sense that does not see the singular term (logaritmic), but small in comparison with other diemsions of the body.
The above analysis seems to indicate then that, under certain conditions, a unique relationship between $J_M$ and $\delta_5$ exists, dependent only on the amount of crack extension.

It is important to note that this relation needs not to be linear, due to the fact that α and β can be changing during growth, simplifications of the model may show here, etc.
The importand point is, though, that resistance curves constructed using $J_M$ or $\delta_5$ appear to be of equal quality as geometry independent descriptions of the crack growth process.

Experimental Results:

McCabe et al. [9] recently tested a A533B pressure vessel steel, with mechanical properties as follow: yield strength = 68 ksi (469 MPa), tensile strength = 90 ksi (620 MPa) and Charpy Energy = 92 ft-lbs (126 J). He produced $\delta_5$- and $J_M$-R curves of Single Edge Notch (SEN) and Double Edge Notch (DEN) specimens, 0.1 in thick, at room temperature.
The results were crossplotted to produce a $\delta_5$-$J_M$ curve for all specimens. The excellency of the results, Figure 1, shows the adequacy of Equation 7.

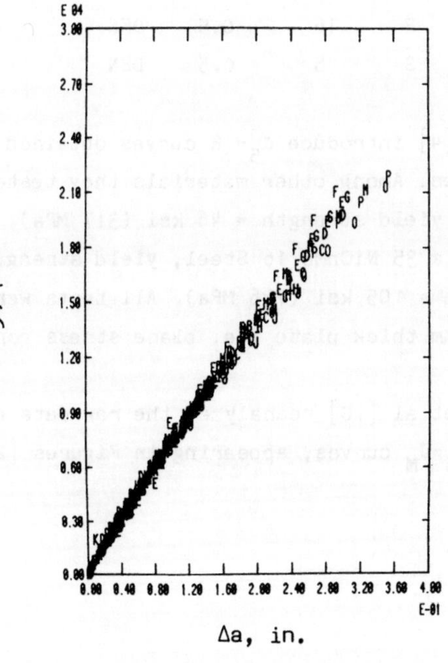


Fig. 1: $J_M$-$\delta_5$ curve A533B at room temperature

LEGEND TO FIG. 1

| SYMBOL | W(IN) | A/W | TYPE |
|---|---|---|---|
| A | 30 | 0.3 | SEN |
| B | 20 | 0.3 | SEN |
| C | 30 | 0.5 | SEN |
| D | 30 | 0.5 | SEN |
| E | 20 | 0.3 | SEN |
| F | 20 | 0.3 | SEN |
| G | 20 | 0.4 | SEN |
| H | 20 | 0.5 | SEN |
| I | 20 | 0.5 | SEN |
| J | 20 | 0.6 | SEN |
| K | 20 | 0.7 | SEN |
| L | 20 | 0.7 | SEN |
| M | 10 | 0.36 | SEN |
| N | 10 | 0.37 | SEN |
| O | 10 | 0.5 | SEN |
| P | 10 | 0.5 | SEN |
| R | 16 | 0.5 | DEN |
| S | 8 | 0.5 | DEN |

Hellmann and Schwalbe [4] introduce $\delta_5$- R curves obtained on different materials and geometries. Among other materials they tested: a 2024-T351 Alu alloy, yield strength = 46 ksi (317 MPa), ultimate strength = 64 ksi (440 MPa) and a 35 NiCrMo 16 Steel, yield strength = 74 ksi (510 MPa), ultimate strength = 105 ksi (726 MPa). All tests were performed on specimens made of a 5 mm thick plate i.e. plane stress conditions.

In a later work Ernst et al [10] reanalyzed the raw data producing $J_M$-R curves and then $\delta_5$-$J_M$ curves, appearing in Figures [2-3].

The results show again the adequacy of Eq(7): i.e. the unique relationship between $J_M$ and $\delta_5$.

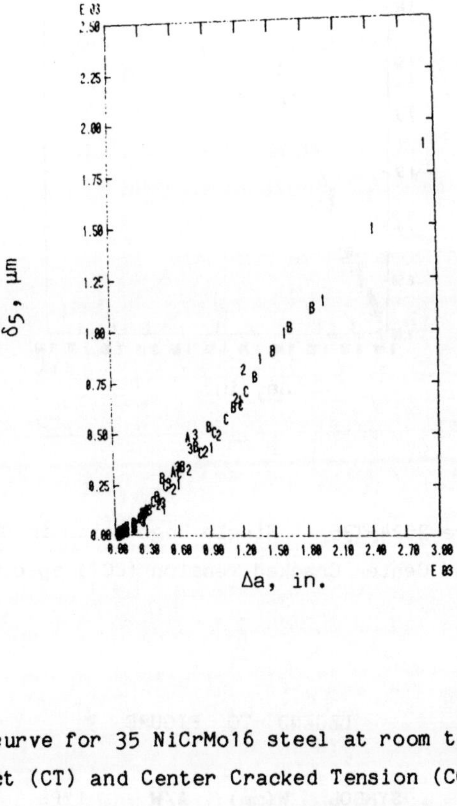


Fig. 2: $J_M$-$\delta_5$ curve for 35 NiCrMo16 steel at room temperature; compact (CT) and Center Cracked Tension (CCT) specimens.

LEGEND TO FIGURE 2

| SYMBOL | W(mm) | A/W | TYPE |
|---|---|---|---|
| A | 50 | 0.3 | CT |
| B | 50 | 0.5 | CT |
| C | 50 | 0.7 | CT |
| 1 | 50 | 0.4 | CCT |
| 2 | 50 | 0.8 | CCT |
| 3 | 50 | 0.9 | CCT |

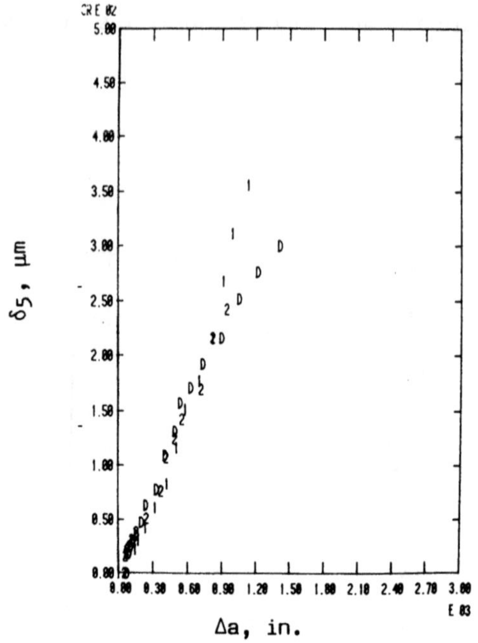


Fig. 3: $J_M$-$\delta_5$ curve 2024-T351. Single Edge Notch in Bending (SENB) and Center Cracked Tension (CCT) specimens.

LEGEND TO FIGURE 3

| SYMBOL | W(mm) | A/W | TYPE |
|---|---|---|---|
| D | 100 | 0.5 | SENB |
| 1 | 50 | 0.585 | CCT |
| 2 | 100 | 0.485 | CCT |

Discussion:

It has been shown that by integrating the expression of Rice et al., Eq(3), and with the identification of $J_x$ as $J_M$, the Hellmann and Schwalbe $\delta_5$ and $J_M$ are directly proportional to each other, under certain circumstances. The coefficient linking the two could be changing with growth and presumably to a smaller degree with the amount of yielding and specimen geometry. At present it is not clearly defined the range of validity of the above statement.

Experimental data on plane stress seem to indicate the general adequacy of the approach, although more work is needed to determine more clearly the mentioned limits (see Figure 3 for example). Finally, it is expected that, the coefficient linking the two parameters is approximately equal to the coefficient linking J and COD (45°) [11] for the case of a stationary crack.

More work is obviously needed in this area, but it is important to see that a particular interpretation of J for growth, i.e. $J_M$, and a particular definition of COD i.e. $\delta_5$, which are proportional to each other, seem to be able to significantly extend the geometry independent range of the resistance curve.

References:

[ 1] Hutchinson, J.W. and Paris, P.C., in ASTM STP 668, American Society for Testing and Materials (1979), 37-64.

[ 2] McCabe, D.E. and Landes, J.D.: $J_R$-Curve Testing of Large Compact Specimens in ASTM STP 803, American Society for Testing and Materials (1983) pp. II-353-II-371.

[ 3] McCabe, D.E. and Ernst, H.A.: Crack Growth Resistance Measurements by Crack Opening Displacement Methods in ASTM STP 833, American Society for Testing and Materials (1984), pp. 648-665.

[ 4] Hellmann, D. and Schwalbe, K.-H.: Geometry and Size Effects on J-R and δ-R curves under Plane Stress Conditions, in ASTM STP 833 American for Testing and Materials, pp. (1984) 577-605.

[ 5] Schwalbe, K.-H. and Hellmann, D.: Correlation of Stable Crack Growth with the J-integral and the Crack Tip Opening Displacement, Effects of Geometry, Size and Material, Progress Report GKSS 84/E/37, GKSS-Forschungszentrum, Geesthacht, West-Germany.

[ 6] Ernst, H.A.: Material Resistance and Instability Beyond J-Controlled Crack Growth, in ASTM STP 803 Ameraican Society for Testing and Materials (1983), pp. I-191-I-213.

[ 7] Rice, J.R., Drugan, W.J. and Sham, T.L.: in ASTM STP 700, American Society for Testing and Materials (1980), pp. 189-221.

[ 8] Ernst, H.A.: Work in progress.

[ 9] McCabe, D.E., Landes, J.D. and Ernst, H.A.: Elastic-Plastic Methodology to Establish R-Curves and Instability Criteria, Final Report to EPRI No. RP 1238-2 (1985).

[10] Ernst, H.A., Schwalbe, K.-H., Hellmann, D. and McCabe, D.E.: Modified J, $J_M$, Resistance Curves under Plane Stress Conditions. To be published.

[11] Shih, C.F.: Relationships Between the J-integral and Crack Opening Displacement for Stationary and Extending Cracks, General Electric Report No. 79CRD075 (1979).

# Application

# A Preliminary Assessment of the Use of the CTOD Criterion in Dynamic Fracture Mechanics

M.F. KANNINEN

Southwest Research Institute
Engineering and Materials Sciences Division
San Antonio, Texas

Summary

The use of dynamic fracture mechanics to predict unstable crack propagation and arrest is now well established for elastodynamic conditions. Advances in the technology are currently being made to cope with ductile and tough materials where the small-scale yielding assumption is not necessarily valid. The key element in this effort lies in identifying a crack tip parameter that controls a rapidly propagating crack. Because of its success in characterizing initiation and stable crack growth, the CTOD is an attractive candidate for use in analyses of unstable crack propagation as well. However, there are difficulties that make this usage more difficult in dynamic fracture mechanics than in the quasi-static regime. This paper first reviews the prior uses of the CTOD in elastic-plastic and dynamic fracture mechanics, with emphasis on the use of the collinear strip yield (Dugdale) model for crack tip plasticity. Then, some preliminary results obtained in a direct determination of this parameter for a fast propagating crack in 4340 steel are reported that illuminate some of the difficulties in dynamic-inelastic analyses.

## Introduction

There are a number of practical situations in which it is not possible to insure the integrity of an engineering structure by precluding crack growth initiation. Some prominent examples of this unfortunate circumstance are for nuclear plant pressure vessels (the thermal shock problem), gas transmission pipelines, ship hulls, LNG storage tanks, and railroad rails. In each of these applications the onset of unstable crack propagation is sufficiently likely that a second line of defense is necessary. It therefore becomes imperative to assess the risk of catastrophic fracture by direct consideration of measures that will arrest unstable crack propagation. The branch of fracture mechanics that encompasses this application area is known as dynamic fracture mechanics [1].

Dynamic fracture mechanics techniques are now well advanced for materials that behave in a linear elastic manner. However, in most of the critical applications of this technology, highly ductile and tough materials are being used. Questions therefore exist on the applicability of the existing elastodynamic technology for such materials. In turn, these questions are spurring fundamental research on dynamic-inelastic material behavior. The central research issue for rapid crack propagation in a material for which a linear elastic description is suspect lies in identifying the proper crack tip characterization parameter.

On the basis of the observed constancy of the crack tip opening displacement (CTOD) parameter in stable crack growth in quasi-static elastic-plastic conditions, coupled with the relative ease of its utilization in finite element models, the CTOD is a prime candidate for dynamic problems also. To set the stage for such an application, a description of previous uses of the CTOD in elastic-plastic and dynamic fracture mechanics is given in this paper. Then, some current research that is addressed to characterizing dynamic crack propagation and arrest problems using the CTOD parameter is described. The background review is drawn from the more complete account given by Kanninen and Popelar [1].

## Use of the CTOD criterion in Elastic-Plastic Fracture Mechanics

While fracture mechanics has been focused exclusively on linear elastic behavior, current applications are to materials for which this assumption is invalid. Initial concern was for conditions in which the cracked structural component itself would likely obey the dictates of linear elastic fracture mechanics (LEFM), but the small-scale laboratory experiments needed to provide the necessary fracture properties would not. Prime examples are materials sufficiently ductile and tough that the extent of plastic yielding accompanying the initiation of the crack growth is comparable to the specimen dimensions.

Current research interest centers on stable crack growth. The purpose is to achieve a better quantitative understanding of structural components where fracture is preceded by significant amounts of stable cracking; e.g., stainless steel nuclear plant piping. This requires direct consideration of crack tip plasticity.

The first quantitative accounting for the effect of the plastic zone at the crack tip was suggested by Irwin et al. [2]. On the basis that a plastically deformed region cannot support the same level of stress that it would if yielding did not intervene, they argued that a cracked body is somewhat weaker than an elastic analysis would suggest. To account for this within the framework of linear elasticity, they supposed that the effect would be the same as if the crack length were slightly enlarged. Thus, a plasticity-modified stress intensity factor for a crack in an infinite medium can be written in terms of a crack length that is mathematically extended by an amount $r_y$. Here $r_y$ is a measure of the plastic zone size; e.g., the radius of a circular yielded region.

Wells [3] advanced an alternative concept to apply even beyond general yielding conditions. This work introduced the crack tip opening displacement (CTOD) as the parameter governing crack initiation. Wells evaluated this parameter using the Irwin et al. plastic zone estimation equation for a center crack in an infinite elastic body. Specifically, taking $\delta$ to be the crack opening at the distance $r_y$ from the crack tip, Wells found that

$$\delta = \alpha \frac{K^2}{E\sigma_Y} \tag{1}$$

where $\sigma_Y$ is the yield stress and $\alpha$ is a numerical factor that in Wells' work was equal to $4/\pi$. Wells recognized that the factor $4/\pi$ is inconsistent with an energy balance approach (which would require a factor of unity) and subsequently adopted $\alpha=1$. Other investigators later found other values of $\alpha$ to be appropriate; see Kanninen and Atkinson [4].

While numerical elastic-plastic analyses to determine the plastic region at a crack tip were available, an explicit relation was needed for $\delta$ to advance the CTOD concept. This was provided by the work of Dugdale [5] who developed a closed form solution applicable for plane stress conditions using methods of the complex variable theory of elasticity. Dugdale supposed that, for a thin sheet loaded in tension, the yielding will be confined to a narrow band lying along the crack line. This idea is identical to placing internal stresses on the portions of the (mathematical) crack faces near its tips; the physical crack being the intermediate stress-free length. The magnitude of the internal stresses are then taken to be equal to the yield stress of the material.

In order to determine the length of the collinear plastic zones at each crack tip, Dugdale postulated that the stress singularity must be abolished. For a crack of length 2a in an infinite medium under uniform tension $\sigma$, this postulate leads to the well-known relation

$$\frac{a}{c} = \cos\left(\frac{\pi}{2}\frac{\sigma}{\sigma_Y}\right) \tag{2}$$

in which c=a+d where d denotes the plastic zone length.

The crack face displacements in the Dugdale model were first worked out by Goodier and Field [6]. But, the form of their result is rather cumbersome. The normal displacement on the crack faces can be written in a more convenient form using the relation developed subsequently by Kanninen et al. [7]. This is

$$v(x) = \frac{2a}{\pi}\frac{\sigma_Y}{E}\left\{\log\left|\frac{\sqrt{c^2-a^2}+\sqrt{c^2-x^2}}{\sqrt{c^2-a^2}-\sqrt{c^2-x^2}}\right| + \frac{x}{a}\log\left|\frac{x\sqrt{c^2-a^2}-a\sqrt{c^2-x^2}}{x\sqrt{c^2-a^2}+a\sqrt{c^2-x^2}}\right|\right\} \quad 0 \le x \le c \tag{3}$$

Of most immediate interest is the displacement at the crack tip given by the Dugdale model. This can be obtained via a limiting procedure applied to Equation (3). Writing $\delta$=2v(a) gives the widely used form first obtained by Goodier and Field. This is

$$\delta = \frac{8}{\pi}\frac{a\sigma_Y}{E}\log\left[\sec\left(\frac{\pi}{2}\frac{\sigma}{\sigma_Y}\right)\right] \tag{4}$$

This key relation was obtained independently by Burdekin and Stone [8] who used it to advance Well's COD concept by demonstrating that fracture could be governed by critical $\delta$ values via experimental results on mild steel in tension and bending. This work provided the basis for the semi-empirical "COD Design Curve" approach evolved by Dawes [9] and coworkers at the UK Welding Institute.

The alternative to the use of the CTOD in elastic-plastic fracture mechanics is one based upon the path-independent contour integral J introduced by Rice [10]. Coupled with the crack characterization studies of Hutchinson [11] and Rice and Rosengren [12], together with the experimental results of Begley and Landes [13] and the estimation procedure developed by Bucci et al. [14], this laid the ground work for the bulk of current applications in elastic-plastic fracture mechanics and for crack tip characterizations in a variety of other applications.

A natural extension of the J parameter was to characterize stable crack growth via the J-resistance curve. Paris et al. [15] introduced the tearing modulus concept to determine fracture instability associated with stable cracking. However, this approach is subject to the limitation of a small amount of crack growth at instability, as noted by Hutchinson and Paris [16]. This limitation has provided the impetus for examining the usefulness of CTOD for a resistance curve criterion; specifically, through the use of a modified version of the CTOD known as the crack tip opening angle (CTOA) parameter. It has been observed that, after an initial transient, the CTOA appears to remain constant for extended amounts of stable crack growth.

Use of the CTOA parameter as a stable crack growth criterion is appealing because of its readily observed physical significance, the opportunity that it offers for direct measurement, and, perhaps most importantly, the fact that after an initial transient, stable crack growth appears to proceed with a virtually constant crack shape. The latter provides a particularly simple criterion. This constancy was reported first in the experimental work of Green and Knott [17], Berry and Brook [18], and in the integrated experimentation/analysis approach of deKoning [19]. Specifically, deKoning found a critical CTOA value of .07 radians in 2024-T3 aluminum, Berry and Brook observed a value of 0.17 radians in 12% Cr steel, while Green and Knott obtained a value of 0.125 radians for a mild steel. Combined experimentation and computational work firmly establishing the practical use of the CTOA for stable crack growth and fracture instability assessments in nuclear plant piping materials was later performed by Kanninen et al. [20] and by Shih et al. [21].

Use of the CTOD Criterion in Dynamic Fracture Mechanics

Use of a criterion based on a crack opening displacement for rapid crack propagation appears to have preceded its use for stable crack growth. Kanninen [22], by incorporating the Dugdale model into the constant length propagating crack (Yoffe) model, found an expression for the CTOD which is just the same as Equation (4) but escalated by a speed-dependent function, L(V). This function is given by the expression (plane stress)

$$L(V) = \frac{1+\nu}{D(V)} \left(\frac{V}{C_2}\right)^2 \left[1-\left(\frac{V}{C_1}\right)^2\right] \tag{5}$$

where $C_1$ and $C_2$ respectively denote the dilatational and shear wave speeds of the material and D(V) is the material-independent function given by

$$D(V) = 4\left[1-\left(\frac{V}{C_1}\right)^2\right]^{1/2} \left[1-\left(\frac{V}{C_2}\right)^2\right]^{1/2} - \left[2-\left(\frac{V}{C_2}\right)^2\right]^2 \tag{6}$$

The root of D(V) is the Rayleigh velocity, $C_R$. Hence, while the L(V) function is unity at V=0, it monotonically increases to infinity at $V=C_R$.

Kanninen [22] may have been the first to suggest that rapid crack propagation occurs at a constant CTOD and to calculate crack speeds under this stipulation. Comparison of CTOD-controlled calculated crack speeds with observed crack speeds in steel foil showed very reasonable agreement.

Atkinson [23] provided a more realistic (expanding) dynamic version of a propagating Dugdale crack that was subsequently extended by Embley and Sih [24]. Using Equation (5) as a multiplicative factor, Equation (4) gives a value of $\delta$ that differs from that obtained by Empley and Sih for an expanding crack. However, for low to moderate applied stresses and crack speeds, the two results are reasonably close whereupon the much simpler form obtained for the constant length crack can be used. Nevertheless, most likely because of the widespread availability of finite element models, little use has been made of this fact.

Another reason for the evident lack of interest in pursuing dynamic Dugdale models is certainly due to the limitations of the Dugdale model itself. It is difficult to incorporate a measure of the strain ahead of the crack tip in other than an artificial way; i.e., because of the zero gage length of a plastically deformed element. Moreover, as Kanninen et al. [7] found, even if a strain measure can be rationalized, a further difficulty arises in incorporating a strain rate dependence of yielding. This can be seen by differentiating Equation (3) which gives

$$\frac{dv}{dx} = \frac{2}{\pi}\frac{\sigma_Y}{E}\log\left|\frac{x\sqrt{c^2-a^2} - a\sqrt{c^2-x^2}}{x\sqrt{c^2-a^2} + a\sqrt{c^2-x^2}}\right| \tag{7}$$

It can readily be seen from setting x=a in Equation (7) that v(x) has a vertical slope at the physical crack tip. Conceptual sketches of the Dugdale model seldom reveal this feature. Nevertheless, it has important consequences even in the stationary crack problem; e.g., the crack tip profile differs from that in an elastic-plastic finite element solution.

An alternative line of attack has also been followed in which a critical CTOD (or CTOA) has been assumed to be the parameter governing crack growth in inelastic conditions with consequences drawn from this assumption. In the stable crack growth situation, Rice and Sorensen [25] and Rice et al. [26] assumed that a critical crack opening displacment is maintained behind a growing crack tip in order to relate a far-field parameter with local microstructural fracture processes. Freund and Douglas [27] have employed a similar assumption for dynamic crack propagation in order to deduce the form of the running fracture toughness as a function of crack speed. Work in progress at the Southwest Research Institute (SwRI) is investigating the efficacy of the CTOA as a criterion for crack propagation in dynamic-viscoplastic conditions.

## The Critical CTOA for Dynamic Crack Propagation in 4340 Steel

Dynamic fracture mechanics capabilities must be extended in several areas for properly based assessments of crack arrest in engineering structures. These include the ability to (1) treat a rapidly applied load, (2) treat a nonsimple flaw-structure geometry, (3) model crack tip plasticity, and (4) compute weld induced residual stresses and crack growth in welded joints. The approach adopted in research led by the author was to examine these complications systematically, beginning with the first. Hence, attention was initially focused on 4340 steel: a well-characterized, high-strength, moderate-toughness material.

The expectation in working with 4340 steel was that the dynamic fracture toughness property, $K_{ID} = K_{ID}(V)$, for a fast propagating crack initiated from quasi-static loading conditions should also characterize crack propagation due to a high rate (impact, blast) loading. Instead, Kanninen et al. [28] found that, while elastodynamic analyses did provide an excellent description of the crack length history in mechanical impact experiments, the apparent $K_{ID}$ values needed to achieve this agreement were substantially different from those given by the quasi-static initiation experiments. The disparity between the fracture toughness values deduced for quasi-static and impact conditions was later explained by Ahmad and Kanninen [29] using an elastic-plastic finite element analysis. Specifically, it was found that, although the plastic zone of the propagating crack remained relatively small, a sizeable plastically deformed region existed at the point of impact. Thus, even though the small scale yielding assumption required for LEFM was not violated, an elastic-plastic analysis was required to account for the energy absorbed remote from the crack tip.

Later work progressed to 4340 steel plates fractured using an explosive detonated at a standoff distance. Elastic-plastic dynamic finite element analyses were then performed to simulate the experiments. As reported by Kanninen [30] these computations demonstrated that, for a uniformly distributed blast load, there is no plasticity remote from the crack tip comparable to that experienced in mechanical impact. This work demonstrated that an elastodynamic approach should therefore be appropriate for blast loading.

The choice of constitutive behavior in applications to tougher and more ductile steels is of obvious importance. Work by Barnes et al. [31] has shown that the crack length histories computed using elastodynamic and elastic-plastic dynamic analyses were virtually indistinguishable for HY-80 steel. Both of these analyses assumed rate-independent behavior. Recent work conducted by a number of different investigators has shown that viscoplastic (rate-dependent plasticity) effects are enormously influencial in determining the local crack tip behavior; see Kanninen and Popelar [1]. This follows from the fact that the strain rate that is imposed on the crack tip by a propagating crack – even in a material that would produce substantial yielding in quasi-static conditions — is so high that yielding is severely constrained.

Work currently in progress at SwRI using a viscoplastic-dynamic analysis model has revealed that the plastic zones attending a running crack in a ductile material are much smaller than the corresponding rate-independent analysis would predict. This strongly suggests that the use of a rate-independent elastic-plastic dynamic computational procedure will provide an unrealistic portrayal of the near-tip region for a propagating crack. But, viscoplastic-dynamic, moving boundary analysis procedures and the supporting material property descriptions are not generally available. Consequently, the practical choice is most often between elastodynamic and elastic-plastic analyses.

The key to deciding upon the more correct constitutive behavior is the nature of the singularity in a dynamic viscoplastic crack propagation analysis. Specifically, asymptotic analyses have revealed that the dynamic viscoplastic singularity is exactly the same as that deduced in an elastodynamic treatment. In rate-independent plasticity, of course, the singularity differs significantly from $r^{-1/2}$. Consequently, to compare the elastodynamic and elastic-plastic dynamic approaches, Cardinal and Kanninen [32] performed an elastodynamic generation phase analysis of a quasi-statically loaded fast fracture three-point bend specimen. This was done in order to contrast it with a previous elastic-plastic analysis [29]. The specimen geometry is given in Figure 1 with the measured crack propagation history shown in Figure 2. A bilinear crack history was used in the analyses, as shown in Figure 2. Figure 3 compares the computational results.

Figure 1

Figure 2

Figure 3

The most prominent feature of the elastic-plastic dynamic analysis shown in Figure 3 was that, after the stabilization of an initial transient, the CTOA remained virtually constant for the remainder of the crack propagation history. The elastodynamic CTOA values are compared to the elastic-plastic CTOA values in Figure 3. Also shown in Figure 3 is an average crack opening angle (COA) which is based upon the displacement at the initial crack tip. It can be seen that the elastodynamic COA agrees more closely with the elastic-plastic dynamic CTOA than with the elastodynamic CTOA. In fact, as the crack propagates, it appears that the elastic COA may provide a lower bound for the critical CTOA. Taken together with the theoretical arguments given in the above, this result leads to the conclusion that, if one is using an elastodynamic approach, it may be most appropriate to use the COA as the critical fracture parameter.

## Discussion

The objective of the research reported herein was to investigate the efficacy of a crack opening angle criterion for dynamic crack propagation in materials that are ductile and tough in quasi-static conditions. Dynamic viscoplastic finite element crack propagation/arrest computations are clearly most appropriate. But, their use requires complex analysis procedures and material constitutive properties that are not generally available. Recent progress in nonlinear dynamic fracture mechanics has clearly shown the inappropriateness of rate-independent plasticity computations for rapidly propagating cracks. Elastodynamic analyses of experiments on 4340 steel suggest the use of an elastodynamic COA as a critical fracture parameter. Future work in this area must nonetheless be structured such that dynamic viscoplastic analyses can be performed to fully validate this conclusion.

## Acknowledgement

This research was conducted under Office of Naval Research Contract Number N-0014-77-C-0576. The author would like to express his appreciation to Dr. Yapa Rajapakse of ONR for his continued support of his work in this problem area. The computational results shown in Figure 3 were obtained by J.W. Cardinal of SwRI.

## References

1. Kanninen, M.F.; Popelar, C.H.: Advanced Fracture Mechanics. Oxford Press, New York, (1985).

2. Irwin, G.R.; Kies, J.A.; and Smith, H.L.: Fracture Strengths Relative to Onset and Arrest of Crack Propagation. Proc. ASTM, 58, (1958) 640-657.

3. Wells, A.A.: Application of Fracture Mechanics at and Beyond General Yielding. British Welding Journal 10 (1963) 563-570.

4. Atkinson, C.; Kanninen, M.F.: A simple representation of crack tip plasticity: the inclined strip-yield superdislocation model. Int. J. Fracture 13 (1977) 151-163.

5. Dugdale, D.S.: Yielding of Steel Sheets Containing Slits. J. Mech. Phys. Solids 4 (1972) 431-442.

6. Goodier, J.N.; Field, F.A.: Fracture of Solids. Wiley, New York (1960) 103-108.

7. Kanninen, M.F.; Mukerjee, A.K.; Rosenfield, A.R.; and Hahn, G.T.: The Speed of Ductile Crack Propagation and the Dynamics of Flow in Metals Mechanical Behavior of Materials Under Dynamic Loads. Springer-Verlag, New York (1969) 96-133.

8. Burdekin, F.M.; Stone, D.E.W.: The Crack Opening Displacement Approach to Fracture Mechanics in Yielding Materials. J. Strain Analysis 1 (1966) 145-153.

9. Dawes, M.G.: Fracture Control in High Yield Strength Weldments. Weld. J. Res. Suppl. 53 (1974) 369S-379S.

10. Rice, J.R.: A path independent integral and the approximate analysis of strain concentrations by notches and cracks. J. Appl. Mech. 35 (1968) 379-386.

11. Hutchinson, J.W.: Fundamentals of the Phenomenological Theory of Nonlinear Fracture Mechanics. J. Appl. Mech. 50 (1983) 1042-1051.

12. Rice, J.R.; Rosengren, G.F.: Plane Strain Deformation Near a Crack Tip in a Power-Law Hardening Material. J. Mech. Phys. Solids 16 (1968) 1-12.

13. Begley, J.A.; Landes, J.D.: The J-Integral as a Fracture Criterion, Fracture Toughness. ASTM STP 514 (1972) 1-20.

14. Bucci, R.J.; Paris, P.C.; Landes, J.D.; Rice, J.R.: J-Integral Estimation Procedures, Fracture Toughness. ASTM STP 514 (1972) 40-69.

15. Paris, P.C.; Tada, H.; Zahoor, A.; and Ernest, H.A.: Instability of the Tearing Mode of Elastic-Plastic Crack Growth, Elastic-Plastic Fracture. ASTM STP 668 (1979) 5-36 and 251-265.

16. Hutchinson, J.W.; Paris, P.C.: Stability Analysis of J-Controlled Crack Growth, Elastic-Plastic Fracture. ASTM STP 668 (1958) 640-657.

17. Green, G.; Knott, J.F.: On Effects of Thickness on Ductile Crack Growth in Mild Steel. J. Mech. Phys. Solids 23 (1975) 167-183.

18. Berry, G.; Brook, R.: On the Measurement of Critical Crack-Opening-Displacement When Slow Crack Growth Proceeds Rapid Fracture. Int. J. Fracture 11 (1975) 933-938.

19. deKoring, A.U., A Contribution to the Analysis of Quasistatic Crack Growth in Sheet Materials. Fracture 1977; Proc. of the Fourth Int. Conf. on Fracture, 3 (1977) 25-31.

20. Kanninen, M.F.; Broek, D.; Hahn, G.T.; Marschall, C.W.; Rybicki, E.F.; Wilkowski, G.M.: Towards an Elastic-Plastic Fracture Mechanics Predictive Capability for Reactor Piping. Nuclear Eng. Des. 48 (1978) 117-134.

21. Shih, C.F.; Delorenzi, H.G.; and Andrews, W.R.; Studies on Crack Initiation and Stable Crack Growth, Elastic-Plastic Fracture. ASTM STP 668 (1979) 65-120.

22. Kanninen, M.F.: An Estimate of the Limiting Speed of a Propagating Ductile Crack. J. Mech. Phys. Solids 16 (1968) 215-228.

23. Atkinson, C.: A Simple Model of Relaxed Expanding Crack. Arkiv for Fysik 26 (1968) 469-476.

24. Embley, G.T.; Sih, G.C.: Plastic Flow Around an Expanding Crack. Eng. Frac. Mech. 4 (1972) 431-442.

25. Rice, J.R.; Sorensen, E.P.: Continuing crack tip deformation and fracture for plane strain crack growth in elastic-plastic solids. J. Mech. Phys. Solids, 26 (1978) 163-186.

26. Rice, J.R.; Drugan, W.J.,; Sham, T.-L.: Elastic-plastic analysis of growing cracks. Fracture Mechanics: Twelfth Conference, ASTM STP 700 (1980) 189-221.

27. Freund, L.B.; Douglas, A.S.: The influence of inertia on elastic-plastic antiplane shear crack growth. J. Mech. Phys. Solids 30 (1982) 59-74.

28. Kanninen, M.F.; Gehlen, P.C.; Barnes, C.R.; Hoagland, R.G.; Hahn, G.T.: Dynamic Crack Propagation Under Impact Loading. Nonlinear and Dynamic Fracture Mechanics ASME (1979).

29. Ahmad, J.; Kanninen, M.F.: Dynamic Elastoplastic Crack Propagation Initiated by Impact, unpublished manuscript (1984).

30. Kanninen, M.F., Dynamic Fracture Mechanics and Its Applications to Material Behavior Under High Stress and Loading Rates. Material Behavior Under High Stress and Loading Rates (1983) 197-210.

31. Barnes, C.R.; Ahmad, J.; Kanninen, M.F.: Dynamic Crack Propagation Through Welded HY80 Plates Under Blast Loading. Fracture Mechanics: Sixteenth Symposium, ASTM STP 886 (1985).

32. Cardinal, J.W.; Kanninen, M.F.: The Analysis of Crack Growth Initiation, Propagation and Arrest in Flawed Ship Structures Subjected to Dynamic Loading SwRI Report, 1985.

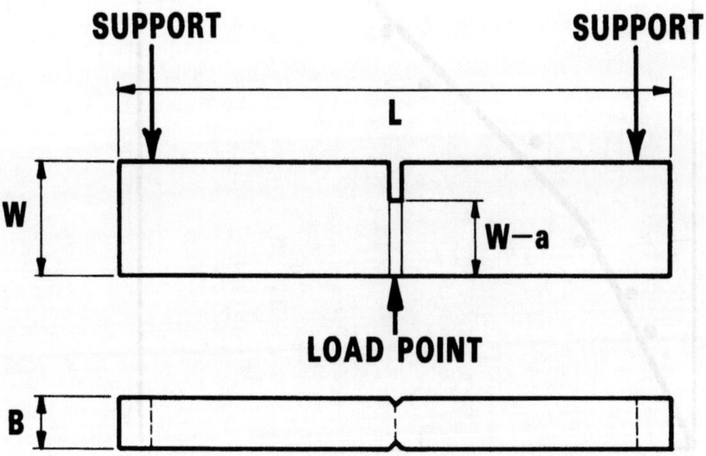


Figure 1. Three Point Bend Specimen Geometry (4340 Steel)
(L=181 mm, W=38 mm, B=15.8 mm, W-a=28.5 mm)

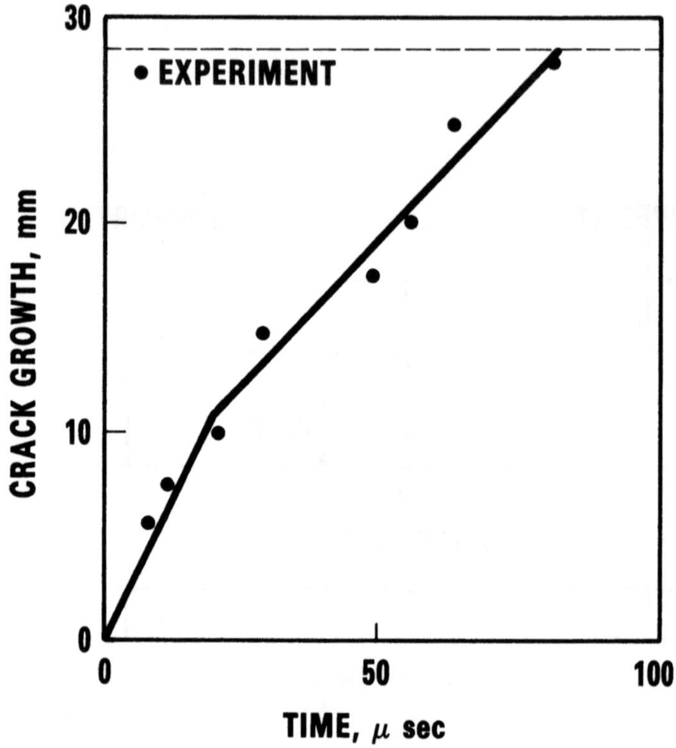


Figure 2. Measured Crack Propagation History for 4340 Three Point Bend Specimen Initiated Under Quasi-Static Loading (Solid line indicates crack history used in elastodynamic generation phase analysis.)

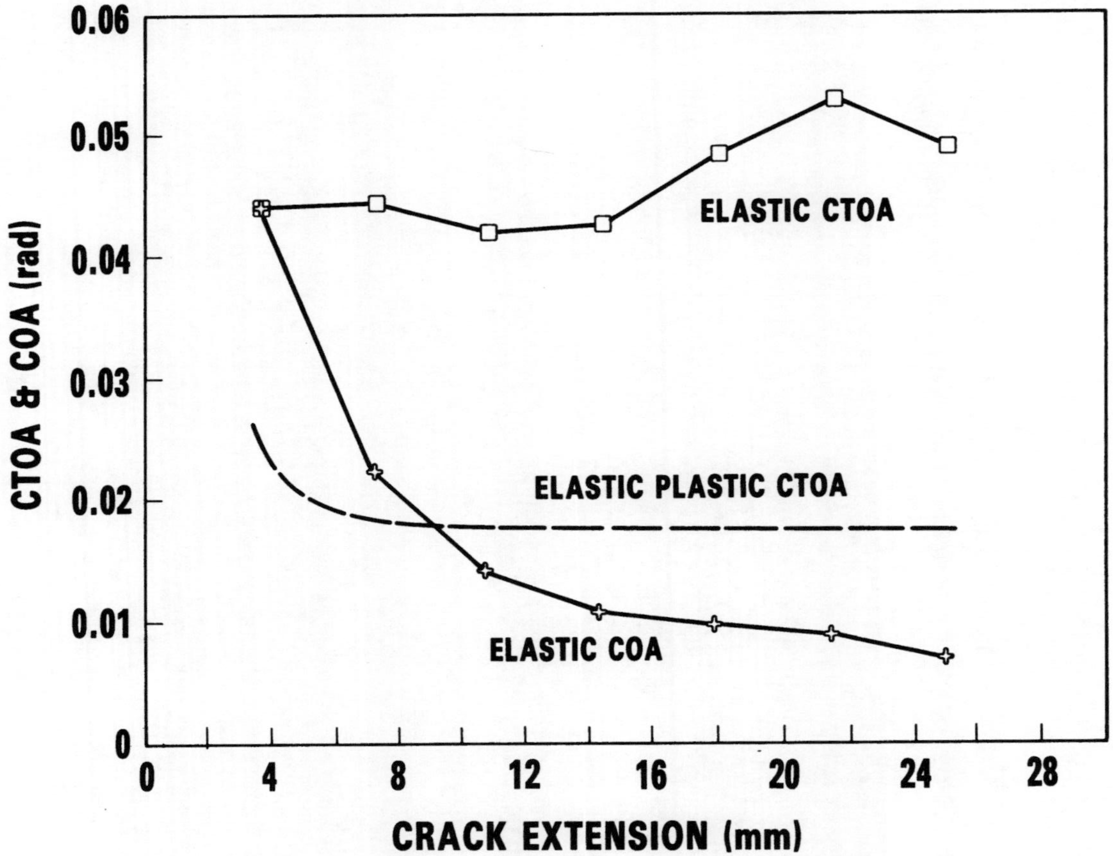


Figure 3. Elastodynamic and Elastic-Plastic Dynamic Results for 4340 Steel Three Point Bend Specimen Initiated Under Quasi-Static Loading

# CTOD Testing of Welds

J.G. BLAUEL and W. BURGET

Fraunhofer-Institut für Werkstoffmechanik, Freiburg FRG

Summary

Experiences and test results of IWM concerning the fracture mechanics characterization of multipass submerged arc and manual metal arc welds in thick components of C-Mn, steel are described. The influences of materials, fabrication parameters and testing variables on the fracture toughness in terms of critical CTOD are discussed.

## Introduction

The CTOD-procedure is widely used not only as a screening test for the fracture toughness of different materials but also as the basis of a quantitative assessment of the structural integrity of load bearing members and components in which crack like defects are found or have to be assumed to exist following non destructive examination arguments. From a photo-macrograph as in Fig. 1 it is immediately obvious that serious difficulties have to be expected if the COD approach is applied to the complex geometry and material conditions of weldments: Since the CTOD is a local crack tip parameter it will depend on the inhomogeneity and anisotropy of the microstructure along the crack front which is determined by the multiple interaction of the materials involved - base material and consumables - and the conditions of fabrication; in addition, details of the test procedure like specimen and crack orientation, crack length, and choice of the critical point have to be defined and measured more carefully than in a bulk material; residual stresses have to be taken into account as well to deliver meaningful values.

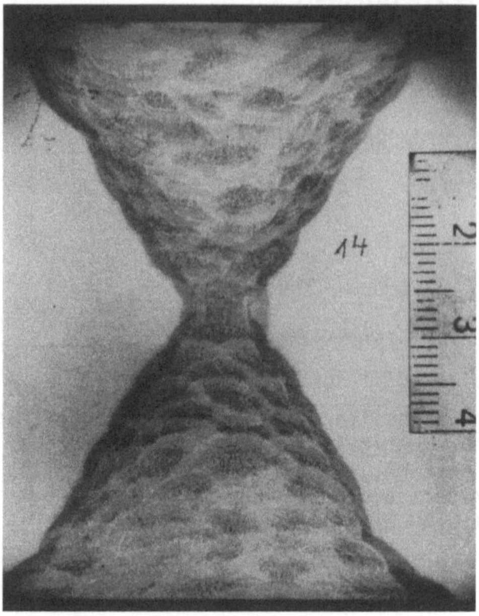

Fig. 1: MMA multipass-weld
(base material: Fe 355, B = 63,5 mm, electrode: EY 4253 MnB, vertical up)

The above mentioned problems are well known and a lot of work has already been done in this field (see for example [1-3]). The available data have shown some of the parameters which control the result of a CTOD-test of a specific weld and they have indicated those parts of the testing procedure which are more important to be followed than others to deliver meaningful and reproducible material characteristics. In the following experiences and test results of IWM in this field are described.

Most of the examples presented are from welding procedure qualifications tests done in connection with offshore projects, that means they concern mainly multipass welds for tubes, nodes, and plates in low alloy structural CMn-steels of the type Fe 355 corresponding to BS 4360 Grade 50D. Submerged arc welding is the preferred procedure which is used for components fabricated in the shop to combine well controlled welding conditions with the advantage of a stress-relief treatment. But manual metal arc welding is considered, too, for joining parts on the site and for repairs; then less favourable welding positions, stronger variations of the process-parameters and welding residual stresses have to be taken into account.

## Characterization of Weld Metal Toughness

### Influence of Notch orientation

When applying the British Standard BS 5762: 1979 [4] to determine critical CTOD-values of the weld material three point bend specimens of component thickness have to be used. Fig. 2 shows different ways of extracting a

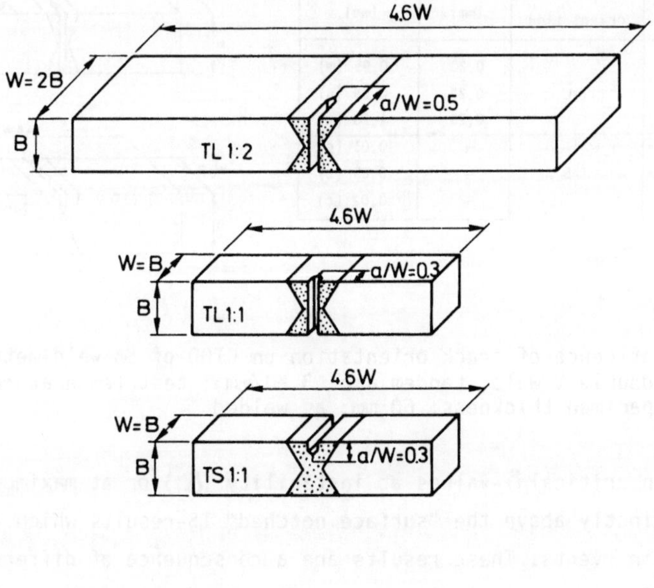


Fig. 2: Test piece geometries for CTOD-testing of weld material

preferred 1:2 test piece or a subsidiary 1:1 test piece from a welded plate of thickness B. To simulate surface crack situations (notch orientation TS) either full-thickness 1:1 or half-thickness 1:2 specimens may be used. Through thickness cracks (TL orientation) could even better be tested with compact (CT) instead of bend specimens as has been proposed by an ASTM task group [5]. The test situation sketched in Fig. 2 is only that for the weld material in a double V joint preparation; other cases are mentioned later especially the K-joint in connection with a fracture mechanics characterization of the heat affected zone (HAZ).

As an example Fig. 3 shows the influence of different notch orientations on the resulting CTOD. The six 1:1 60 mm thick specimens have been extracted from a double V tandem submerged arc weld; they have been tested in the as welded condition at -10°C with crack depths between 0.3 and 0.5 a/W. The "through thickness notched" TL-specimens showed stable crack growth re-

| specimen | crack orientation | $\delta_i$ [mm] | $\delta$ [mm] |
|---|---|---|---|
| T1 | T-L | 0,20 | 0,95 (m) |
| T2 | | 0,22 | 0,39 (u) |
| T3 | | 0,23 | 1,20 (m) |
| B1 | T-S | – | 0,01 (c) |
| B2 | | – | 0,05 (c) |
| B3 | | – | 0,02 (c) |

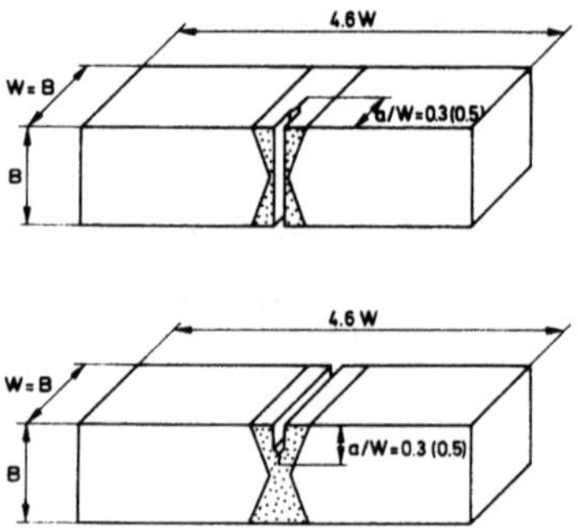


Fig. 3: Influence of crack orientation on CTOD of SA weld-metal (double V weld, tandem arc: 3 KJ/mm); test temperature: - 10°C; specimen thickness: 60 mm; as welded

sulting in critical $\delta$-values at instability ($\delta_u$) or at maximum load ($\delta_m$) distinctly above the "surface notched" TS-results which are derived from pop-in events. These results are a consequence of different distributions of weld metal microstructure along and around the crackfront which are shown in Fig. 4: The front of the through-thickness crack extending in the welding direction cuts through many layers of less tough dendritic and tougher recrystallized material; the crack front of a TS-specimen on the other hand, if it is straight, sees more uniform conditions of only as deposited or only reheated material depending on the crack length.

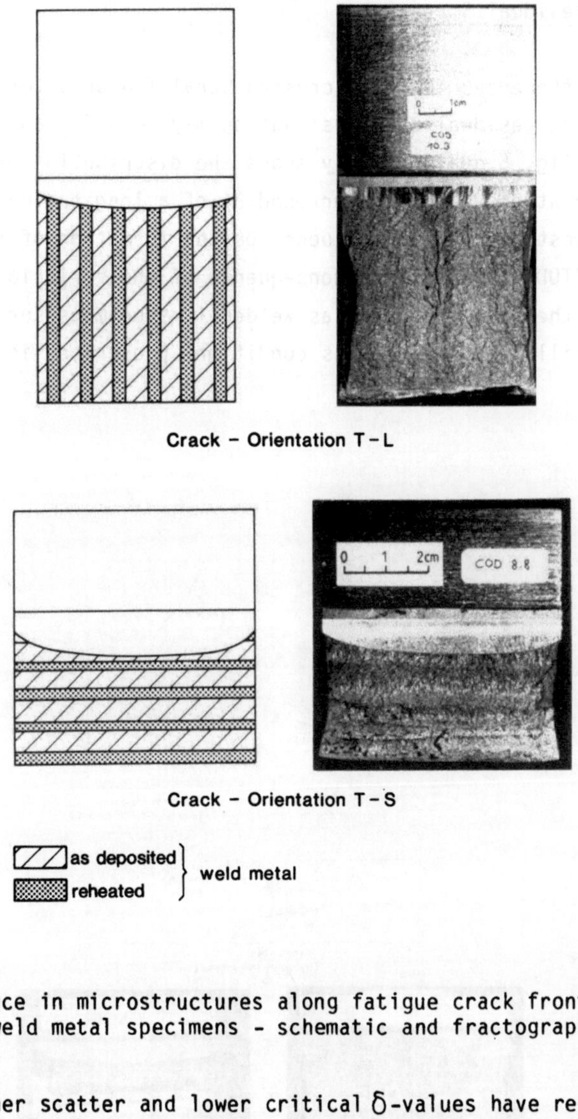


Fig. 4: Difference in microstructures along fatigue crack front for TL- and TS-weld metal specimens - schematic and fractographic phenomena

Therefore a higher scatter and lower critical $\delta$-values have resulted for the TS-specimens as compared to the TL-specimens, which have delivered something like an average value over the different microstructures.

## Influence of Residual Stresses

Besides the differences in the microstructural features for TL- and TS-orientations the residual stress situation may be different for both types of specimens. Fig. 5 qualitatively shows the distribution of longitudinal and transverse stresses in the surrounding of a longitudinal butt welded plate and demonstrates their influence on the formation of the fatigue precrack for CTOD-testing. As a consequence of the variation of residual stresses over the crack front in as welded TL-specimens serious problems arise to fullfill the straightness conditions prescribed in the COD test

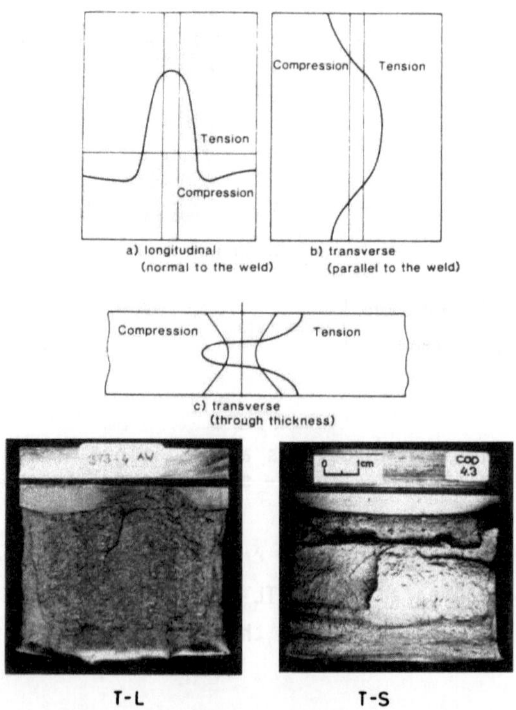


Fig. 5: Qualitative sketch of distribution of welding residual stresses for a longitudinal butt welded plate and influence on straightness of fatigue crack front in TL- and TS-COD-specimens

standard. In TS-specimens the residual stresses may influence the overall fatigue crack growth rate but since being constant along the crack front they do not effect its straightness.

The magnitude of residual stresses developing in a welded joint - besides other parameters - depends on the restraint conditions imposed during welding. Fig. 6 shows a real size construction tube for offshore applications:

Fig. 6: Tubular offshore component and welding qualification test plate welded under high constraint

diameter 1600 mm, wall thickness 63 mm with a longitudinal weld. As a welding qualification test for such a component usually a relatively small plate - like test piece is produced and in addition to other tests is investigated for its fracture toughness in terms of critical CTOD. To be conservative concerning the residual stresses it is important to induce as high a constraint on the weld as possible to simulate the high stiffness of the tube; a possible solution is also indicated in Fig. 6. To quantify this argumentation a program has been started at IWM using the available X-ray equipment - see fig. 7 [6] - to measure the residual stresses on tubular components, test pieces, and specimens.

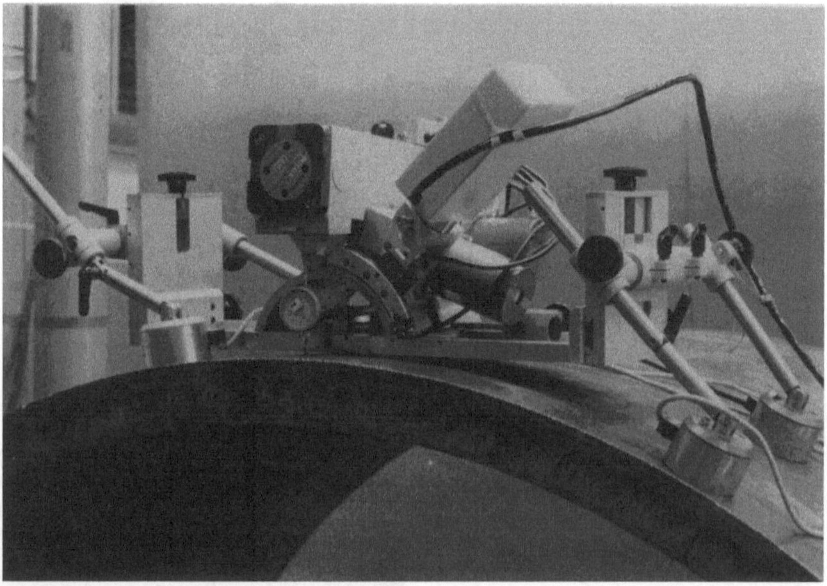

Fig. 7: Measurement of welding residual stresses on tubular components using a mobile X-ray diffractometer

A thermal stress-relief treatment can be used to reduce the residual stresses in welded specimens and by that straight crack fronts for CTOD-testing can be achieved. Fig. 8 as an example shows the result of a test on a MMA weld in a 25 mm thick 9% nickel steel plate: Fatigue precracking the 3 point bend specimen in the as welded condition resulted in an extremely curved crack front; after a thermal stress-relief treatment by further fatigue loading a more or less straight crack front could be developed which then allowed for a valid $\delta$-determination.

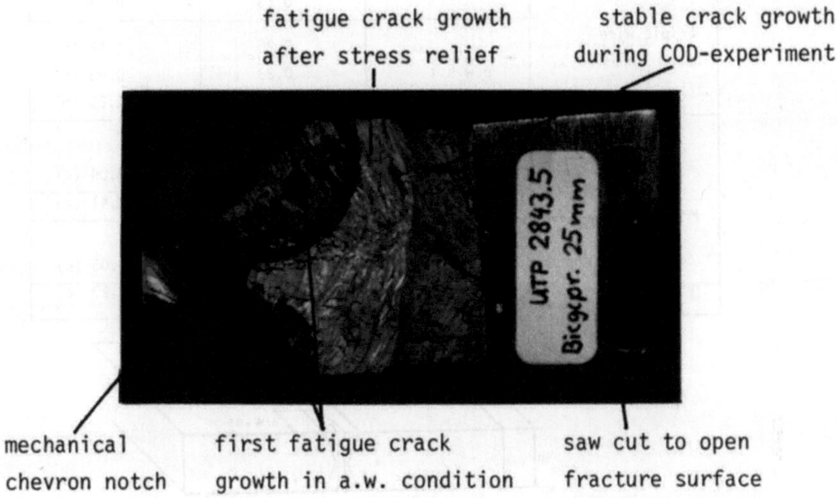


Fig. 8: Fatigue crack growth before and after thermal stress-relief treatment in a welded X8Ni9-COD specimen

In general a thermal stress-relief treatment seems to improve the fracture toughness of the class of offshore weld materials as used nowadays, but can be detrimental in some cases for the base metal. The set of results in Fig. 9 for three wire SA welds and for MMA welds clearly shows the improvement of CTOD in the weld material, the effect is especially large for the manual welds - although on a relatively low level of toughness when compared with that of the submerged arc welds; the improvement by post weld heat treatment effects the toughness values at initiation $\delta_i$, at instability $\delta_u$, and at maximum load $\delta_m$.

| Process | | Condition | $\delta_i$ [mm] | $\delta$ |
|---|---|---|---|---|
| SA triple wire | | as welded | 0,16 | 0,38 (u) |
| | | | 0,13 | 0,86 (m) |
| | | | 0,13 | 0,24 (u) |
| | | p.w.h.t | 0,20 | 1,12 (m) |
| | | | 0,23 | 0,98 (m) |
| | | | 0,21 | 1,25 (m) |
| MMA | Electrode A | as welded | - | 0,04 (c) |
| | | p.w.h.t | - | 0,41 (u) |
| | Electrode B | as welded | - | 0,05 (c) |
| | | p.w.h.t | - | 0,23 (c) |

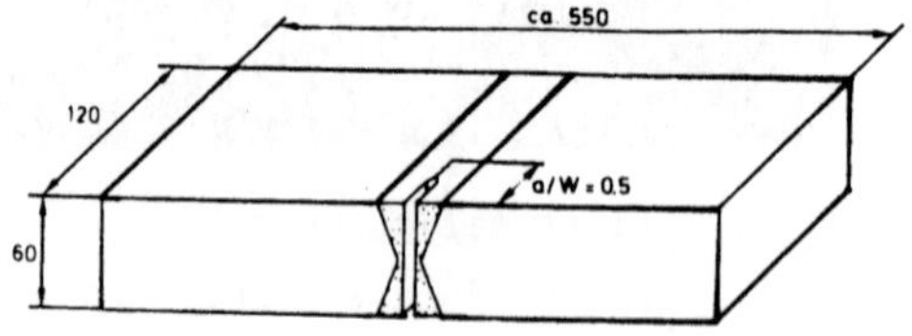


Fig. 9: Influence of thermal stress-relief treatment on weld metal toughness (steel Fe 355; electrodes E 7018)

In many practical cases a stress relief treatment is not feasible or too expensive. Then an alternative method, the method of plastic precompression as developed by Dawes [7] can be employed to produce straight crack fronts in as welded specimens and by this enable valid COD-testing. Fig. 10 demonstrates the principle of the method and the following two figures show results of such a mechanical stress-relief treatment.

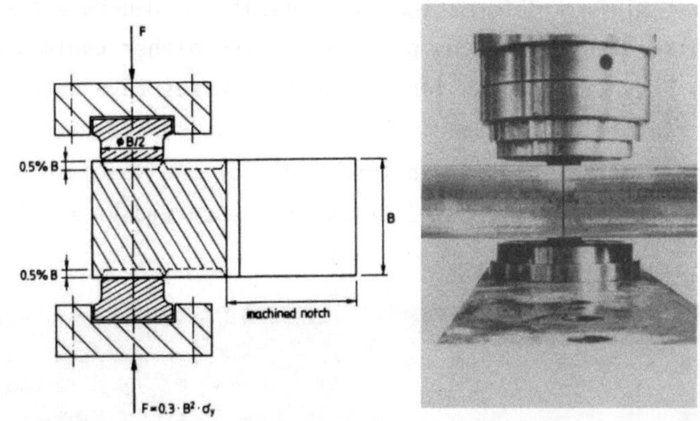


Fig. 10: Principle of mechanical stress relief treatment by local plastic precompression after [7]

In Fig. 11 two TL-type SAW-specimens of 60 mm thickness have been compressed to 0,3 and 0,6 % plastic deformation of their original thickness B.

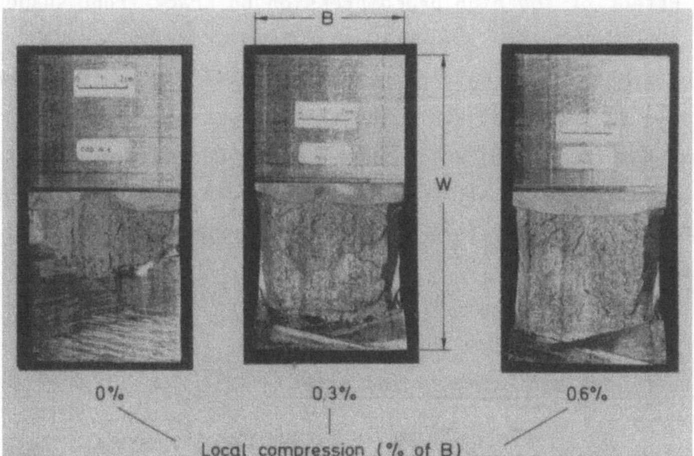


Fig. 11: Influence of local compression on the relief of welding residual stresses and the formation of straight fatigue crack fronts.

Obviously some minimum deformation is necessary to generate the desired straight fronts. Increasing the preloading even higher could produce an effect as in Fig. 12: Compression residual stresses have been induced in the surface regions of the specimen by the plastic deformation and the equilibrating tension stresses in the middle of the specimen have accelerated the fatigue crack growth. Crack fronts curved as in Fig. 12

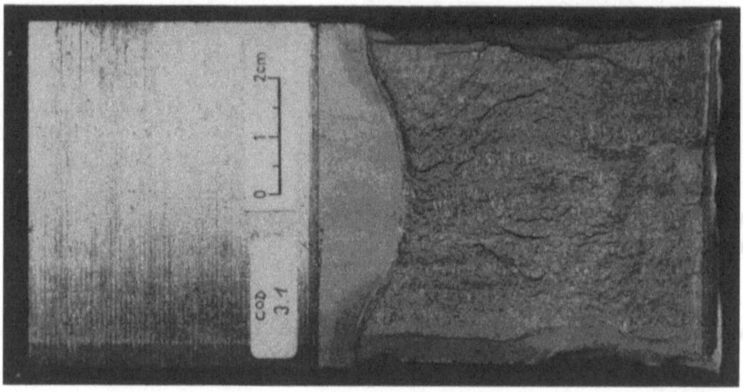


Fig. 12: Effect of too high precompression on crack front shape

have to be avoided because from stress analysis arguments higher critical values could be enforced. Excluding this effect our results seem to support other findings (e.g. [8]) which show a minor decrease of the critical values - if a change at all - by this procedure of mechanical stress-relief treatment and it is therefore applied throughout all testing of as welded specimens.

## Influence of Specimen Geometry and Size

Effects of specimen geometry and specimen size have to be considered when assessing the relevance of a CTOD value determined in the laboratory for a construction. This is the case for any plain material as well as for the compound-system welded joint. An increase of critical CTOD's with decreasing specimen size and crack length is to be expected. This follows from the argument that due to this change of the geometrical parameters the crack tip constraint will be diminished and therefore more plastic deformation takes place. In the case of cleavage type initiation, this increase

| specimen | Condition | crack orientation | B,W [mm] | δ [mm] |
|---|---|---|---|---|
| S1 | as welded | T-S | 60,60 | 0,01 (c) |
| S2 | | | | 0,05 (c) |
| S3 | | | | 0,02 (c) |
| P1 | | | 30,60 | 1,14 (m) |
| P2 | | | | 0,99 (m) |
| P3 | | | | 0,96 (m) |

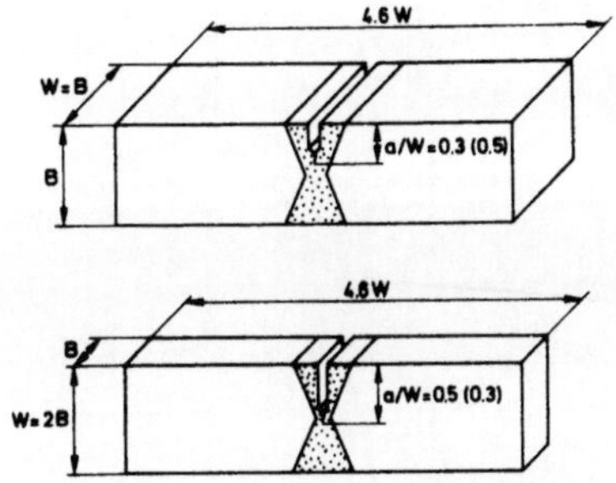


Fig. 13: SA-weld metal toughness as a function of specimen thickness (double-V-joints, tandem-wire)

in CTOD critical values is mostly expected due to a statistical argument e.g. the smaller the thickness B, the lower the initiation points number. The table in Fig. 13 is one set of results to show this trend for the weld metal toughness of TS-specimens of two thicknesses taken from one double V tandem-wire SA joint. The half-thickness specimens exhibited maximum load type behaviour; whereas the higher constrained 60 x 60 mm specimens exhibited pop-in-behaviour indicated by a noise during the test and a distinct load drop in the force-displacement diagram. Figure 14 shows an example together with a picture of the corresponding fracture surface on which the limitated, but predominantly brittle instability can be seen. In this case clearly the pop-in has to be quoted as a critical event in the meaning of the COD-standard.

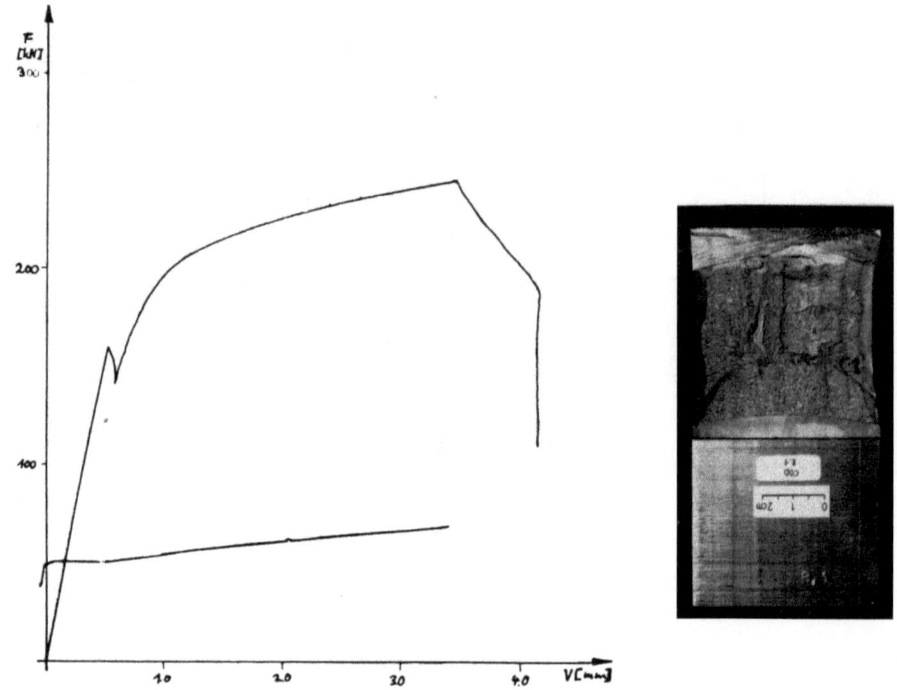


Fig. 14: Pop-in event during the COD-test of a TL oriented 60 x 120 mm 3 point bend specimen from a three-wire SA-weld (double V-joint)

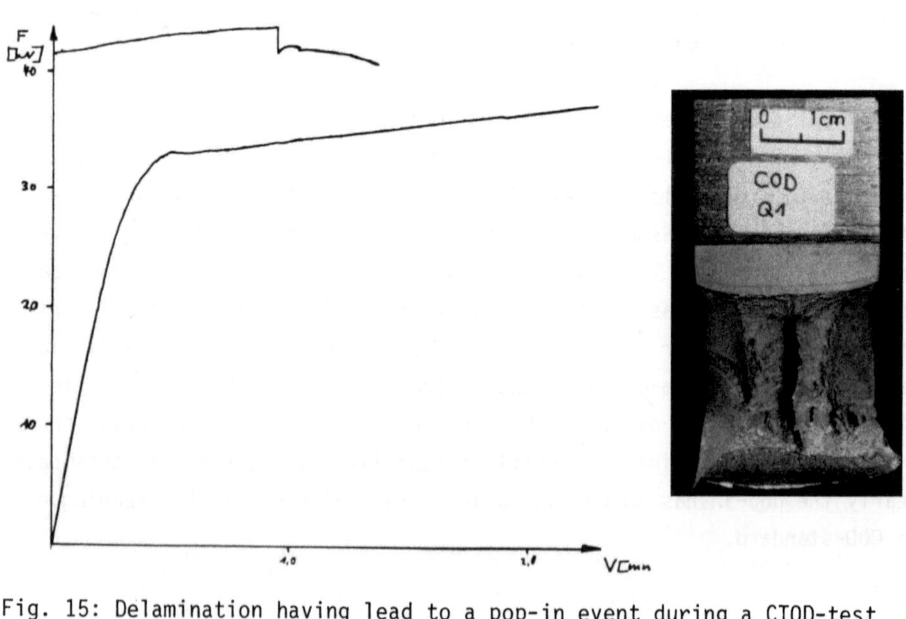


Fig. 15: Delamination having lead to a pop-in event during a CTOD-test (material: API LX Gr. 60, B = 22 mm)

There are other cases as in Fig. 15, where the load drop is caused by the initiation of secondary cracks in planes normal to the main crack. Such "separations" or "delaminations" result in an increased crack resistance by internal shear lip formation. Then the pop-in is no longer a critical event.

## Influence of Welding Procedure and Process Parameters

In Figs. 16 through 19 examples are given for the complex interaction of procedures and parameters in producing SA multipass welds.

In Fig. 16 results are compared for 2 and 3-wire submerged arc weldments.

| Process | Condition | $\delta_i$ | $\delta_i$ [mm] | $\delta$ |
|---|---|---|---|---|
| SA tandem wire | a.w | 0,13 | 0,13 | 0,21 (u) |
| | | 0,16 | | 1,10 (m) |
| | | 0,11 | | 1,03 (m) |
| | p.w.h.t | 0,22 | 0,21 | 1,14 (m) |
| | | 0,21 | | 1,10 (m) |
| | | 0,19 | | 1,08 (m) |
| SA three wire | a.w | 0,16 | 0,14 | 0,38 (u) |
| | | 0,13 | | 0,86 (m) |
| | | 0,13 | | 0,24 (u) |
| | p.w.h.t | 0,20 | 0,21 | 0,98 (m) |
| | | 0,23 | | 1,12 (m) |
| | | 0,21 | | 1,25 (m) |

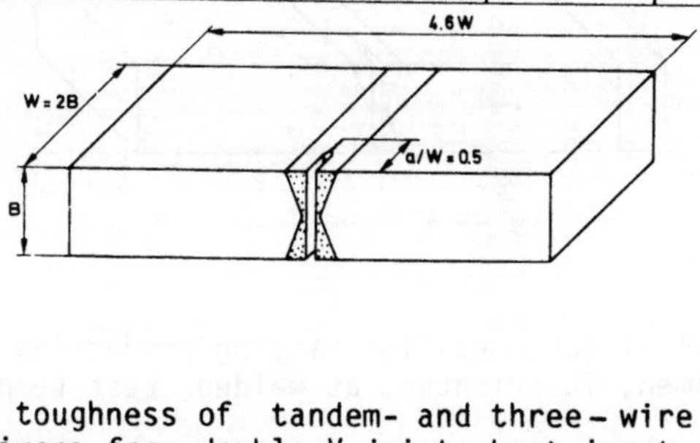


Fig. 16: Weld metal toughness of tandem- and three- wire SAW joints (Bx2B specimens from double V-joint; heat input = 30 KJ/cm)

Besides using the same combination of wire and flux (UPY 42 55 53 BFB 1 65) the processes were carefully controlled to deliver a constant heat input of 30 KJ/cm. Then for TL-specimens the same initiation toughness $\delta_i$ was found with again a 50% improvement by post weld heat treatment.

In Fig. 17 electrodes A, B, and C of the same type (E 7018) but from different manufacturers have been used by the manufacturers themselves and by a welding research institute to produce best workmanship manual metal arc X- and K-joints which were then COD-tested at -10$^{o}$C with 60 x 120 three point bend specimens. The results look quite disappointing concerning the chance to reproduce the quality of a weld if different manufacturers are

| Electrode | Joint geometrie | δ [mm] | welded by electrode manu-facturer |
|---|---|---|---|
| A | X | 0,11 | A |
| B | X | 0,07 (c) | B |
| C | K | 0,26 | C |
| A | X | 0,04 (c) | welded by SLV |
| B | X | 0,05 (c) | " |
| C | K | 0,06 (c) | " |

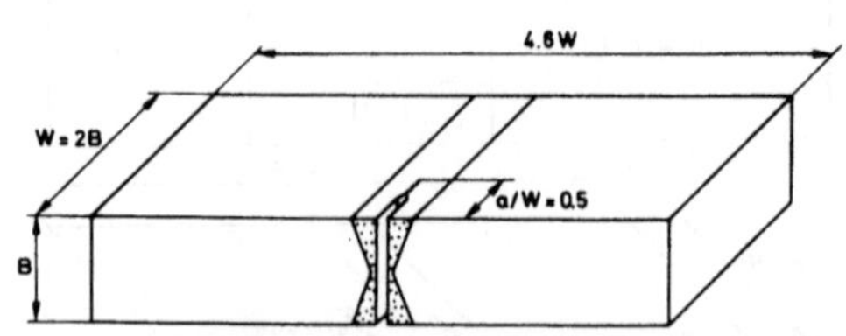


Fig. 17: MMA weld metal toughness for varying production parameters (1:2 specimen, TL oriented, as welded, test temperature: -10$^{o}$C)

involved. But in Fig. 17 K-joints beeing welded in transverse position are compared to vertical up welded X-joints. The different positions certainly change the microstructural configuration as it is shown in Fig. 18 for MMA-welds. When testing through-thickness notched specimens different orientations of the weld beads with the crack plane result in differences of fracture toughness: whereas for the K-joint the crack plane intersects the main orientation of the dendrites with an angle of 30 to 40 degree causing a "high" $\delta_u$-value in the double V joint the dendrites are more or less parallel with the crack extension direction and the crack can follow easily the weak planes of the microstructure causing a"low" $\delta_c$-value. A corresponding tendency has been described in [9].

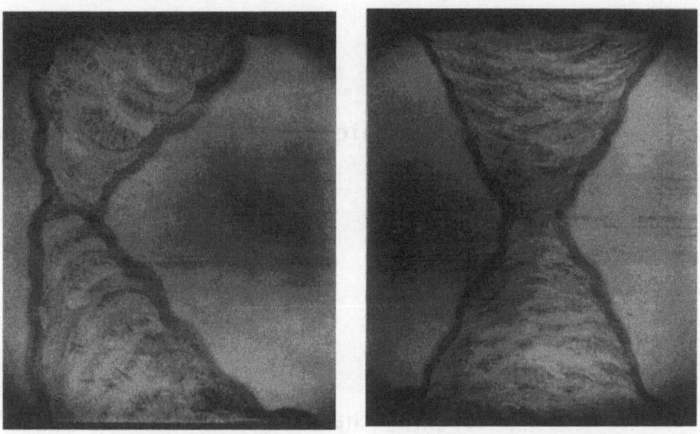

$\delta_u = 0.7\,mm$ $\delta_c = 0.1\,mm$

Fig. 18: Influence of welding position in MMA-joints on formation of microstructure and weld metal toughness

Fig. 19 demonstrates the influence of two different ways of path control on the weld metal fracture toughness. The same electrode has been used in the MMA process and both double V-joints have been welded in vertical-up position: The weaver bead joint gives much better toughness than the stringer bead because of its higher content of grain-refined interpass HAZ microstructure at the crack front.

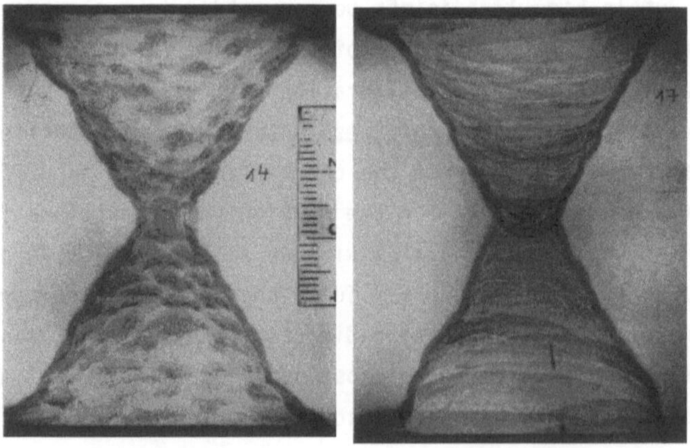


stringer bead $\delta_c = 0.04$ mm　　weave bead $\delta_c = 0.11$ mm

Fig. 19: Influence of welding technique on MMA weld metal toughness

## Influence of Testing Technique

As a supplement to the preceeding chapters some details of IWM's experimental techniques in CTOD-testing are described in the following. Multiple clip gages over the crack depth are used as shown in Fig. 20. An optical technique, applied in parallel, allows the measurement of the opening of the crack which for each experiment separately can be extrapolated to the crack tip to deliver a CTOD-value.

Fig. 20: 3 clip-gage arrangement mounted on a 60 mm preferred COD-testpiece

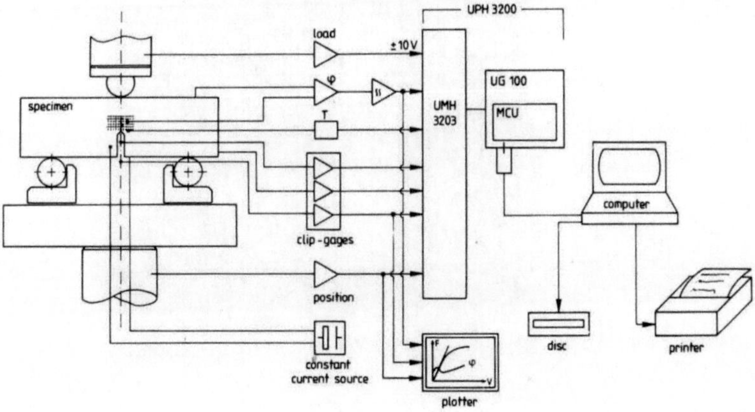


Fig. 21: Test set-up and data acquisition system

Fig. 21 schematically shows the test set-up and the data acquisition system. Fig. 22 shows one example of the optical crack opening measurement at the surface of the specimen.

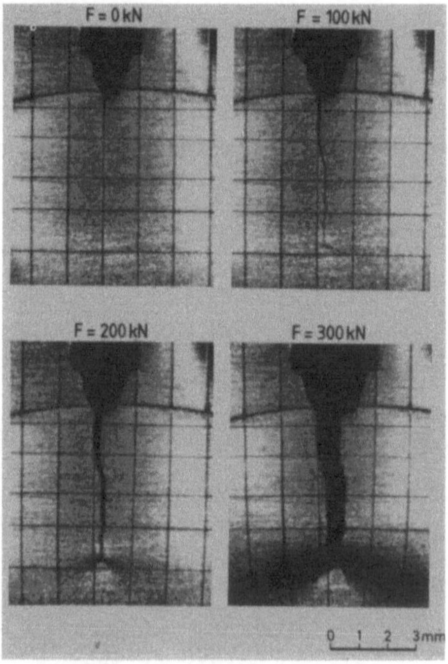


Fig. 22: Optical crack opening measurement

With the help of multiple crack opening measurements it is possible to calculate individual rotational factors r and to check the conservatism of the current evaluation following BS 5762 using a constant factor r = 0.4. Fig. 23 is an evaluation of some weld metal tests. It shows that the BSI-values are only slightly conservative compared to the individual CTOD data.

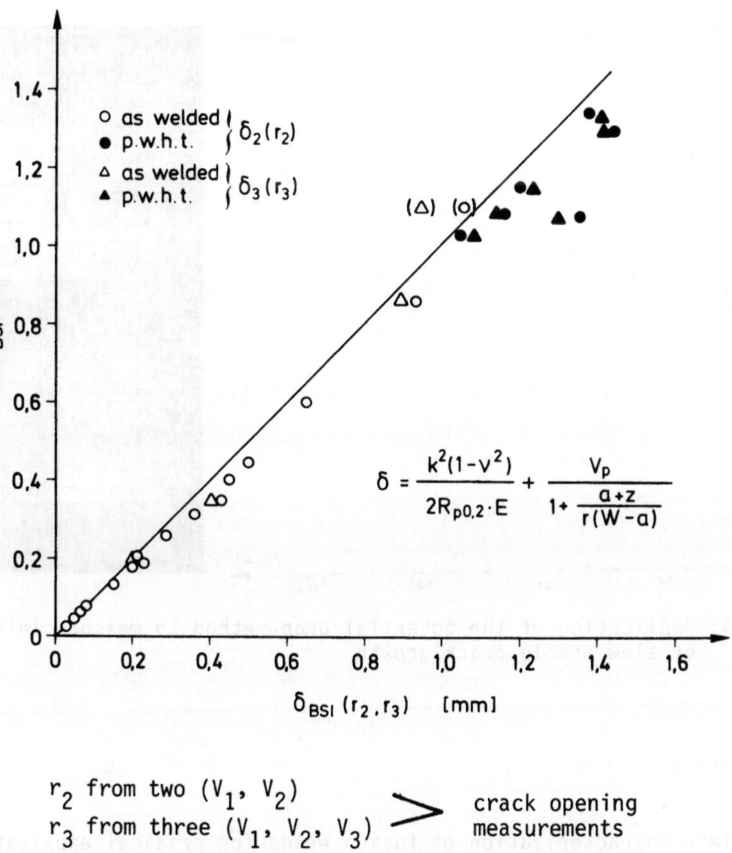


Fig. 23: Evaluation of CTOD comparing constant and experimental hinge point ratio determination

To detect initiation of slow stable crack growth DC- or AC-potential drop methods are used. The change of slope in the $\varphi$-V-diagram (see Fig. 24) is a clear indication for the onset of stable crack growth. Critical CTOD-values calculated for onset of stable crack growth ($\delta_i$) are a conservative but compared to $\delta_u$ and $\delta_m$ a more reliable material toughness parameter, because they show less scatter.

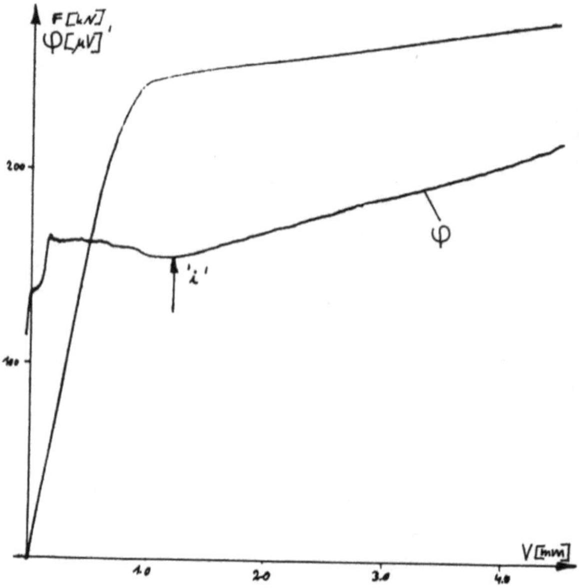


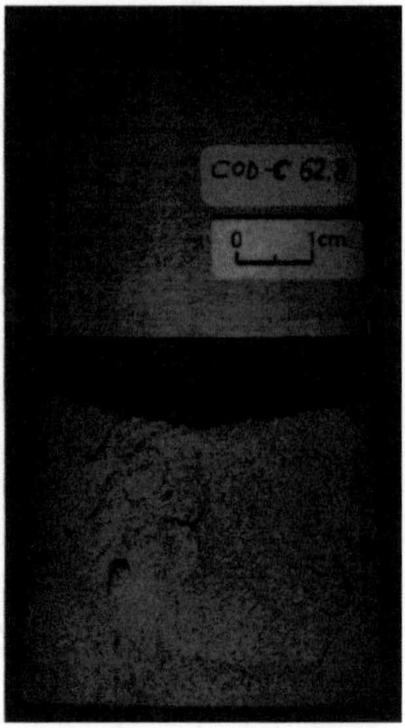


Fig. 24: Application of the potential drop-method to measure initiation of slow stable crack growth

HAZ-Testing

A complete characterization of fusion welds for critical applications requires also the determination of the fracture toughness of the heat affected zone. Depending on the welding process, welding parameters, joint geometry, and materials the extension and form of the critical coarse grained zone in the HAZ can differ significantly and therefore cause several practical problems in connection with CTOD-testing. Fig. 25 shows examples for different joint geometries and notch orientations to determine full thickness HAZ-CTOD values.

From Fig. 25 it is evident that machining of the notch and fatigue precracking is difficult with respect to a precise and reproducible location of the crack-front in the material zone of interest. Therefore the real position of the crack-front and the point of initiation of unstable or stable crack propagation in the coarse grained material has to be controlled and documented in photomacrographs for each specimen tested. Fig. 26 shows an example, which has been accepted as a valid HAZ-test based on a specific offshore specification.

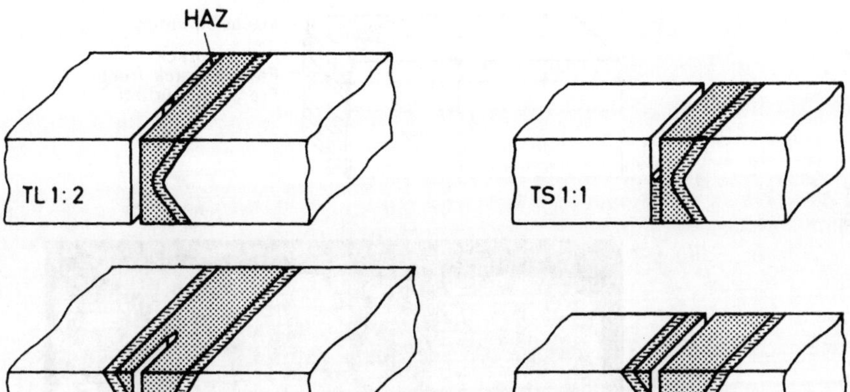


Fig. 25: HAZ-testing on double V- or K-joints

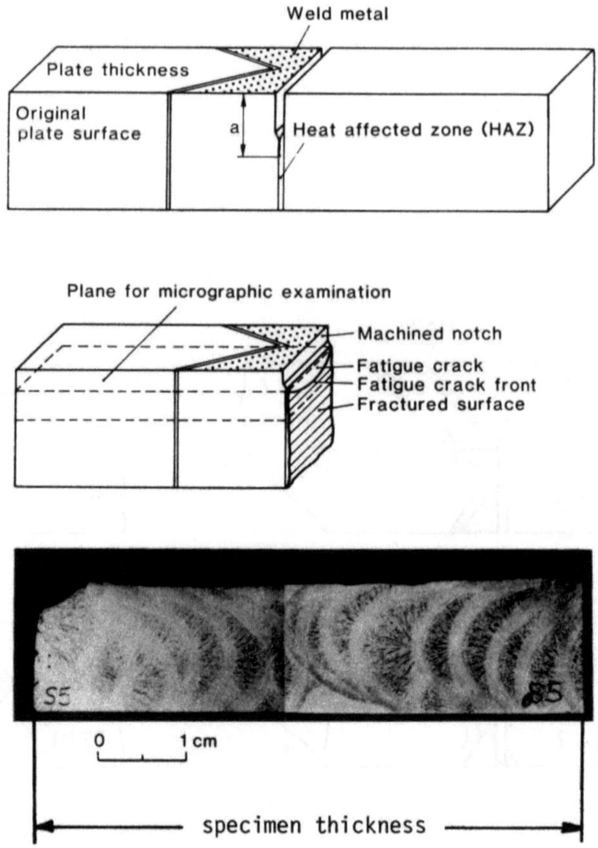


Heat affected zone microstructure along the fatigue crack front (through thickness notch)

Fig. 26: Location of the crack front in the heat affected zone

## Conclusions

Based on our experience in CTOD-testing of welded joints only a very strict control and a complete documentation of production parameters, like

- welding procedure and parameters
- welding position
- joint preparation (geometry)
- welding consumables
- restraint conditions (residual stresses)
- post weld heat treatment

and of testing parameters, like

- specimen geometry (SENB, CT) and thickness
- notch orientation
- evaluation of CTOD

will enable to generate well defined and reproducible CTOD test-results for welds with an acceptable scatter. Only such values may be used as a basis for a systematic investigation into the improvement of strength and toughness of multipass welds (weld metal and heat affected zone) and for structural applications in material selection, design, or defect assessment.

References

[1] Pisarski, H.G; Dawes, M.G.
"Measurement of crack opening displacement in weldments with particular reference to offshore structures"; ECSC Working Group 2 Meeting, Genova 11.10.1979

[2] Harrison, J.D.; Anderson,T.L.
"The application of fracture mechanics to welded construction"; Eighteenth ASTM National Symposium on Fracture Mechanics, Boulder, Colorado, USA, 24-27.6.85

[3] Satoh, K.; Toyoda, M.
"Applicability of bending COD test of welds for securing required fracture performance of welded structure"; IIW Doc. X-1063-84

[4] BS 5762, 1979 "Methods for crack opening displacement (COD) testing"; British Standards Institution, London

[5] Draft ASTM test method for crack tip opening displacement (CTOD) testing; Working document E 24 commitee, Feb. 1984 (unveröffentlicht)

[6] Christian, H.; Elfinger F.-X.; Klug, W.
"Ein Röntgengoniometer für den mobilen Einsatz"; Materialprüfung 18(1976) Nr. 10

[7] Towers, O.L.; Dawes, M.G.
"Welding Institute research on the fatigue precracking of fracture toughness specimens"; Elastic-plastic Fracture Test Methods: The User's Experience, ASTM STP 856, E.T. Wessel and F.J. Loss

[8] (see [1])

[9] Ruge, J.; Lee, B.-Y.; Wösle, H.
"Einfluß der Kerblage auf die Kerbschlagarbeit unterpulvergeschweißter Nähte"; Schweißen und Schneiden 38 (1984), H.4

# Statistical Analysis of Critical CTOD and J Values in SAW C-Mn Steel Welds

A. FONTAINE*, J.A. LE DUFF**, B. MARANDET*, Y. SIXOU*

* Institut de Recherches de la Sidérurgie Française (IRSID) St Germain-en-Laye, France
** Institut de Soudure (IS), Ennery, France

## ABSTRACT

A statistical analysis of fracture toughness test results obtained in the case of the weld metal of a structural steel is reported. It is shown that, when the standard J and CTOD tests methods are used, no significant differences between different laboratories are observed. CT specimens are as well suited as SENB for CTOD determination. The scatter of the fracture toughness values at initiation is correctly fitted by a log-normal distribution. The level of fracture toughness at a given temperature depends on the in-plane dimensions of the specimens, when initiation takes place by ductile tearing. The relationship between J and CTOD critical values is reanalysed.

## INTRODUCTION

Steels and steel weldments usually exhibit a large scatter in fracture toughness values in the brittle - ductile transition temperature range [1-3]. This scatter generally increases with temperature up to the ductile regime. Experience shows moreover that the larger the dimensions of the fracture mechanics specimen (especially the ligament width), the higher the temperature of the brittle-ductile transition [4-5].

In the present study, a statistical analysis of fracture toughness results was applied to tests carried out on the weld metal of welded joints prepared using a fine grained structural steel. The purpose of this work was to evaluate the effects of experimental parameters (specimen type and dimensions, temperature, laboratory) on the variability of fracture toughness results and on fracture initiation mechanism, and to analyse the statistical relationship between the critical values of J and CTOD.

## EXPERIMENTS AND RESULTS

This work was carried out on a bipass-submerged arc weld metal. The weldments were prepared at IS using a 20 mm thick C-Mn steel plate and were stress relieved at 580°C for 2h30. Conventional mechanical properties and chemical compositions of base plate and weld metal are given in Table I.

The microstructure of the weld metal consists of proeutectoid ferrite veins or lath-like ferrite grains nucleated at the prior austenite grain boundaries, and of acicular ferrite grains interspersed with carbide-ferrite aggregates.

The fracture mechanics specimens were positioned so that the fatigue precrack would grow along the centerline of the weld, parallel to the welding direction and only within the weld metal (Figure 1). These specimens were taken at random positions from several welded joints in order to avoid any sampling effect [6].

Two types of SENB and four types of CTJ specimens were used. The thickness B of all the specimens was kept constant and the width W was varied from 17,5 mm to 750 mm (Figure 2). The fatigue precrack was extended from the mechanical notch according to the method of ASTM E399 and an original crack length/ specimen width ratio $a_o/W$ of about 0.5 was achieved.

The critical values of J and CTOD were estimated for each test following the standards BS 5762 and ASTM E 813. In the case of the CTJ specimens, the critical CTOD values were estimated, using the same formula as for SENB specimens, from the load-line displacement. The critical values of the Stress Intensity Factor (called $K_J$) were calculated by the usual relationship :

$$K_J = \left( \frac{EJ}{1 - \nu^2} \right)^{\frac{1}{2}}$$

with E = 206000 MPa and $\nu$ = 0.3 (Temperature variations of E and $\nu$ are neglected).

The terminology of BS 5762 was used for the critical fracture toughness values : depending on the type of fracture initiation, different subscripts are used ($CTOD_C$ - $K_{Jc}$, $CTOD_u$ - $K_{Ju}$, $CTOD_i$ - $K_{Ji}$, $CTOD_m$ - $K_{Jm}$).

Except for the case of the largest specimens (CR30), the onset of ductile tearing was detected using the AC potential drop method developped at IRSID [7].

In order to study the effect of different laboratories practices on the variability of the results, 88 specimens were broken at IS and 100 at IRSID. Five temperatures within the brittle-ductile transition range were selected (-120°C, -80°C, -60°C, -40°C and -20°C).

As expected, the scatter of the fracture toughness values in the weld metal increases markedly with temperature, whether the results are expressed in terms of $K_J$ or of CTOD (Figure 3). Furthermore, the critical fracture toughness values at and after ductile tearing ($CTOD_i$ and $CTOD_u$) measured with the largest specimens (CTJ100 and CR30) are much greater than those measured with smaller ones. The critical fracture toughness values at cleavage initiation seem to be independent of the specimen type (SENB or CTJ) and of the in-plane dimensions of the specimens.

## Statistical analysis

According to different authors [1,2,8], the fracture toughness values of steel and steel weldments follow a lognormal distribution in the ductile to brittle transition range. The statistical analysis was thus carried out on the lognormal transformation of the experimental values.

The ability of this distribution to represent the experimental scatter was checked by statistical and visual tests. At each temperature, the effect of specimen type and dimensions was analysed by plotting the critical values of CTOD and $K_J$ on lognormal graphs. Figure 4 shows that at -40°C, for example, whatever the specimen type, all the fracture toughness values measured at IRSID follow a lognormal distribution defined by the following cumulative probability of failure :

$$f(\mathrm{CTOD}) = \frac{1}{\sigma\sqrt{2\pi}}\,\frac{1}{\mathrm{CTOD}}\,\exp\left[\frac{(\ln \mathrm{CTOD} - \mu)^2}{2\sigma^2}\right] ,$$

where the mean value $\mu$ is equal to 1,844, and the standard deviation $\sigma$ is equal to 0,2618 (CTOD in µm).

These statistical parameters are given in Table II for all the distributions of both fracture parameters (CTOD and $K_J$) analysed during this study.

Surprisingly, except for the case of the largest specimens (CTJ100 and CR30), the statistical analysis has shown that there was no significant difference between the fracture toughness values measured at the onset of cleavage or at initiation of ductile tearing.

In the case of cleavage initiation, the fracture toughness values measured at each temperature in both laboratories with different specimen types and in-plane dimensions were shown to belong to the same population, always correctly fitted by a lognormal distribution, whatever the fracture parameter used ($K_J$ or CTOD) [6].

The cumulative probability of the critical $K_J$ values measured in this study are plotted Figure 5 on a lognormal graph. It can be seen that the distributions are shifted with the temperature and that the standard deviation of the results, which is related to the slope of the plots, remains approximately constant, except for the lowest temperature. Moreover, this figure shows that there is no difference between the fracture toughness values measured in both laboratories.

## FRACTURE INITIATION MODES

Two fracture initiation modes have been encountered during this study : cleavage and ductile tearing (followed by brittle fracture or by plastic collapse).

It is generally expected that the specimen type and the in-plane dimensions influence :

- the measured fracture toughness values;
- the probability of initiating fracture by one of the two possible mechanisms. This probability was evaluated here by the percentage of fractures initiated by ductile tearing.

The plot of the CTOD values at -20°C versus the specimen width W (Figure 6) indicates that, at constant thickness B :

- the fracture toughness values at cleavage initiation ($CTOD_c$) are independent of specimen width;
- the fracture toughness values at and after ductile tearing initiation ($CTOD_i$, $CTOD_u$ and $CTOD_m$) increase strongly with specimen width;
- cleavage initiates or reinitiates by pop-ins in the case of all CR30 specimens and only half CTJ100 specimens. In the case of SENB BxB specimens no pop-in was encountered.

The probability of fracture by one or the other mechanism depends strongly on the width of the specimen W, as diagrammaticaly presented on Figure 7. This diagram shows practically that at a temperature where most of the small specimens break by ductile tearing, only a few large specimens break by this mechanism. Moreover, the proportion of fractures initiated by ductile tearing increases more steeply with temperature in the case of the largest specimens.

In this study, ductile tearing initiation appears above -80°C for SENB BxB specimens and only above -20°C for CR30 specimens. At -20°C, 80 % of SENB BxB specimens still initiate by ductile tearing. More generally, small fracture toughness specimens tend to yield conservative CTOD or $K_J$ values for ductile tearing initiation and correct values for cleavage initiation. Nevertheless it must be noticed that the initiation mechanism is often different, at the same temperature, for small and large specimens.

## CTOD AND J CRITICAL VALUES RELATIONSHIP

The theorical relationship between the fracture parameters is :

$$J = m\,\sigma_y\ \mathrm{CTOD},$$

where $\sigma_y$ is the yield stress at the given temperature.

As shown previously, the distribution of the critical J and CTOD values is lognormal. Consequently the m values also follow a lognormal distribution. This means that, instead of estimating the mean value of the coefficient m by a linear regression between the critical values of $J/\sigma_y$ and CTOD, it should be better to describe first the scatter of the m values by a lognormal distribution and then to use as a mean value of m the mean value of its statistical distribution [6].

In order to take into account the influence of the constraint factor on m, the experimental values obtained at different temperatures have been separated into three groups corresponding to different constraint factors. Figure 8 shows that the three groups of results plotted on a lognormal graph are correctly fitted by this distribution and that their mean values are slightly different. It is also confirmed that the larger the specimen in-plane dimensions, the smaller the m value (the plane stress theoretical value is equal to 1).

It can be seen on Figure 9 and Table III that the mean m values estimated by this statistical analysis (see Figure 8) are different from those, which correspond to regression lines drawn directly through the experimental points.

## CONCLUDING REMARKS

The statistical analysis of CTOD and $K_J$ critical values measured in the brittle-ductile transition range for a SAW C-Mn steel weld metal shows that the variability of fracture toughness results can be correctly fitted by a lognormal distribution, whithout any significant influence of the testing laboratory.

The fracture toughness seems to be independent of specimen type (SENB or CT) and in-plane dimensions at the onset of cleavage fracture but a strong increase of the critical values measured at or after ductile crack extension is observed when the specimens become larger.

Similarly, it appears that the increase of the specimen in-plane dimensions tends to promote cleavage initiation at higher temperatures.

The m values, which relate the critical values of J and COD, also follow a lognormal distribution and decrease significantly when larger specimens are used.

ACKNOWLEDGEMENTS

The authors are indebted to MM REGNIER (IRSID) and DOUCET (IS) for their participation to this study.

Sincere thanks are expressed here to Miss THOMAS and M. LALOI who carried out the experimental part of the project.

This work was partly carried out with the financial support of the DGRST (Délégation Générale à la Recherche Scientifique et Technique).

REFERENCES


1. O.L. TOWERS et al "ECSC Collaborative Elastic-Plastic Fracture Toughness Testing and Assessment Methods". Agreement Number 7210/KE/805 - F5. 1/78, Final Report Welding Institute N° 3571/10M/84, March 1984.

2. G.O. JOHNSTON "Toughness Distributions in two Steels" Welding Institute Report N° 106/1979.

3. G.O. JOHNSTON "Toughness Distributions in Submerged Arc Welds" Welding Institute Report 7304.04/80/1971.

4. I. MILNE and G.G. CHELL "Effect of Size on the J Fracture Criterion", ASTM STP 668, 1979 pp 358-377.

5. P. DE ROO, B. MARANDET, G. PHELIPPEAU and G. ROUSSELIER "Effect of Specimen Dimensions on J1C at Initiation of Crack Growth by Ductile Tearing", ASTM STP 833, 1983 pp 606-621.

6. A. FONTAINE, B. MARANDET, Y. SIXOU, J.A. LE DUFF "Analyse statistique des mesures de ténacité dans le métal fondu d'un joint soudé", Rapport IRSID N° RE 1184, Avril 1985.

7. B. MARANDET, G. SANZ "Experimental Verification of the J1C and Equivalent Energy Methods for the Evaluation of the Fracture Toughness of Steels", ASTM STP 631, 1977 pp 462-476.

8. F. BASTENAIRE "Etude statistique de l'influence des divers facteurs sur les résultats des mesures de COD dans les soudures", Note IRSID STAT 77/24, August 1977.

CHEMICAL COMPOSITION (W %)

| | C | Mn | Si | S | P | Cr | Nb | Al | Cu | Ni | Ti | V | $N_2$ | $O_2$ |
|---|---|---|---|---|---|---|---|---|---|---|---|---|---|---|
| Base plate | 0,17 | 1,44 | 0,27 | 0,010 | 0,011 | 0,007 | 0,003 | 0,056 | 0,022 | - | - | - | 0,010 | 0,002 |
| Weld metal | 0,11 | 1,55 | 0,35 | 0,011 | 0,019 | <0,05 | 0,017 | 0,036 | - | <0,05 | <0,017 | <0,005 | - | 0,007 |

MECHANICAL PROPERTIES

| | Tension Tests (20°C) | | | Charpy V Impact Tests | |
|---|---|---|---|---|---|
| | y (MPa) | UTS (MPa) | Elongation (%) | TK 28J (°C) | Upper Shelf Kv (J) |
| Base plate | 395 | 558 | 34 | - 60 | 155 |
| Weld metal | 569 | 654 | 23 | - 42 | 150 |

**TABLE I** : Chemical compositions and mechanical properties of studied materials

Fracture Parameter : CTOD

| Température (°C) | IRSID | IS | Both Laboratories | Total number of specimens |
|---|---|---|---|---|
| -20 | 1,947/0,1946 | 1,680/0,1630 | * | 17 |
| -40 | 1,844/0,2618 | 1,802/0,1497 | 1,827/0,2211 | 37 |
| -60 | 1,596/0,2127 | 1,683/0,1316 | 1,622/0,1962 | 30 |
| -80 | 1,321/0,1792 | 1,262/0,2518 | 1,296/0,2120 | 38 |
| -120 | 0,817/0,0887 | 1,026/0,1571 | * | 26 |

Fracture Parameter : $K_J$

| Température (°C) | IRSID | IS | Both Laboratories | Total number of specimens |
|---|---|---|---|---|
| -20 | 2,132/0,0999 | 2,022/0,0645 | 2,093/0,1028 | 17 |
| -40 | 2,100/0,1210 | 2,042/0,0856 | 2,076/0,1142 | 37 |
| -60 | 1,978/0,0993 | 2,022/0,0728 | 1,996/0,0971 | 30 |
| -80 | 1,865/0,0816 | 1,773/0,1209 | 1,826/0,1092 | 38 |
| -120 | 1,661/0,0519 | 1,640/0,0862 | 1,650/0,0658 | 26 |

* distributions statistically different between the two laboratories

**TABLE II** : Estimated means and standard deviations of the lognormal distribution of CTOD (µm) and $K_J$ (MPa$\sqrt{m}$) critical values.

| Specimens | CR30,CTJ100 | CTJ50,CTJ25,SENB Bx2B | SENB BxB |
|---|---|---|---|
| Linear Regression | 1,20 | 1,39 | 2,16 |
| Statistical Analysis | 1,35 | 1,68 | 1,92 |

**TABLE III** : Mean values of the m coefficient for the IRSID results estimated by the usual linear regression method and by the proposed statistical analysis.

Fig.1. Diagrammatic SENB Bx2B specimen positionning in the welded joint.

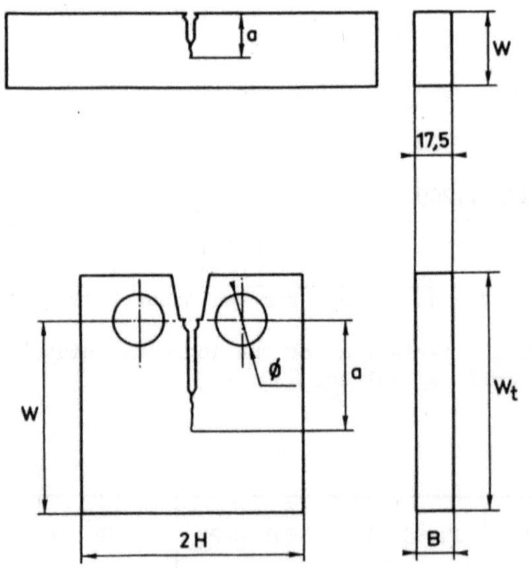


SENB specimens

| | B × B | B × 2B |
|---|---|---|
| W(mm) | 17,5 | 35 |

CTJ specimens

| | CTJ 25 | CTJ 50 | CTJ 100 | CR 30 |
|---|---|---|---|---|
| W(mm) | 50 | 100 | 200 | 750 |
| $W_t$ (mm) | 62,5 | 125 | 250 | 938 |
| 2H(mm) | 60 | 120 | 240 | 900 |
| ø (mm) | 12,5 | 25 | 50 | 188 |

Fig.2. Geometries of the six types of fracture specimens used. Note that the thickness B is kept constant.

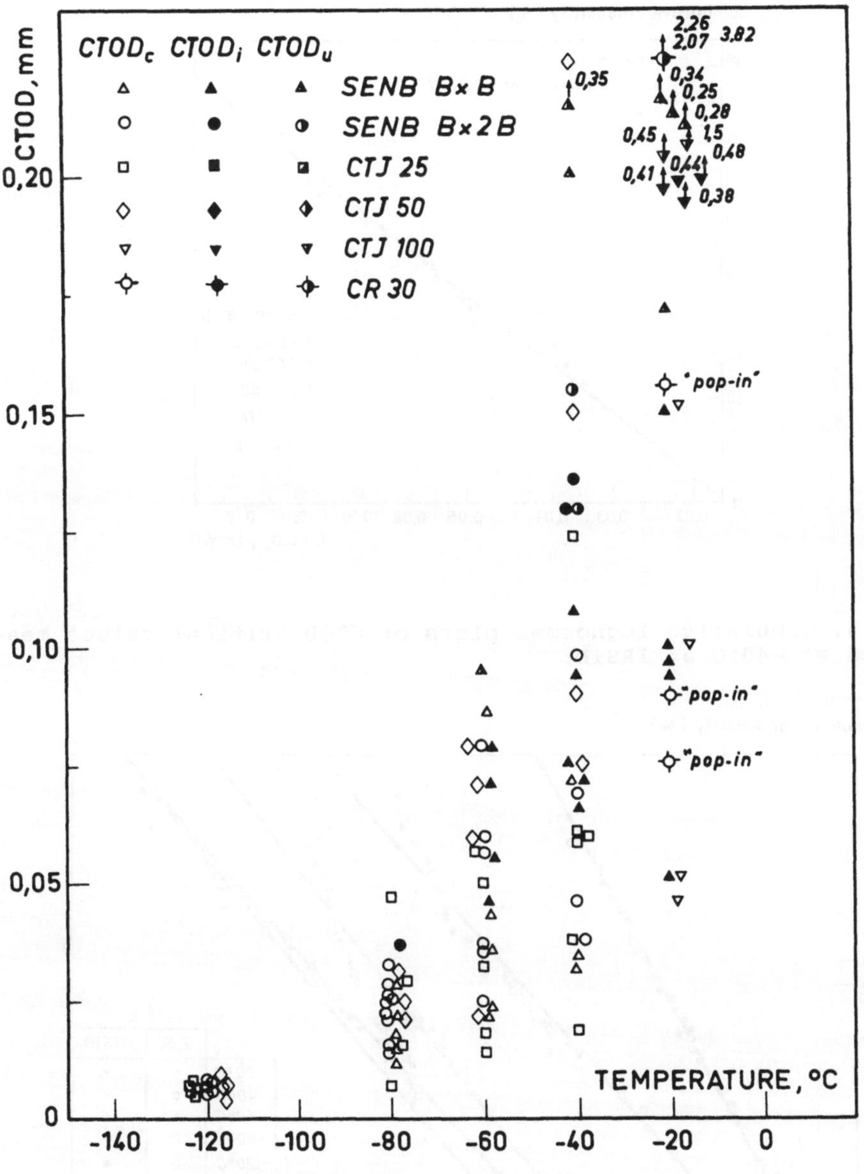


Fig.3. Critical CTOD values for all the specimens tested at IRSID as a function of temperature.

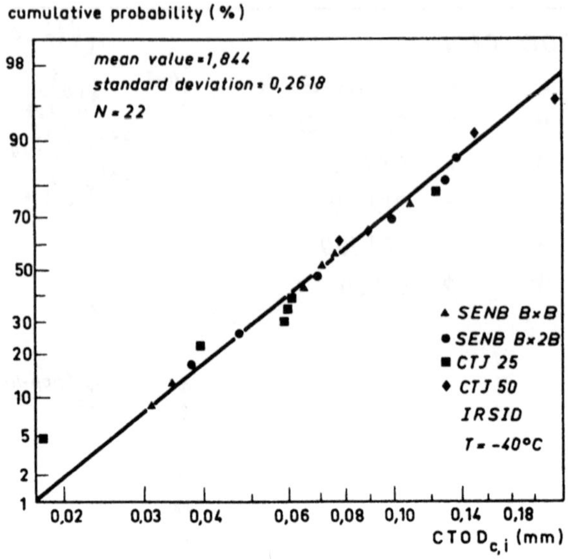


Fig.4. Cumulative lognormal plots of CTOD critical values measured at -40°C at IRSID.

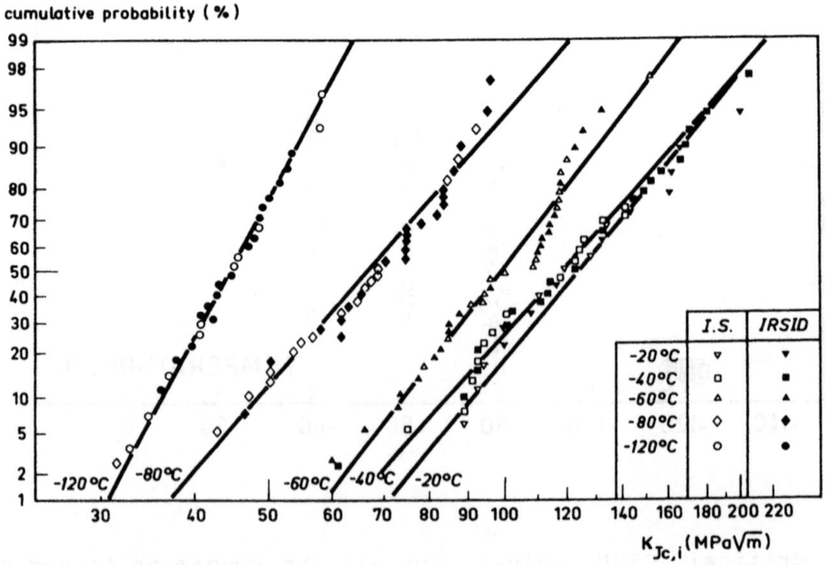


Fig.5. Cumulative lognormal plots of all KJ critical values measured during this study.

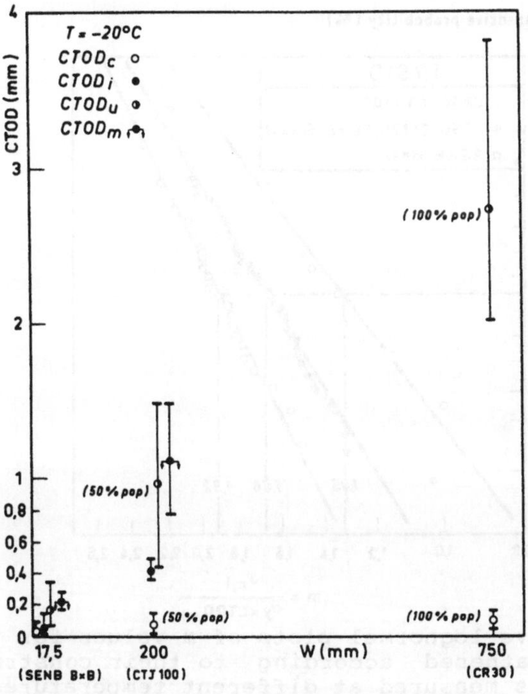


Fig.6. Effect of specimen width W on the critical CTOD values measured at -20°C. Note the percentage of pop-ins in the case of cleavage ruptures.

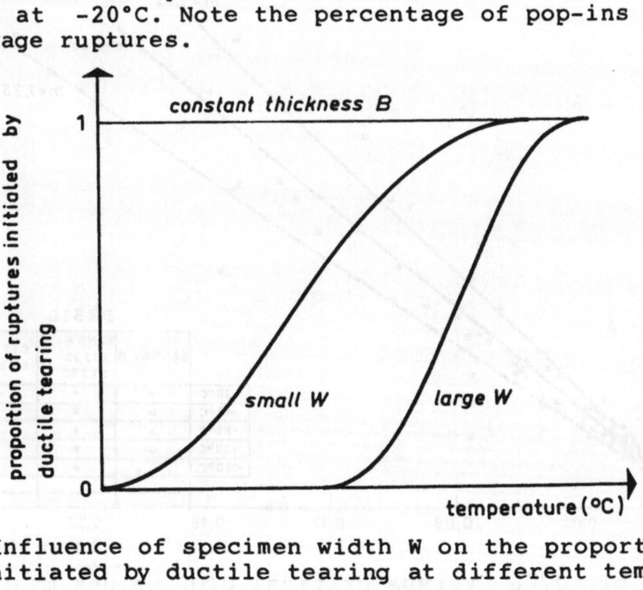


Fig.7. Influence of specimen width W on the proportion of ruptures initiated by ductile tearing at different temperatures.

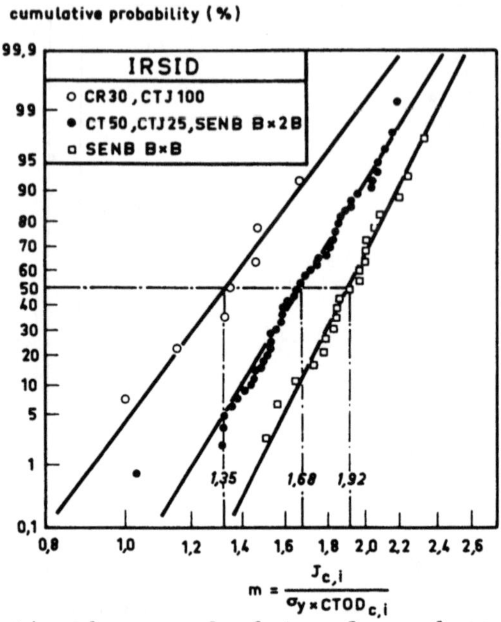


Fig.8. Cumulative lognormal plots of m values for three groups of specimens gathered according to their constraint factor. Toughness values measured at different temperatures.

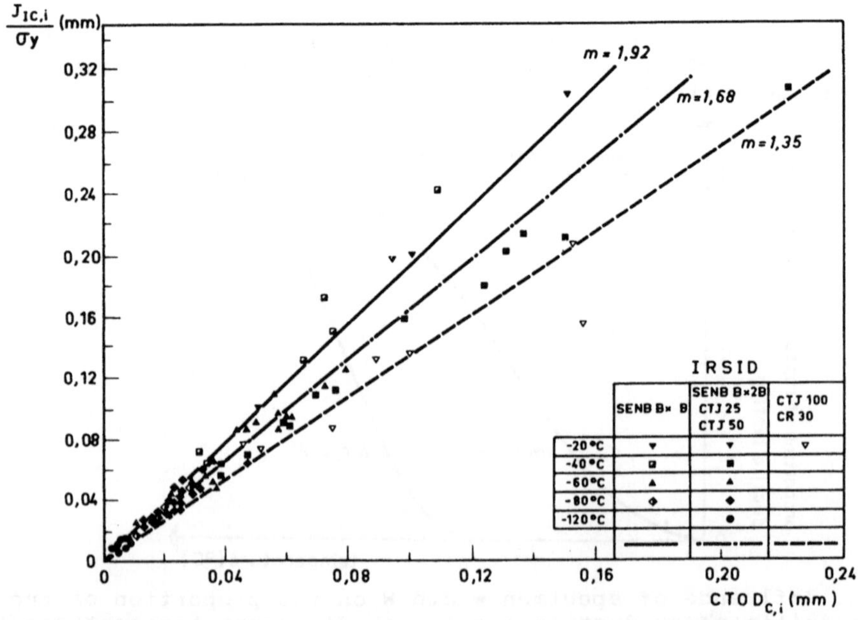


Fig.9. Plot of $J/\sigma_y$ versus critical CTOD values measured at IRSID. The slopes of the three lines drawn on the diagram correspond to the mean values of m estimated by the statistical analysis.

# On the Toughness Evaluation of Sulphide-Controlled Fe 510 Steel by Charpy-V-Impact, Fracture Mechanics, and Wide Plate Tests

D. DORMAGEN, A. HALIM, H. EHRHARDT AND W. DAHL

Institute of Ferrous Metallurgy, RWTH Aachen

## Abstract

Four qualities of Fe 510 steel with different sulphur content and/or sulphur shape control are investigated in Charpy-V-impact, fracture mechanics and wide plate tests in order to evaluate the toughness behaviour of sulphide-controlled steels. Different transition temperatures are compared amongst each others and are related to the fracture of wide plates, for which quantitative fracture mechanics analysis were carried out.

## Introduction

It is common practice to improve the toughness behaviour of structural steels in the upper shelf regime by lowering the sulphur content [1,2] and/or performing a sulphide shape control [3,4]. Moreover, the anisotropy in the mechanical properties is reduced [1,5]. But there is also experimental evidence, that in Charpy-V-impact tests the fracture appearance transition temperature FATT 50 is shifted to higher values [6-8], although the impact energy is improved, when the sulphur content is reduced and a sulphide shape control is performed. This is known as the "sulphide effect" and leads to a contradictory toughness evalution of the steels, especially in the ductile/brittle transition region. Thus the question arises, how relevant is a fracture appearance criterion for the assessment of the failure behaviour of a structural component? In the following this problem shall be discussed by comparing different test techniques such as Charpy-V-impact-, fracture mechanics- and wide plate tests.

## Material

The materials investigated were four types of Fe 510 steel with basically indentical chemical composition, but different sulphur content. Besides a typical Fe 510 steel, a Cer-treated, a CaSi-treated quality and a pearlite-reduced version were tested in 15 mm thick, normalized plates. The chemical composition and the microstructure of the four steels are shown in fig. 1.

Chemical composition

| Steel | C | Si | Mn | P | S | Al | Cu | Cr | Ni | V | Nb |
|---|---|---|---|---|---|---|---|---|---|---|---|
| Fe 510 | 0,15 | 0,28 | 1,39 | 0,021 | 0,019 | 0,032 | 0,08 | 0,03 | 0,02 | — | — |
| Fe 510+Cer | 0,16 | 0,49 | 1,21 | 0,016 | 0,011 | 0,035 | 0,03 | 0,07 | 0,02 | — | — |
| Fe 510+CaSi | 0,20 | 0,47 | 1,56 | 0,021 | 0,002 | 0,033 | 0,07 | 0,05 | 0,03 | — | — |
| Fe 510 pearlite-red. | 0,11 | 0,37 | 1,53 | 0,011 | 0,002 | 0,031 | 0,05 | 0,04 | 0,06 | 0,07 | 0,04 |

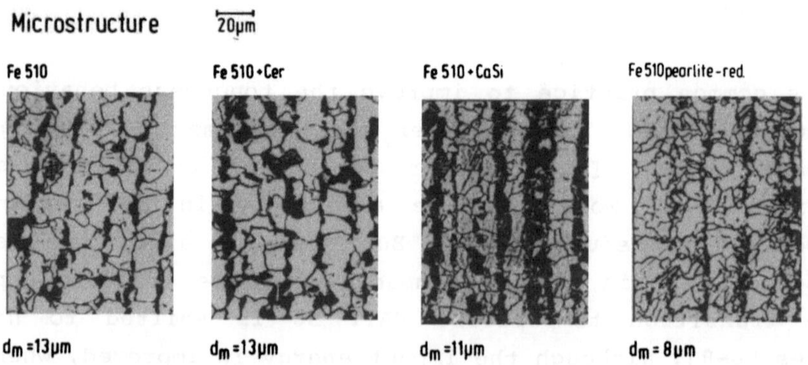


Figure 1: Chemical composition and microstructure of the four different types of Fe 510 steel

The sulpher content varies from 0.019% for Fe 510 steel to 0.011% for the Cer-treated quality, to 0.002% for the CaSi added steel and to 0.002% for the pearlite-reduced quality. As for the latter steel the carbon content is reduced, micro-alloying elements such as Vanadium and Niobium are added to assure similar strength levels. For all steels the microstructure is characterized by ferrite with bands of pearlite, being less pronounced in the pearlite-reduced quality. The mean ferrite grain size is similar for the Fe 510 steel and the Cer- and CaSi-treated quality being 12 µm and 11 µm respectively and reduced to 8 µm for the pearlite-reduced steel due to the addition of the micro-alloying elements.

Test techniques

The toughness behaviour of the four qualities was investigated in Charpy-V-Impact, fracture toughness- and wide plate tests. The Charpy-V-notched specimens were tested at an impact velocity of 5 m/s and the impact energy and the ductile fracture appearance was determined. The fracture mechanics tests were carried out in displacement-control using plain sided 13 mm thick CT-specimens with a width of 50 mm. The specimens were prefatigued for an a/W-ratio of 0.6 according to ASTM E 813 [9] and the critical CTOD-values were determined after BS 5762 [10]. The initiation of stable crack growth was measured using the direct current potential drop technique (DCPD) [11].

In order to evaluate the toughness behaviour in a structural component wide plates of CNT-types were tested with a thickness of B = 15 mm and a width of W = 300 mm and a flaw size of a = 15 mm. The notch acuity was varied by introducing notches with a radius of $\rho$ = 0.1 mm and prefatigued cracks. The tests were carried out in ram-displacement control of $V_R$ = 2 mm/min and $V_R$ = 1000 mm/min. The overall elongation was measured by inductive transducers at three positions over a gage length of 300 mm. Besides characteristic stresses and strains three different transition temperatures due to the fracture behaviour of wide plates [12] were determined. All specimens were tested in the T-L-orientation.

## Results

The tensile test properties at room temperature are listed in tab. 1, and show similar strength and toughness properties for all steels. Due to the smaller ferrite grain size the yield stress is slightly increased for the pearlite-reduced quality.

| Steel | $\sigma_{YS}$ / MPa | $\sigma_{UTS}$ / MPa | EF / % | RA / % |
|---|---|---|---|---|
| Fe510 | 371 | 530 | 27 | 63 |
| Fe 510 + Cer | 356 | 522 | 27 | 69 |
| Fe510+CaSi | 410 | 598 | 23 | 72 |
| Fe510pearlite-red. | 430 | 562 | 28 | 76 |

EF = elongation at fracture
RA = reduction of area at fracture

Table 1: Tensile test properties at room temperature of the Fe 510 steels

In fig. 2 the results of the Charpy-V-impact test are presented. In the upper part the Charpy-V-impact energy (CVE) and in the lower part the ductile fracture appearance (DFA) is plotted as a function of temperature. It is to be seen, that in the upper shelf regime the toughness is highly improved by lowering the sulpher content and/or sulphide shape control. The upper shelf energy (CVUSE) is increased from 56 J for the Fe 510 steel to 121 J for the Cer-treated quality, to 176 J for the CaSi-added type and to 200 J for the pearlite-reduced steel.

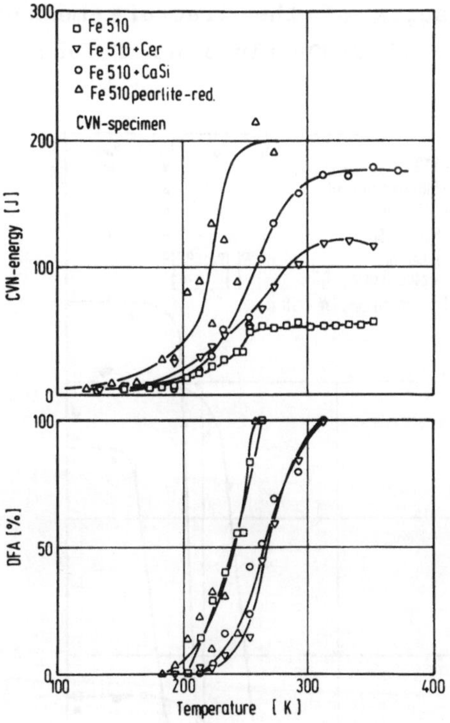


Figure 2: Charpy-V-impact energy and percentage of ductile fracture appearance as a function of temperature for the Fe 510 steels

Within the ductile/brittle transition region the CVE-values are also improved for the sulphide-controlled steels, whereas a slight increase of the transition temperature $T_{0.5\ USE}$ defined by 50% of USE is observed for the Cer- and CaSi-treated steel. A more pronounced shift in the transition temperatures to higher values and with that a detoriated toughness assessment is observed when the ductile fracture appearance using the FATT 50-criterion is regarded. For the Cer- and CaSi-treated quality an increase of FATT 50-values of 29 K and 25 K, respectively is observed. Thus the Charpy-V-impact test leads to an ambiguous toughness evalution for steels with sulphide-control.

In fig. 3 the results of the fracture mechanics tests are presented in terms of CTOD-values as a function of temperature.

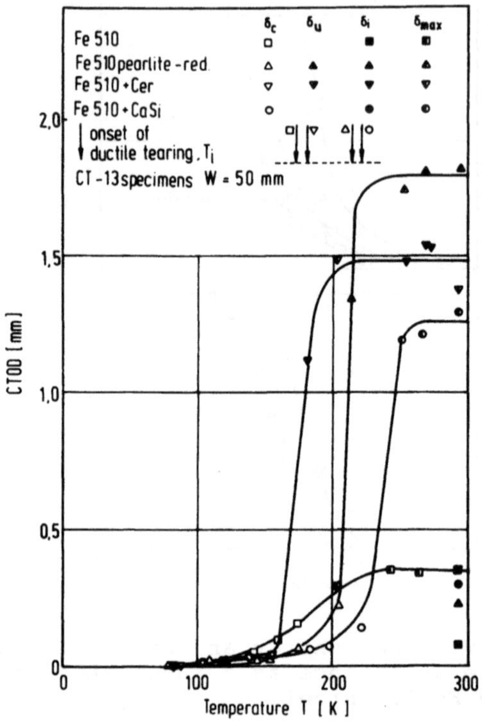


Figure 3: CTOD-values as a function of temperature for the Fe 510 steels

Compared with the Charpy-V-impact data it is confirmed by fracture mechanics that in the upper shelf regime the toughness of steels with sulphide-control is highly improved. This is to be seen for CTOD-values at maximum load but also for initiation-values for ductile crack growth. In the brittle regime there is no influence of sulphur content and/or sulphide shape control on the toughness behaviour of the steels.

Concerning the ductile/brittle transition region one should emphasize some differences with regard to the Charpy-V-impact results. Compared with the sulphide-controlled steels showing a steep ductile/brittle transition behaviour the Fe 510 steel is characterized by a very flat transition, which leads to an intersecting of the curves. Thus above a characteristic temperature higher CTOD-values are measured for steels with sulphide-control, and below this intersection point higher CTOD-values are determined for the basic quality. The consequences of this behaviour will be discussed in fig. 5. This kind of intersecting was not observed in the Charpy test-results; in the transition region the CVE-values were always improved for steels with lower sulphur content and for sulphide shape control.

The arrows marking the first appearence of ductile tearing indicate that compared with the Fe 510 steel the transition from brittle to ductile fracture is shifted to higher values especially for the pearlite-reduced quality and the CaSi-treated version. Thus the improved transition behaviour of the pearlite-reduced quality measured by impact values is not confirmed by the fracture mechanics tests. Further investigations on the influence of sulphur content and/or sulphide shape control revealed that the ductile/brittle transition strongly depends on notch acuity, stress state and loading rate [13].

Fig.4 shows as an example the results of the wide plate tests. For the CaSi-treated quality the yield stress and the net section stresses and strains at maximum load and at fracture are plotted as a function of temperature. By three characteristic transition temperatures four different regions can be defined. Below $T_{gy}$ fracture occurs before the yield stress is reached in the net section (low stress brittle fracture). Between $T_{gy}$ and $T_i$ the fracture surface is again characterized by cleavage fracture but fracture occurs at general yielding conditions. Above $T_i$ the crack first propagates by ductile tearing and then after a certain amount of stable crack growth in a cleavage type of fracture until the temperature $T_a$, above which only ductile tearing is observed on the fracture surface. Above $T_{gy}$ a slight decrease of the net section stresses is observed as a function of temperature according to the temperature dependence of the uni-axial yield strength. The

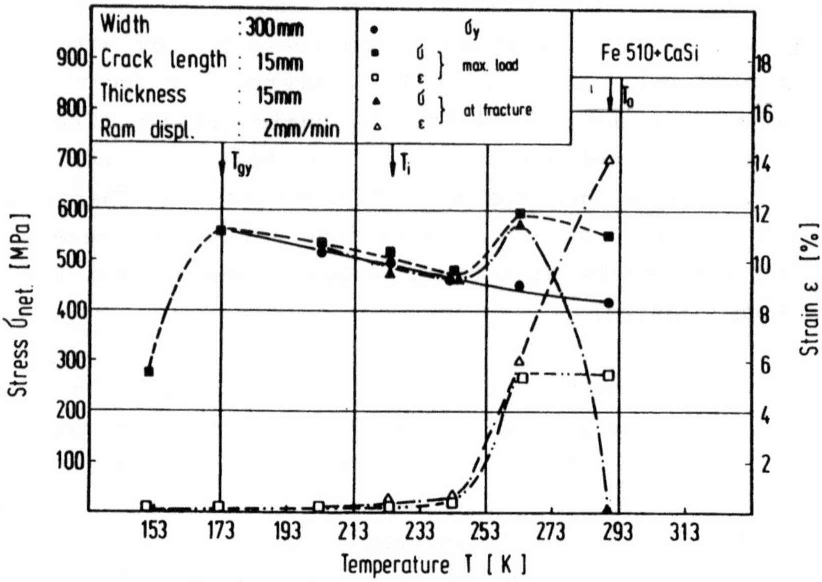


Figure 4: Net-section stresses and strains as a function of temperature for the CaSi-treated Fe 510 steel

stresses at maximum load values are slightly higher than the yield stresses. Between $T_i$ and $T_a$ a steep increase in strains at maximum load and at fracture is observed. The strain hardening causes the net section stress at maximum load to rise between $T_i$ and $T_a$ whereas the fracture stress decreases continuously due to stable crack growth.

By comparing the aforementioned transition temperatures $T_{gy}$, $T_i$ and $T_a$ in tab. 2 one can see that under quasi-static conditions the temperatures $T_i$ and $T_a$ which are indicating the ductile failure of the wide plate are shifted to higher values for sulphide-controlled steels. There is no influence of notch acuity on $T_i$ and $T_a$, whereas an increased loading rate results in an embrittlement for the Cer-and CaSi-treated steels. This embrittlement was not observed in the case of the pearlite-reduced quality, an effect which was tried to be explained by the different adiabatic testing conditions [13]. Thus the tendency to fracilitate cleavage fracture by lowering the sulpher content and/or sulphide shape control is affirmed by wide plate tests.

| Steel | Ram displacement mm/min | Notch length mm | Crack length mm | $T_{gy}$ K | $T_i$ K | $T_a$ K |
|---|---|---|---|---|---|---|
| Fe 510 | 2 | 15 | | <153 | 170 | 243 |
| Fe 510 pearlite-red. | 2 | 15 | | <150 | 203 | 293 |
| | 2 | | 15 | 170 | 203 | 290 |
| | 1000 | 15 | | <173 | <173 | 243 |
| Fe 510 + Cer | 2 | 15 | | <153 | 201 | 243 |
| | 2 | | 15 | <150 | 203 | 243 |
| | 1000 | 15 | | <173 | 223 | 242 |
| Fe 510 + CaSi | 2 | 15 | | <154 | 223 | 293 |
| | 2 | | 15 | 173 | 224 | 290 |
| | 1000 | 15 | | <174 | 233 | 293 |

Table 2: Comparison of the transition temperatures $T_{gy}$, $T_i$ and $T_a$ from wide plate tests of the Fe 510 steels

In reality it is generally avoided that components operate at loading conditions compared with those reached in specimens during a laboratory test. General yielding conditions are usually avoided and for the toughness behaviour of a structure tolerable defect sizes or allowable stresses in the present of a flaw are of more importance. Thus, based upon the CTOD-data presented in fig. 3, a failure assessment of the wide plates was carried out using the procedure given by PD 6493 [14]. Within the limits of application the results of wide plates analyses i.e. only contained and gross-section yielding conditions were considered are presented in tab. 3. The tolerable defect size $a_{tol}$ of the wide plate was calculated for the loading conditions given by $\varepsilon_{max}/\varepsilon_y$ and the ratio of $a_{specimen}$ $a_{tol}$ is a measure of safety against failure. Comparing the four steels one can see that within the limits of application the tolerable defect size is bigger in steels with reduced sulphur content and/or sulphide shape control, the mean safety factor being 1.5.

| steel | ram displacement mm/min | notch length mm | crack length mm | temperature K | CTOD mm | $\varepsilon_{max}/\varepsilon_y$ | $a_{spec}$ mm | $a_{tolerable}$ |
|---|---|---|---|---|---|---|---|---|
| Fe 510 | 2 | 15 | | 293 | 0.34 | 17 | 15 | 4 |
| | 2 | 15 | | 263 | 0.33 | 16 | 14 | 4 |
| Fe 510 + Cer | 2 | 15 | | 290 | 1.5 | 29 | 15 | 10 |
| | 2 | 15 | | 261 | 1.5 | 28 | 14 | 10 |
| | 2 | 15 | | 243 | 1.5 | 25 | 16 | 10 |
| | 1000 | 15 | | 290 | 1.5 | 29 | 16 | 10 |
| | 1000 | 15 | | 261 | 1.5 | 27 | 15 | 10 |
| | 1000 | 15 | | 242 | 1.5 | 25 | 14 | 10 |
| | 2 | | 15 | 290 | 1.5 | 25 | 18 | 11 |
| | 2 | | 15 | 260 | 1.5 | 25 | 16 | 10 |
| | 2 | | 15 | 243 | 1.5 | 24 | 16 | 10 |
| Fe 510 + CaSi | 2 | 15 | | 293 | 1.26 | 29 | 16 | 7 |
| | 2 | 15 | | 263 | 1.26 | 26 | 15 | 8 |
| | 1000 | 15 | | 293 | 1.26 | 33 | 14 | 6 |
| | 1000 | 15 | | 261 | 1.26 | 18 | 16 | 11 |
| | 2 | | 15 | 290 | 1.26 | 27 | 17 | 7 |
| | 2 | | 15 | 263 | 1.26 | 26 | 17 | 7 |
| | 2 | | 15 | 151 | 0.05 | 0.5 | 17 | 12 |
| Fe 510 pearlite-reduced | 2 | 15 | | 293 | 1.8 | 31 | 15 | 9 |
| | 2 | 15 | | 263 | 1.8 | 32 | 14 | 9 |
| | 2 | 15 | | 243 | 1.8 | 27 | 15 | 10 |
| | 1000 | 15 | | 290 | 1.8 | 32 | 15 | 9 |
| | 1000 | 15 | | 263 | 1.8 | 31 | 14 | 9 |
| | 1000 | 15 | | 243 | 1.8 | 31 | 15 | 9 |
| | 2 | | 15 | 290 | 1.8 | 29 | 17 | 10 |
| | 2 | | 15 | 263 | 1.8 | 27 | 17 | 10 |
| | 2 | | 15 | 243 | 1.8 | 24 | 18 | 11 |
| | 2 | | 15 | 153 | 0.03 | 0.8 | 21 | 2 |

Table 3: Comparison between actual and tolerable defect sizes according to PD 6493 for wide plates of the Fe 510 steels

Thus, at least above T = 243 K the Cer-treated and the pearlite-reduced quality and above T = 263 K the CaSi-added steel could be used preferably. Below these temperatures the wide plates failed under net section yielding conditions which lead to gage-length dependent strain values, so that the loading conditions of the wide plates can't be expressed uniquely. This results in an inappropriate failure analysis of those wide plates.

Therefore a failure assessment of any component whatever was carried out using the PD 6493 procedure assuming equal loading conditions for all steels.

The ratio

$$\frac{a_{tol}}{a_{tol\ (Fe\ 510)}} = \frac{c\left(\frac{\delta_{crit}}{\varepsilon_y}\right)}{c\left(\frac{\delta_{crit\ Fe\ 510}}{\varepsilon_y}\right)} \qquad (1)$$

with $$c = \frac{1}{2\pi\left(\frac{\sigma}{\sigma_y}\right)^2}$$

or $$c = \frac{1}{2\pi\left(\frac{\varepsilon}{\varepsilon_y} - 0.25\right)}$$

depending on the ratio of applied load to yield strength reduces for equal loading conditions $\sigma/\sigma_y$ and equal yield strengths to

$$\frac{a_{tol}}{a_{tol(Fe\ 510)}} = \frac{\sigma_{crit}}{\sigma_{crit(Fe\ 510)}} \qquad (2)$$

Calculating the ratio $a_{tol}/a_{tol\ (Fe\ 510)}$ of eq.(2) as a function of temperature the intersecting of the CTOD-temperature curves of sulphide-controlled steels with that of the Fe 510 steel results in a characteristic temperature, above which bigger tolerable defect sizes would be determined for the sulphide-controlled (SC) steels than for the conventionally melted (CM) steels, due to improved CTOD-values ($T_{a_{tol\ SC} \leq a_{tol\ CM}}$). Below this temperature bigger defect sizes would be tolerated for the basic Fe 510 quality. This temperature is compared in fig. 5 with the FATT 50-criterion and the transition temperatures $T_i$ and $T_a$ of the wide plate tests under quasi-static conditions, where the strains are plotted as a function of temperature. By comparing the different transition temperatures among each others and with regard to the failure behaviour of the wide plates, some

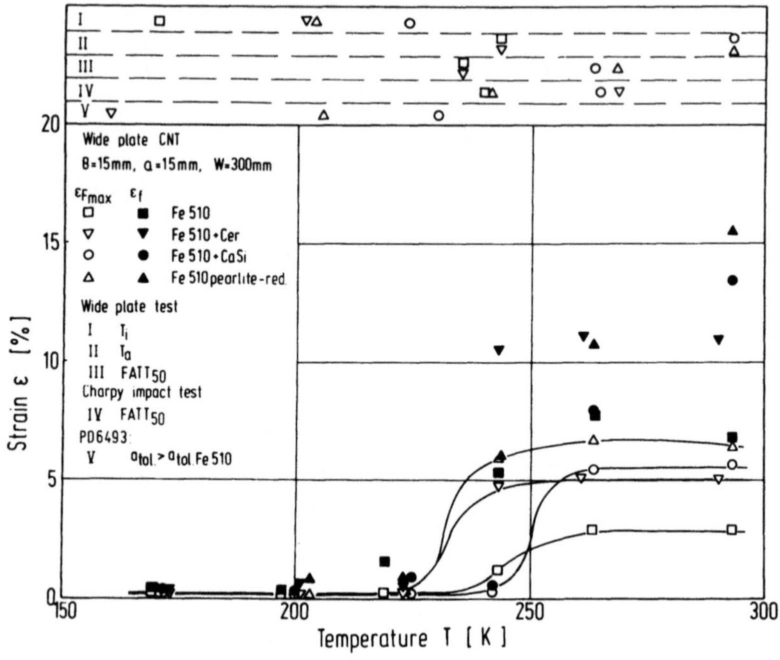


Figure 5: Different transition temperatures in relation to the failure behaviour of wide plates for the Fe 510 steels

important conclusions can be drawn concerning the transferability of transition temperatures, defined by the fracture appearance, to a component.

Firstly, establishing the service-temperature of a component above FATT 50 does not garantuee a fully ductile failure ($T_a$ of wide plates above FATT 50). Secondly, even below temperatures such as FATT 50 a component can be operated with bigger tolerable flaw sizes or stresses when sulphide-controlled (SC) steels are used down to temperatures where CTOD-values are higher than those of conventionally melted (CM) qualities ($T_{a_{tol\,SC} \leq a_{tol\,CM}} \leq T_{a_{tol\,SC} \geq a_{tol\,CM}} <$ FATT 50).

Therefore service-temperatures shouldn't be established on the basis of transition temperatures such as FATT 50 but based upon fracture mechanics analysis.

## Conclusions

Four qualities of Fe 510 steel with different sulphur content and/or sulphide shape control were investigated in Charpy-V-impact, fracture mechanics and wide plate tests in order to evaluate the toughness behaviour of sulphide-controlled steels. This investigation demonstrated that sulphide-controlled steels show an improved toughness behaviour in the upper shelf regime by higher CV-upper shelf energies, higher CTOD-values for crack initiation and maximum load and also increased strains at fracture in wide plate tests. Concerning the ductile/brittle transition behaviour, there is a more pronounced tendency of sulphide-controlled steel to fail by cleavage. Thus transition temperatures defined by the fracture appearance are shifted to higher values compared to the conventionally melted Fe 510 steel.

It could be demonstrated by the comparison of different transition temperatures and fracture mechanics analysis that FAT-temperatures are inappropriate for the safety assessment of a component and that lowest service-temperatures should be established by quantitative fracture mechanics analysis.

## Acknowledgement

The project entitled "Auswirkungen der Sulfidkontrolle auf das Zähigkeitsverhalten von Baustählen" was sponsored under BMFT-FB S 246 by the Bundesminister für Forschung und Technologie (BMFT), Bonn, Federal Republic of Germany.

References

[1] Dahl, W.
in: Werkstoffkunde Stahl, Vol. 1, Ed: VDEh, Springer-Verlag Berlin, Heidelberg, New York, Tokyo (1984), p. 223/400, Verlag Stahleisen mbH Düsseldorf

[2] Dahl, W., El Gammal, T. and L. Lorenz
Arch. Eisenhüttenwesen 44(1973) 11, p. 243/6

[3] Meyer, L.
Stahl und Eisen 97 (1977) 6, p. 410/6

[4] Fischer, W.A. and H. Bertram
Arch. Eisenhüttenwesen 44 (1973), p. 97/109

[5] Wilson, A.D.
in: ASTM STP 668 (1979), p. 469/93

[6] Fuchs, A., Täffner, K.; Krisch, A. and A. Kochendörfer
Arch. Eisenhüttenwesen 46 (1975), p. 127/36

[7] Gunn, K.W.
Research Report No PK/TSR/75/D 20, (1975) Australien Iron and Steel PTY Ltd., Australien

[8] Nilserik, H.
pres. at: Conference: "HSLA steels - Experiences in Application", Versailles 22.-24.Jan. (1970)

[9] ASTM Annual Book of ASTM Standards Metals Test-Methodes and Analytical Procedures, Sect. 3, Vol. 03.01

[10] Methods for Crack Opening Displacements Testing, British Standards 5762/1979, British Standard Institution, London 1979

[11] Sun, D.Z., Dormagen, D. and W. Dahl
Steel Research 56 (1985) 8, p. 445/49

[12] Dahl, W. and H.J. Rosezin
Stahl und Eisen 104 (1984) 18, p. 909/12

[13] Dahl, W. und D. Dormagen
Final Report for the Bundesministerium für Forschung und Technologie (BMFT), BMFT-FB S 246 (1985)

[14] PD 6493
Guidance on some methods for the derivation of aceptance levels for defects in fusion welded joints, British Standard Institution

# The Use of CTOD Methods in Fitness for Purpose Analysis

**T.L. ANDERSON,* R.H. LEGGATT and S.J. GARWOOD**

**The Welding Institute,**
**Abington Hall**
**Abington,**
**Cambridge, CB1 6AL**
**England**

**SUMMARY**

**A three tier fracture assessment procedure is proposed which allows flaws to be assessed at various levels of complexity, depending on the application. All three levels utilise crack tip opening displacement (CTOD) as the fracture toughness input. The CTOD design curve, with minor modifications, occupies the first (simplest) level. Several restrictions to the application of the CTOD design curve are recommended. The second level utilises a strip yield model, which is reasonably accurate below net section yield. Level 3, the most complex and accurate method, utilises a reference stress approach which allows fracture assessments to be performed on work hardening materials above net section yield. The proposed three tier procedure also allows for some flexibility in the way stress gradients are treated. The proposed options for treating stress gradients range from simple and conservative to complex and accurate. Some guidelines for choosing the appropriate assessment level are given.**

**INTRODUCTION**

**A number of different methods for calculating allowable or critical flaw sizes under post-yield conditions are currently in use in the UK and other countries throughout the world. These methods include the British Standards Institute (BSI) PD6493 procedure (1), the Central Electricity Generating Board (CEGB) R6 method (2), and Turner's EnJ method (3). A survey has been conducted by Burdekin under the auspices of Commission X of the International Institute of Welding (4). It was found that BSI PD6493 was the most widely used post-yield assessment method among respondents representing 32 companies and institutions in 15 countries. The CTOD design curve as used in BSI PD6493 does, however, have a number**

*** Now at Mechanical Engineering Department, Texas A & M University, College Station, Texas, U.S.A.**

of limitations which may in some cases produce an excessive degree of conservatism in the calculated allowable flaw size. In particular:

- no allowance is made for stress variation through the thickness under post-yield conditions;
- for applied stress to yield strength ratios greater than 0.5, the CTOD design curve is partly empirical;
- a variable safety factor is inherent in the approach, making probabilistic studies and critical analyses difficult;
- plastic collapse solutions are not an integral part of the analysis.

The present paper has two principal aims. The first is to propose alternative assessment methods that are more accurate and less conservative than the present PD6493 approach. This is not to say that PD6493 should be completely discarded in favour of more advanced methods. In many cases a simple approach is adequate and the extra complexity of the proposed methods is not warranted. Hence, the second aim of this paper is to present a unified framework within which flaws can be assessed at a level of complexity appropriate to the problem under consideration.

A three tier approach is proposed which is sufficiently flexible to allow the use of different design equations and different assumptions for stress distributions within one overall framework. The approach is based entirely on crack tip opening displacement (CTOD) as the fracture toughness input to the analysis. The advantage of CTOD is that it is currently the best documented method for measuring the toughness of weldments as well as parent steels in the ductile-to-brittle transition region.

Several new stress parameters are defined below. This is followed by a discussion of the proposed three tier approach.

The three levels of assessment are presented in increasing order of complexity and accuracy. The CTOD design curve, as applied in PD6493, occupies the first level, although several modifications to the existing PD 6493 method are proposed. The second level utilises a strip yield model, which also forms the basis of the second revision of the CEGB R6 method (2). The reference stress approach is on the third and highest level. The CEGB has incorporated a similar reference stress approach into proposals for the revision of the R6 method (5).

## DEFINITION OF EFFECTIVE STRESSES

Before the three levels of assessment are discussed, it is necessary to define three stress parameters that are used at each level, namely the effective primary, secondary and net section stresses. The purpose of these parameters will become evident in subsequent sections.

### Effective Primary Stress

Primary stresses are the result of the net force and moment acting on the structure. When these stresses are calculated, the effect of stress concentrations and crack-like flaws are ignored. Primary stresses are generally caused by loads applied to the structure, and are thus usually load controlled. However, certain types of displacement-controlled stresses, such as fit-up stresses acting in one component of a multiply-connected structure, may cause unbalanced net forces or moments in the cross section and should be treated as primary stresses.

Primary stresses often vary in the cross section of a structure. For example, if a plate is loaded in bending, the stress will vary from tension to compression through the thickness. If a structure contains a flaw, the fracture resistance of the structure will be affected by stress gradients. Thus an accurate fracture mechanics analysis must take these effects into account.

An effective value of primary stress has been defined which allows stress gradient effects to be incorporated into a fracture assessment. The effective primary stress, $\sigma_p$, can be expressed as a function of the stress intensity due to primary stresses, $K_I^P$ :

$$\sigma_p = \frac{K_I^P}{\sqrt{\pi a}} \qquad [1]$$

The dimension a corresponds to the flaw dimension of interest, i.e. the half length of a through-thickness flaw, the depth of a surface flaw, or the half height of an embedded flaw. $\sigma_p$ has the units of stress but, in general, is not equal to the actual primary stress. (The exception is an infinite plate in uniform tension with a through-thickness flaw of length 2a). The effective primary stress is simply a convenient parameter that contains the stress intensity and flaw size. Simplified methods for calculating $\sigma_p$ are given in a later section.

**Effective Secondary Stress**

Secondary stresses are defined as those stresses which are self-equilibrating within the structure or component being analysed. They may include residual stresses, thermal stresses and local stress concentrations at geometric discontinuities. Residual and thermal stresses are calculated ignoring the presence of the flaw.

In some cases, where the residual stress distribution is unknown, it is necessary to assume that it is uniform, and equal in magnitude to some constant value such as the material yield stress. In such cases, the residual stress is treated as a secondary stress, despite the fact that the assumed distribution is not self-equilibrating.

The effective secondary stress, $\sigma_s$, is defined in a manner similar to $\sigma_p$:

$$\sigma_s = \frac{K_I^s}{\sqrt{\pi a}} \qquad [2]$$

where $K_I^s$ is the stress intensity due to the secondary stress distribution. Simplified methods for calculating $\sigma_s$ are given in a later section.

**Effective Net Section Stress**

The effective net section stress, $\sigma_n$, characterises the stress in the uncracked area of the section under consideration. Net section yield occurs when $\sigma_n$ is equal to the yield stress. Expressions for $\sigma_n$ for various configurations can be obtained by considering force and moment equilibrium between a fully plastic stress distribution in the net section and the nominal primary stresses at collapse (see Appendix).

**Table 1 Formulae for effective net section stress** (See Appendix)

| FLAW TYPE | MEMBRANE STRESS | BENDING STRESS | COMBINED STRESSES |
|---|---|---|---|
| Through-wall flaw | $\frac{\sigma_m}{(1 - 2a/W)}$ | $\frac{2\sigma_b}{3(1 - 2a/w)}$ | $\frac{\sigma_b + \sqrt{\sigma_b^2 + 9\sigma_m^2}}{3(1 - 2a/W)}$ |
| Long surface flaw | $\frac{\sigma_m}{(1 - a/t)}$ | $\frac{2\sigma_b}{3(1 - a/t)^2}$ | $\frac{\sigma_b + \sqrt{\sigma_b^2 + 9\sigma_m^2 (1 - a/t)^2}}{3(1 - a/t)^2}$ |
| Embedded flaw | $\frac{\sigma_m}{(1 - 2a/t)}$ | $\frac{2\sigma_b}{3\left[(1 - 2a/t)^2 + 8ap/t^2\right]}$ | See below |

For an embedded flaw,

$$\sigma_n = \frac{\sigma_b + 6\sigma_m a/t + \sqrt{(\sigma_b + 6\sigma_m a/t)^2 + 9\sigma_m^2 (1 - 4a^2/t^2)}}{3(1 - 4a^2/t)^2} \quad \text{if} \quad \frac{\sigma_m}{\sigma_b} > \frac{\left[1 - 2(p + a)/t\right]}{6(1 - p/t)\,p/t}$$

$$\sigma_n = \frac{\sigma_b + \sqrt{\sigma_b^2 + 9\sigma_m^2 \left[(1 - 2a/t)^2 + 8ap/t^2\right]}}{3\left[(1 - 2a/t)^2 + 8ap/t^2\right]} \quad \text{if} \quad \frac{\sigma_m}{\sigma_b} < \frac{\left[1 - 2(p + a)/t\right]}{6(1 - p/t)\,p/t}$$

In the case of a section subjected to a uniform membrane stress, $\sigma_m$, consideration of force equilibrium leads to the following simple expression for effective net section stress.

$$\sigma_n = \sigma_m A_g/A_n \qquad [3]$$

where $A_g$ is the area of the gross section and $A_n$ is the area of the net section.

Formulae for the effective net section stress under membrane, bending and combined loading are given in Table 1 for through-wall and long part-wall flaws. These expressions are derived in the Appendix. Similar expressions can be derived for short part-wall flaws (i.e. flaws whose length is of the same order of magnitude as their depth) but these expressions are rather complex for bending and combined loading. For practical applications, it is recommended that short part-wall flaws be treated as through-wall flaws of the same length when computing $\sigma_n$. (However, the actual flaw dimensions should be used to compute $\sigma_p$ and $\sigma_s$). This procedure would give a conservative estimate of $\sigma_n$ but the degree of conservatism would be small provided that the flaw length is small compared to the width of the section.

## LEVEL 1: THE CTOD DESIGN CURVE

The Level 1 approach is the simplest and most conservative of the three proposed methods. This approach is based on the existing PD 6493 method, which is outlined below.

### Background

The CTOD design curve (6) forms the basis of the fracture assessment in PD6493 (1). In terms of non-dimensional CTOD ($\Phi$), the design curve equations can be stated as:

$$\Phi = \frac{\delta_{crit} E}{2\pi\ \sigma_{YS}\bar{a}_m} = \left(\frac{\sigma_1}{\sigma_{YS}}\right)^2 \qquad \text{for } \frac{\sigma_1}{\sigma_{YS}} < 0.5 \qquad [4a]$$

$$\Phi = \frac{\sigma_1}{\sigma_{YS}} - 0.25 \quad \text{for } \frac{\sigma_1}{\sigma_{YS}} > 0.5 \qquad [4b]$$

where $\delta_{crit}$ is the critical CTOD value,

$E$ is Young's modulus,

$\bar{a}_m$ is the allowable equivalent flaw size, equal to the half length of a through thickness flaw in an infinite plate in tension,

$\sigma_{YS}$ is the yield strength,

$\sigma_1$ is the peak stress, defined as:

$$\sigma_1 = K_t \left(\sigma_m + \sigma_b\right) + \sigma_Q$$

where $K_t$ is the elastic stress concentration factor,

$\sigma_m$ is the primary membrane stress,

$\sigma_b$ is the bending stress,

$\sigma_Q$ is the secondary stress.

Equation [4a] is based on a linear elastic fracture mechanics (LEFM) analysis with a safety factor of 2. Equation [4b] was obtained from an empirical correlation between CTOD data and wide plate tests. Kamath (7) has validated this correlation with additional wide plate and CTOD data.

Primary and secondary stresses and a critical CTOD value are inserted into Eq. [4a] or [4b] to obtain an allowable flaw size, $\bar{a}_m$. This value can then be converted to allowable dimensions of surface or embedded flaws by means of Figs. 12 and 13 in PD6493. The inverse of this procedure can be used to assess a known flaw.

It is necessary to adhere to certain restrictions when using the CTOD design curve, namely:

i) The crack size to component width ratio ($\bar{a}/W$) should not exceed 0.1.

ii) The primary stresses should not exceed yield.

iii) A separate check for the attainment of the flow stress on the remaining ligament (ie. plastic collapse) is required.

Unfortunately the above restrictions are not explicitly stated in PD6493.

Recently Dawes (8) has reiterated the finite size and stress limitations of the design curve. Also, he has proposed a finite size correction to the design curve which allows it to be used for $\bar{a}/W$ ratios up to 0.5.

The plastic collapse check can be performed with the aid of Fig. 6 in PD6493 (1). Alternatively, other published collapse solutions (9) can be used. PD6493 describes a procedure where surface flaws are recategorised as through-thickness flaws and embedded flaws are recategorised as surface flaws (or in some cases through-thickness flaws) when the stress on the remaining ligament behind the flaw exceeds the flow stress. However, this recategorisation procedure is unnecessary if a plastic collapse check has been performed.

## Proposed Level 1 Assessment Method

The CTOD design curve can be reformulated in terms of the effective stresses in order to provide a means for incorporating stress gradient effects and to make this approach consistent with Level 2 and Level 3 assessement methods. The reformulated design curve is given by

$$\frac{\delta_{crit} E}{2\pi \, \sigma_{YS} \, a_m} = \left(\frac{\sigma_p + \sigma_s}{\sigma_{YS}}\right)^2 \quad \text{for} \quad \frac{\sigma_p + \sigma_s}{\sigma_{YS}} < 0.5 \qquad [5a]$$

$$\frac{\delta_{crit} E}{2\pi \, \sigma_{YS} \, a_m} = \frac{\sigma_p + \sigma_s}{\sigma_{YS}} - 0.25 \quad \text{for} \quad \frac{\sigma_p + \sigma_s}{\sigma_{YS}} > 0.5 \qquad [5b]$$

subject to the restriction

$$\sigma_n/\sigma_{YS} < 0.85$$

The main difference between Eq. [5] and the original design curve (Eq. [4]) is that in the above equations, the LEFM corrections for flaw shape are included in the effective stresses, $\sigma_p$ and $\sigma_s$. Note that the allowable effective flaw dimension, $\bar{a}_m$, has been replaced with the actual allowable flaw size, $a_m$.

The above formulation of the CTOD design curve makes it possible to include stress gradient effects. It should be noted that previous authors (10,11) have also proposed modifications to the design curve to incorporate stress gradients, although their proposals are slightly different from Eq. [5]. The present proposal allows substantial flexibility in the way in which stress gradients are treated. For example, the stress distributions can be treated in the same manner as in Eq. [4] by assuming primary and secondary stresses are uniform when calculating $\sigma_p$ and $\sigma_s$. Alternatively, a more sophisticated and accurate approach can be used. Various options for treating stress gradients are given in a later section.

Another important difference between Eq. [5] and the present PD 6493 approach is that an upper limit has been placed on the net section stress, above which Eq. [5] cannot be used. This restriction is based on recent studies (8, 12) which have indicated that the CTOD design curve is potentially unsafe at high net section stresses. As long as the user adheres to this restriction, he can apply Eq. [5] to finite width sections or structures with large cracks, i.e. cases where $\bar{a}/W > 0.1$. With this restriction, there is no longer a need for a separate plastic collapse check. However, a similar calculation to determine $\sigma_n$ must be performed as part of the assessment.

The proposed modifications to the CTOD design curve outlined above eliminate some of the limitations mentioned at the beginning of this article. However, the CTOD design curve still suffers from the disadvantage of being primarily an empirical equation with a variable

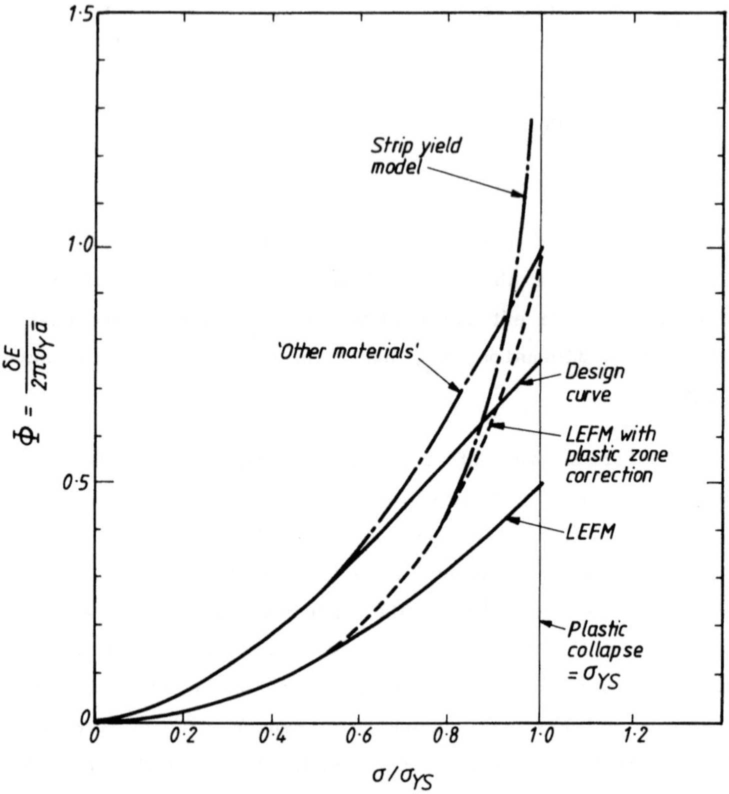


*Non-dimensional CTOD* $\Phi = \dfrac{\delta E}{2\pi\sigma_{YS}\bar{a}}$

*LEFM*

$$\Phi = \frac{1}{2}\left(\frac{\sigma}{\sigma_{YS}}\right)^2$$

*LEFM + plastic zone correction*

$$\Phi = (\sigma/\sigma_{YS})^2 / (2-(\sigma/\sigma_{YS})^2)$$

*Design curve*

$$\Phi = \left(\frac{\sigma}{\sigma_{YS}}\right)^2 \qquad \text{for } \frac{\sigma}{\sigma_Y} \leqslant 0{\cdot}5$$

$$\Phi = \frac{\sigma}{\sigma_{YS}} - 0{\cdot}25 \qquad \text{for } \frac{\sigma}{\sigma_Y} > 0{\cdot}5$$

*'Other materials'*

$$\Phi = \left(\frac{\sigma}{\sigma_{YS}}\right)^2$$

*Strip yield model*

$$\Phi = \frac{4}{\pi^2} \ln \sec\left(\frac{\pi\sigma}{2\sigma_{YS}}\right)$$

Fig. 1 Comparison of the CTOD design curve, strip yield model and LEFM equations

safety factor. In order to overcome this limitation, it is necessary to replace the design curve with a more accurate and less empirical equation, such as those which form the basis of the Level 2 and Level 3 approaches described below.

## LEVEL 2: STRIP YIELD MODEL

Although the proposed Level 2 approach is slightly more complex than Level 1, it has the advantage of being considerably more accurate. This approach is based on the strip yield model which is described below.

### Background

In 1960 Dugdale (13) formulated a strip yield model for the plastic zone ahead of a notch in a thin plate under tensile loading. Several investigations (14-16) have used this model to derive equations for K or CTOD as a function of applied stress in a cracked plate. These equations, which assume plane stress conditions and elastic-perfectly plastic behaviour, all have a similar form.

Garwood (17) used a collapse modified strip yield model to assess ductile fracture in wide plate specimens. The collapse modified equation is given by:

$$\delta = \frac{8\,\sigma_{pc}^{2}\,\bar{a}}{\pi\,E\,\sigma_{YS}}\,\ell n\,\sec\left(\frac{\pi}{2}\frac{\sigma}{\sigma_{pc}}\right) \qquad [6]$$

where $\sigma$ is the remotely applied stress and $\sigma_{pc}$ is the assumed value of $\sigma$ at plastic collapse. $\sigma_{pc}$ was calculated by assuming that collapse would occur when the stress on the remaining cross section reached the flow stress. Since Garwood (17) analysed plain material specimens in tension, there was no need to incorporate secondary stresses or stress gradient effects into Eq. [6].

In the case of an infinite sheet and a non-work hardening material (ie. $\bar{a} \ll W$ and $\sigma_{pc} = \sigma_{YS}$) Eq. [6] reduces to the original strip yield equation derived by several authors (14-16). This equation is compared with the CTOD design curve and LEFM analyses in Fig. 1. At stresses below 90% of yield, the strip yield model agrees closely with the LEFM analysis that includes a plastic zone correction. There is no inherent safety factor in the strip yield model, as is the case with the CTOD design curve, and it is also applicable to materials other than ferritic steels. Note that at high stresses the design curve becomes non-conservative compared with the strip yield model.

Equation [6] can be expressed in terms of the ratio of applied CTOD to critical CTOD, $\delta_r$:

$$\delta_r = \frac{\pi^2}{8} S_r^2 \left(\ell n \sec \frac{\pi}{2} S_r\right)^{-1} \qquad [7]$$

where $S_r$ $(= \sigma/\sigma_{pc})$ is the stress ratio. Equation [7], which is plotted in Fig. 2, is identical to the original CEGB R6 failure assessment diagram, except that the R6 diagram is formulated in terms of the critical K which is computed from a critical value of the J integral.

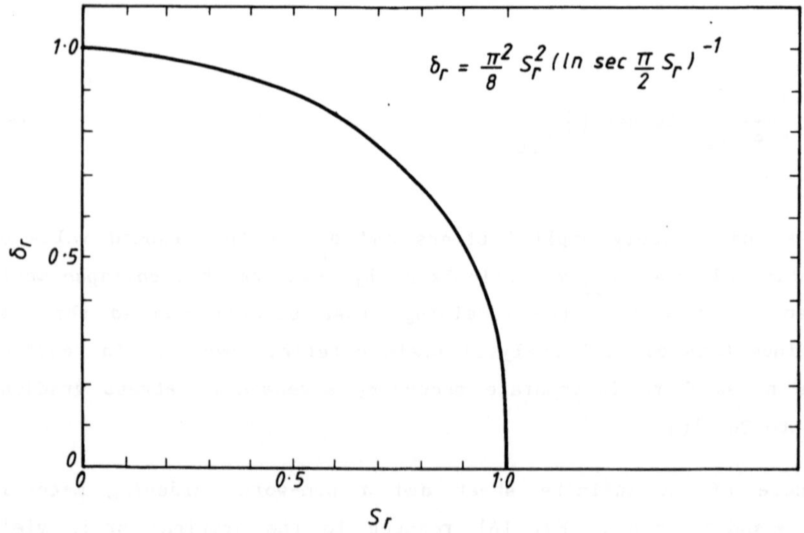


Fig. 2 CTOD failure assessment diagram

In Ref. (17) the modified strip yield model (Eq. [6]) was used in conjunction with crack growth resistance (R) curves to predict the behaviour of wide plate and pressurised burst tests. Reasonably accurate predictions were obtained for a pipeline steel but predictions for an austenitic stainless steel were not particulary good. It was concluded that the accuracy of this procedure did not justify its complexity. Instead, a simplified strip yield assessment procedure, which uses maximum load toughness ($\delta_m$) as input, was recommended. However, it was found that a conservative estimate of the austenitic stainless steel's flow stress was required in order to ensure safe predictions for this material.

## Proposed Level 2 Assessment method

Equation [6] applies to uniform membrane loading with no secondary stresses. In order to apply the strip yield model to the general case, Eq. [6] was modified to incorporate $\sigma_p$, $\sigma_s$, and $\sigma_n$. The final result is as follows:

$$\frac{\delta E}{\pi \sigma_{YS} a} = \left[\frac{\sigma_p}{\sigma_n}\left(\frac{8}{\pi^2}\ \ell n \sec\left(\frac{\pi}{2}\frac{\sigma_n}{\sigma_{YS}}\right)\right)^{\frac{1}{2}} + \frac{\sigma_s}{\sigma_{YS}}\right]^2 \qquad [8]$$

In order to understand how Eq. [8] was obtained, it is necessary to examine the '$\ell n$ sec' term at small stresses:

$$\ell n \sec(x) = \frac{x^2}{2} \text{ for small } x$$

Therefore, for small $\sigma_n$, Eq. [8] becomes

$$\frac{\delta E}{\pi \sigma_{YS} a} = \left[\frac{\sigma_p}{\sigma_n}\left(\frac{8}{\pi^2}\ \frac{1}{2}\ \frac{\pi^2}{4}\ \frac{\sigma_n^2}{\sigma_{YS}^2}\right)^{\frac{1}{2}} + \frac{\sigma_s}{\sigma_{YS}}\right]^2$$

$$= \left(\frac{\sigma_p + \sigma_s}{\sigma_{YS}}\right)^2 \qquad [9]$$

Substituting Eqs. [1] and [2] into the right hand side of Eq. [9] gives

$$\delta E = \frac{(K_I^p + K_I^s)^2}{\sigma_{YS}} \qquad [10]$$

For small-scale yielding under plane stress conditions,

$$\delta = \frac{K_I^2}{E\ \sigma_{YS}} \qquad [11]$$

Therefore at low stresses Eq. [8] reduces to

$$K_I = K_I^p + K_I^s \qquad [12]$$

which is simply the linear elastic summation of stress intensities due to primary and secondary stresses. Eq. [8] was formulated in such a way that it would reduce to the LEFM solution at low stresses while applying a strip yield plasticity correction to the primary stress at intermediate and high net section stresses.

The treatment of primary stresses in Eq. [8] is essentially identical to that in the second revision of the CEGB R6 method (2). However the treatment of secondary stresses is slightly different for the two approaches. In Eq. [8] the secondary stress term is outside of the ' $\ell n$ sec' term which defines the collapse limit. This is believed to be a more logical formulation because secondary stresses cannot, by definition, contribute to collapse.

The procedure for fitness for purpose assessments with Eq. [8] can be summarised as follows:

i) Measure fracture toughness ($\delta_{crit}$).

ii) Assume a flaw size and shape.

iii) Compute $\sigma_p$ and $\sigma_s$ (from a procedure described later) and $\sigma_n$ (Table 1).

iv) Compute the applied CTOD, $\delta$, from Eq. [8].

v) Repeat steps ii to iv for various flaw sizes.

vi) Plot $\delta$ versus flaw size. Fracture is predicted when $\delta$ reaches or exceeds $\delta_{crit}$.

There is no need for a separate collapse check because CTOD increases to infinity in Eq. [8] at net section yield. Equation [8] is conservative at high stresses because it does not allow for work hardening. The reference stress approach which follows takes work hardening into account and allows fracture assessments to be extended beyond net section yield.

## LEVEL 3: THE REFERENCE STRESS METHOD

The CTOD design curve (Level 1) is a simple and conservative approach for determining flaw acceptance levels. The strip yield approach (Level 2) provides a higher degree of accuracy and less conservatism. However, both methods rely on simple assumptions which are not valid at high stresses and strains. This is especially true for high work-hardening materials.

A proposed method for extending fracture analyses beyond net section yield is described below. This approach is more accurate than those discussed previously, but it is also more complex. Moreover, more input data are required for this method. The accuracy of the method is limited by the accuracy of these input data. Thus, the approach which follows should only be used in cases where: 1) simpler approaches (e.g. the CTOD design curve and strip yield model) prove to be inadequate; and 2) the required input data are available.

### Background

Kumar, et. al. (18) developed a set of fracture analysis equations based on the Hutchinson, Rice and Rosengren (HRR) stress-strain singularity (19,20). These equations assume that a material's flow behaviour obeys a Ramberg-Osgood power law relationship:

$$\frac{\varepsilon}{\varepsilon_o} = \frac{\sigma}{\sigma_o} + \alpha \left(\frac{\sigma}{\sigma_o}\right)^n \qquad [13]$$

where $\sigma_o$ is a reference point (usually $\sigma_{YS}$) on the flow curve, $\varepsilon$ is strain, $\varepsilon_o = \sigma_o/E$, and $\alpha$ and n are dimensionless constants.

Many materials, such as austenitic stainless steel, do not obey a power law hardening relationship. For this reason, Ainsworth (21) modified the Kumar, et. al. equation for J to reflect the actual shape of a materials stress-strain curve, rather than a power law approximation of it. He defined a reference stress, $\sigma_{ref}$, as follows.

$$\sigma_{ref} = (P/P_o)\ \sigma_{YS} \qquad [14]$$

where P is the remotely applied load and $P_o$ is the load at net section yield. The reference stress is in fact equal to the effective net section stress, $\sigma_n$. The strain at $\sigma_{ref}$ in a uniaxial tensile test was designated the reference strain, $\varepsilon_{ref}$. By substituting the reference stress and strain into the Kumar et. al. equation for J, Ainsworth was able to eliminate the power law dependence of this equation. The new equation utilises a material's stress-strain curve.

Ainsworth (21) made a further modification which simplified the analysis. Plastic finite element solutions are required to solve for boundary conditions in the Kumar et. al. equations. Ainsworth made an approximation which eliminated the need for plastic finite element analysis. He also formulated his approach in terms of a failure assessment diagram (FAD). This reference stress FAD is included in the proposals for revision of the CEGB R6 assessment method (5).

Anderson (12) used Ainsworth's approach to obtain a reference stress equation for CTOD.

For a uniform membrane stress and no secondary stresses, the non-dimensional CTOD is given by:

$$\frac{\delta E}{\pi \sigma_{YS} \bar{a}} = \left(\frac{\sigma_m}{\sigma_{YS}}\right)^2 \left[\frac{\bar{a}_e}{\bar{a}} + \frac{E \varepsilon_{ref}}{\sigma_{ref}} - 1\right] \quad [15]$$

where $\bar{a}_e$ is the effective $\bar{a}$, including a plastic zone correction. That is, the actual flaw size is increased to an effective value (18):

$$a_e = a + \frac{1}{2\pi} \frac{(K_I^p)^2}{\sigma_{YS}^2} \frac{1}{1 + (\sigma_{ref}/\sigma_{YS})^2} \quad [16]$$

Equation [15] was used to predict applied CTOD versus applied stress curves in a number of wide plate specimens (12). In most cases, the agreement between the predictions and experimental data was very good.

**Proposed Level 3 Assessment Method**

Equation [15] was modified to take account of stress gradients and secondary stresses. The resulting equation is given by:

$$\frac{\delta E}{\pi \sigma_{YS} a} = \left[\frac{\sigma_p}{\sigma_{YS}} \left(\frac{a_e}{a} + \frac{E \varepsilon_{ref}}{\sigma_{ref}} - 1\right)^{\frac{1}{2}} + \frac{\sigma_s}{\sigma_{YS}}\right]^2 \quad [17]$$

As in the Level 2 assessment equation. Eq. [17] can be viewed as the CTOD equivalent of the LEFM summation of primary and secondary stress intensities, with a plasticity correction factor applied to the primary stress. The correction factor on Level 3 is based on the actual material stress-strain curve.

As mentioned previously, $\sigma_{ref} = \sigma_n$. However the term 'reference stress' has been retained in Level 3 in order to avoid confusion between the general concept of an effective net section stress, as used in any assessment method, and its application in the Level 3 method in which actual flow curves are used to relate reference stress to reference strain.

The procedure for using Eq. [17] to assess brittle fracture is summarised below.

i) Obtain the appropriate fracture toughness value, $\delta_{crit}$, and the material's true stress-true strain curve. This procedure applies only to $\delta_c$ and $\delta_u$ values. Some guidelines for ductile fracture assessment are given in the next section. The stress-strain curve should be relevant to the microstructure being assessed. In cases where this is not practical (eg. the heat affected zone) a lower bound estimate of the flow curve should be obtained.

ii) Assume a flaw of a given size and shape.

iii) Compute $\sigma_{ref}$ (Table 1), $\sigma_p$, $\sigma_s$, and $a_e$.

iv) Obtain $\varepsilon_{ref}$ (at $\sigma_{ref}$) from the stress-strain curve.

v) Calculate $\delta$ from Eq. [17].

vi) Repeat steps iii to v for various flaw sizes.

vii) Plot $\delta$ versus flaw size. Fracture is predicted when $\delta$ reaches or exceeds $\delta_{crit}$.

There is no need for a separate plastic collapse check as in PD 6493. However, special consideration should be given to cases where materials with different flow properties are subjected to the same stress, such as in a weldment. For example, consider an axially loaded structure where the weld metal over-matches the yield strength of the parent plate and a flaw in the weld metal is to be assessed. If the flaw is small compared to the overall cross section, collapse will be controlled by the weaker parent metal. Thus a separate check for yielding in the parent plate is necessary. The interaction between materials with different flow properties may be quite complex. In such cases, elastic-plastic finite element analysis may be the only alternative if an accurate assessment is required.

## DUCTILE FRACTURE ASSESSMENT

The three levels of assessment outlined above are intended mainly for cleavage fracture analysis, i.e. cases where $\delta_c$ and $\delta_u$ values are measured in laboratory tests. However, ductile fracture can also be assessed with these methods. In the case of Levels 1 and 2, maximum load toughness ($\delta_m$) values can be applied to the assessments as outlined above. Garwood (17) has shown that using $\delta_m$ values in conjunction with the CTOD design curve or the strip yield model will result in conservative predictions. If an accurate ductile fracture assessment is required, a modified version of the Level 3 approach can be used.

The Level 3 procedure as described in the previous section is applicable to stationary cracks; i.e. the amount of stable crack growth prior to fracture is negligible. However this assumption is generally not valid for ductile fracture. An analysis which takes crack growth into account is required for accurate predictions. Such an analysis is described in detail in Ref. (22) and is outlined briefly below. Figure 3 illustrates the principle of the method. Equation [17] is used to plot a series of applied CTOD versus remote stress curves for stationary cracks of various sizes. The material's resistance (R) curve is used to construct CTOD-stress curves for a growing crack (solid line, Fig. 3b). In the case of a load controlled structure, instability occurs when the maximum stress is reached.

For simple geometries and loading conditions, it is possible to solve for the instability point analytically. This was done in Ref. (12) for a uniformly loaded panel with a centrally located through-thickness crack. Since stress gradients and secondary stresses were not considered, Eq. [15] was applicable. Imposing the condition $d\sigma_m/d\delta = 0$ (cf. Fig. 3b) and assuming power-law hardening resulted in

$$\frac{d\delta}{da} = \frac{2\ \delta_m\ (n + 1)}{W - 2a_{crit}} \qquad [18]$$

where $a_{crit}$ is the half crack length (including crack growth) at instability (or point of maximum stress) and n is the hardening exponent as defined by Eq. [13]. The derivative $d\delta/da$ is the slope of the R curve.

In Ref. (22), Eq. [18] is used in conjunction with Eq. [15] to solve for critical crack size, stress and tear length.

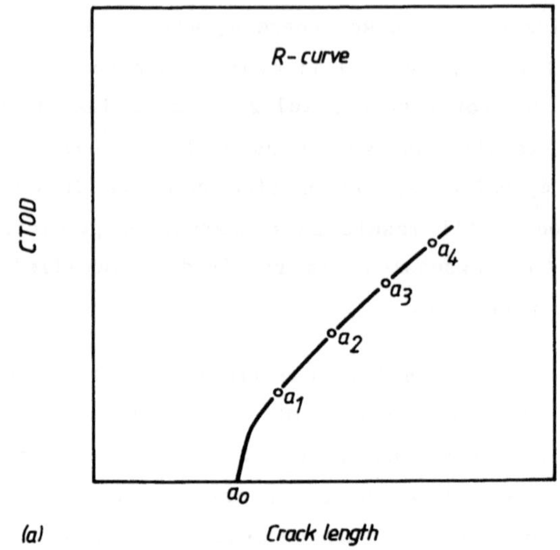


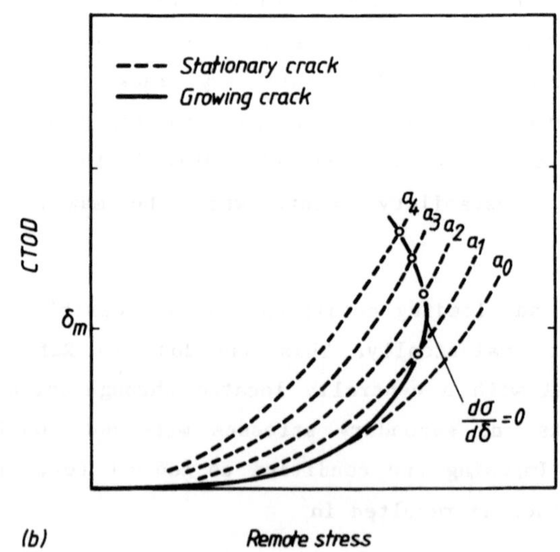


Fig. 3 Incorporation of crack growth into the reference stress approach

## CALCULATION OF $\sigma_p$ AND $\sigma_s$

Effective primary and secondary stresses, as defined in Eqs. [1] and [2], are related to primary and secondary stress intensity factors as well as flaw size. Various assumptions can be used to compute $\sigma_p$ and $\sigma_s$, depending on available information and required accuracy of the analysis. Three approaches are described below in increasing order of complexity.

**Option 1: Treat Stresses as Uniform**

The most conservative approach is to assume that stress is uniform across the section. This is the approach recommended in PD6493 for elastic-plastic assessments. The maximum stress in the cross-section is assumed to act uniformly across the entire section. For this assumption the effective primary and secondary stresses are as follows:

$$\sigma_p = \frac{M_m}{Q_o\sqrt{\pi}} (\sigma_m + \sigma_b) \quad [19]$$

$$\sigma_s = \frac{M_m}{Q_o\sqrt{\pi}} (\sigma_{r\,max} + \sigma_{t\,max}) + (K_T - 1)(\sigma_m + \sigma_b) \quad [20]$$

where $Q_o$ and $M_m$ (together with $M_b$ which is required for Option 2) are the flaw shape and stress intensity correction factors which are used in PD6493 to calculate stress intensity factors in LEFM analyses. For a through-thickness flaw where a << W , $M_m/(Q_o \sqrt{\pi}) = 1$ . $\sigma_r$ and $\sigma_t$ are the residual and thermal stresses acting in the section, excluding any components giving rise to net forces and moments, which are included in the primary stresses. However, as stated previously, if the residual stresses are known to be balanced within the section, but are assumed to be uniform due to lack of information on their distribution, they are treated as secondary stresses. The values of $\sigma_r$ and $\sigma_t$ are obtained ignoring the presence of the flaw. The maximum value within the area of the flaw is taken.

$K_T$ is the stress intensity magnification factor due to geometric stress concentration.

**Option 2: Treat Stresses as Linear**

If the variation of primary and secondary stresses are approximated as linear, these stresses may be resolved into bending and axial components. The effective stresses would then be defined as:

$$\sigma_p = \frac{1}{Q_o\sqrt{\pi}} (M_m\sigma_m + M_b\sigma_b) \quad [21]$$

$$\sigma_s = \frac{1}{Q_o\sqrt{\pi}} (M_m\sigma_{sm} + M_b\sigma_{sb}) + (K_T - 1)(\sigma_m + \sigma_b) \quad [22]$$

where $\sigma_{sm}$ and $\sigma_{sb}$ are the secondary membrane and bending stresses obtained by linearising the secondary stresses (excluding geometric stress concentration effects) as shown in Fig. 3 of PD6493.

PD6493 recommends the above approach when performing an LEFM analysis. However, stresses are usually assumed to be uniform (Option 1) for an elastic plastic analysis to PD6493.

## Option 3: Use Stress Intensity Solutions

In most structures, primary stresses can be resolved into bending and axial components, in which case Eq. [21] is adequate. However, there are cases where more complex stress distributions are present. Also, secondary stresses seldom vary linearly through the thickness. In such instances it may be desirable to use accurate stress intensity solutions to determine $\sigma_p$ and $\sigma_s$. These may be obtained from stress intensity solution handbooks, weight function techniques, finite element analysis, or other numerical methods. The effective primary and secondary stresses would then be computed from Eqs. [1] and [2].

## Selecting the Appropriate Options

The complexity and accuracy of the stress analysis would not necessarily correspond to the complexity and accuracy of the assessment equation. That is, Option 1 need not be applied to Level 1, Option 2 need not be applied to Level 2, and so on.

In general, the stress analysis option should be chosen based on the following factors.

1) **Required accuracy.** If a simple and conservative analysis is all that is required, Option 1 may be sufficient. However, Options 2 and 3 would tend to give more realistic results.

2) **Available data.** Obviously, Option 3 cannot be applied to secondary stresses in a weldment if the residual stress distribution is not known. If accurate input data are not available, a sophisticated analysis would be of little value.

3) **Severity of stress gradients.** If stresses are reasonably uniform in the section of interest, Option 1 may be adequate. However, Option 1 would be over conservative if steep stress gradients were present near the flaw being assessed.

4) **Sensitivity of the analysis.** For example, if a Level 3 analysis is applied at high plastic strains, the secondary stress term in Eq. [17] would be negligible. Thus, a sophisticated secondary stress analysis would be unnecessary.

## COMPARISON OF ASSESSMENT EQUATIONS

The CTOD design curve, strip yield model and reference stress approach (Levels 1, 2 and 3, respectively) were compared to experimental data from Ref. (12). A structure (i.e. a wide plate specimen) in uniform tension with no secondary stresses was considered for this comparison. This way, the three assessment equations could be compared, independent of considerations for secondary stresses and stress gradients.

The material used for the comparison was 51mm thick A533B steel plate which had a yield strength of $460N/mm^2$ and a tensile strength of $595N/mm^2$ at 70°C. The test specimen was 510mm wide and contained a semi-elliptical surface crack 240mm long and 38.7mm deep.

Figure 4 is a plot of applied CTOD versus remote stress ($\sigma_m$). The reference stress equation agrees very well with experiment. The strip yield equation gives a reasonable approximation of applied conditions but the CTOD design curve bears little resemblance to experimental data. At high stresses the CTOD design curve appears to be non-conservative.

Figure 5 shows the safety factor on applied CTOD in the wide plate specimen. As indicated in Fig. 4, the safety factor is approximately equal to 1.0 for the reference stress equation. This equation is slightly non-conservative at high stresses and high CTOD values but this is probably because crack growth has been neglected in this example. The safety factor of the strip yield equation varies between 1.15 and 1.5 below net section yield. The CTOD design curve has a highly variable safety factor which falls below 1.0 at a net section stress of approximately 0.9 $\sigma_{YS}$. These results, along with previous investigations (8,12) have led to the recommendations of the upper limit of 0.85 $\sigma_{YS}$ on net section stress for the CTOD design curve.

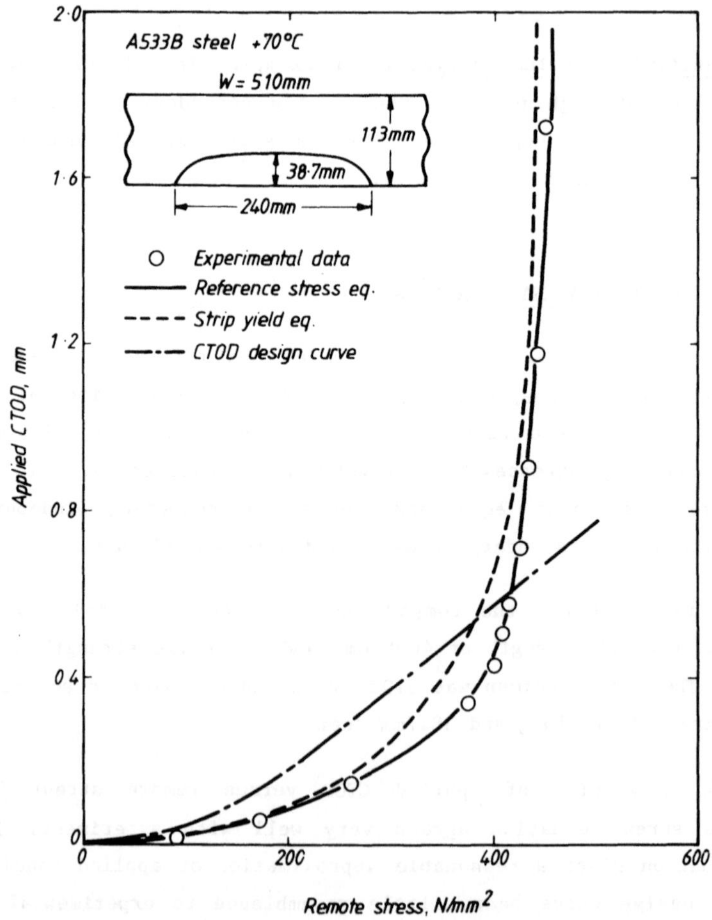


Fig. 4 Comparison of predicted and experimental applied CTOD in an A533 B steel wide plate

It should be emphasised that Fig. 5 merely represents the safety factor on applied CTOD. The overall safety factor in a fracture analysis would depend on additional factors. For example, the Option 1 treatment of stress gradients would add an extra degree of conservatism to the analysis if bending stresses were present. Also, lower bound estimates of toughness are usually used in a fracture assessment, thereby making the assessment more conservative.

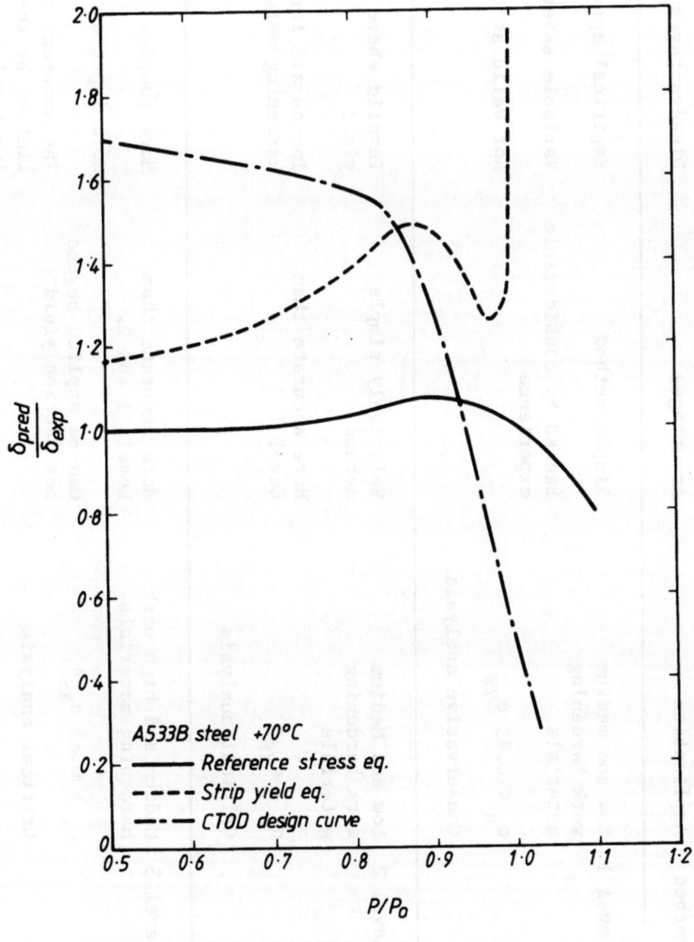


Fig. 5 Safety factor on applied CTOD for the three assessment equations

**Table 2. Summary of the 3 assessment levels**

| Method | Application | Advantages | Disadvantages |
|---|---|---|---|
| Level 1 | Low and medium work hardening materials<br>$\sigma_n \leq 0.85\ \sigma_{YS}$<br>Conservative analysis | Simple method<br>Backed by considerable experience | Empirical approach<br>Variable safety factor<br>Not valid at high stresses |
| Level 2 | Low and Medium work hardening materials<br>$\sigma_n < \sigma_{YS}$<br>Critical analysis | Relatively simple method<br>More accurate than Level 1 | Invalid above net section yield<br>Inaccurate for high work hardening materials |
| Level 3 | Medium and high work hardening materials<br>$\sigma_n > 0.85\ \sigma_{YS}$*<br>Critical analysis | More accurate than Levels 1 and 2.<br>Can be applied beyond net section yield | More complex than Levels 1 and 2<br>More input data are required<br>The accuracy of the analysis is limited by the accuracy of the input data |

* Although the Level 3 approach is valid at all stresses, its complexity is not warranted at lower stress levels

## DISCUSSION

A three tier approach to flaw assessment has been presented. The level of complexity which is most appropriate for a given application depends on a number of factors. The advantages and disadvantages of each level are summarised in Table 2.

The CTOD design curve approach in PD6493 has the advantage of being relatively simple but it has a variable safety factor (Fig. 5) which makes critical analyses difficult. At high net section stresses, the CTOD design curve under-estimates the applied CTOD. Consequently the design curve should be restricted to net section stresses below 0.85 $\sigma_{YS}$. Above this stress, Level 2 or Level 3 assessment equations should be used.

The strip yield approach is slightly more complex than the CTOD design curve but can be considerably more accurate below net section yield. The strip yield model assumes elastic-perfectly plastic behaviour and is therefore inaccurate for work hardening materials above net section yield.

The reference stress method allows fracture assessments to be peformed above net section yield. However, this method is more complex than the Level 1 and 2 approaches because the reference stress assessment curve (Fig. 4) is material dependent. Thus, an accurate material flow curve is required each time an assessment is performed. According to Fig. 5 the strip yield model based on the yield stress is only slightly conservative up to net section yield. Therefore, there is no real advantage in using a complex Level 3 approach to perform assessments at low and medium stress levels. However, for high stresses and strains, the reference stress approach is the only one of the three methods that gives reasonably accurate results.

The way in which stress gradients and secondary stresses are treated has a significant effect on the overall accuracy of a fracture assessment. In fact, in some cases these considerations dominate the analysis and the assessment equation is of secondary importance.

Finally, it must be stated that the accuracy of any fracture analysis, irrespective of the assessment method, is likely to be controlled by the amount of scatter in fracture toughness values. Cleavage fracture toughness values are subject to wide variations with slight changes in microstructure. Since engineering materials are seldom homogeneous, it is nearly impossible to predict accurately the critical conditions for fracture in a given structure. However, the authors believe that probabilistic approaches, used in conjunction with sophisticated fracture analyses, will soon be able to predict accurately the risk of failure in a given situation.

## CONCLUSIONS

1. The CTOD design curve approach is a simple method that gives conservative predictions, provided the user adheres to certain restrictions.

2. The strip yield method is relatively simple but can also be reasonably accurate below net section yield.

3. The reference stress method is the most accurate of the three approaches but its complexity is warranted only for work hardening materials at high stresses and strains.

4. The assumptions for secondary stresses and stress gradients are as important as the assessment equation to the accuracy of a fracture analysis.

## REFERENCES


1. PD. 6493. "Guidance on some methods for the derivation of acceptance levels for defects in fusion welded joints". British Standards Institution, March 1980.

2. HARRISON, R.P., LOOSEMOORE, K., MILNE, I., and DOWLING, A.R. "Assessment of the integrity of structures containing defects". CEGB Report R/H/R6-Rev.2, 1980.

3. TURNER, C.E. "The J-estimation curve, R-curve, and tearing resistance concepts leading to a proposal for a J-based design curve against fracture". proc. conf. "Fitness for purpose validation of welded constructions." The Welding Institute, Paper 17, 1982.

4. BURDEKIN, F.M. "Final report on questionare on the use of fracture mechanics methods for the assessment of the significance of weld defects". IIW Document X-1076-84, 1984.

5. MILNE, I., DOWLING, A.R., AINSWORTH, R.A., and STEWART, A.T. "Assessment of the integrity of structures containing defects - addendum for structures made with a high capacity for work hardening". CEGB Report R/H/R6-Rev.2, Addendum 1, December 1984.

6. DAWES, M.G. "The COD design curve". 2nd Advanced Seminar on Fracture Mechanics, Ispra, Italy, April 1979. (Published in 'Advances in Elastro-Plastic Fracture Mechanics', Edit. L.H. Larsson, Applied Science Publishers, London, 1979).

7. KAMATH, M.S. "The COD design curve: an assessment of validity using wide plate tests." Welding Institute Report 71/1978/E, September 1979.

8. DAWES, M.G. "The CTOD design curve approach: limitations and finite size". Welding Institute Report 278/1985, June 1985.

9. WILLOUGHBY, A.A. "A survey of plastic collapse solutions used in the failure assessment of part wall defects". Welding Institute Report 191/1982, September 1982.

10. KAMATH, M.S. "The crack tip opening displacement (CTOD) design curve: some proposals for incorporating stress gradient effects". proc. conf. "Fitness for purpose validation of welded constructions". The Welding Institute, Paper 23, 1982.

11. HARRISON, J.D. "Fracture mechanics developments related to weld defect acceptance methods given in British Standard PD 6493". Presented at "Fracture 84, 2nd National Conference on Fracture". Johannesburg, South Africa, November 1984.

12. ANDERSON, T.L. "Elastic-plastic fracture assessments based on CTOD". Welding Institute Report 276/1985, June 1985.

13. DUGDALE, D.S. "Yielding in steel sheets containing slits". J. Mech. Phys. Solids, 8, 1960. p. 100.

14. BURDEKIN, F.M. and STONE, D.E.W. "The crack opening displacement approach to fracture mechanics in yielding materials". Journal of Strain Analysis, 1(2), 1966, p. 144.

15. BILBY, B.A., COTTRELL, A.H. and SWINDEN, K.H. "The spread of plastic yield from a notch". Proc. Royal Society, A 272, 1963, p. 304.

16. HEALD, P.T., SPINK, G.M. and WORTHINGTON, P.J. "Post yield fracture mechanics". Mat. Sci. Eng., 10, 1972, p. 129.

17. GARWOOD, S.J. "A crack tip opening displacement (CTOD) method for the analysis of ductile materials". Presented at the Eighteenth National Symposium on Fracture Mechanics, 25-27 June 1985, Boulder, Colorado, U.S.A.

18. KUMAR, V., GERMAN, M.D. and SHIH, C.F. "An engineering approach for elastic plastic fracture analysis". EPRI Report NP 1931, July 1981.

19. HUTCHINSON, J.W. "Singular behaviour at the end of a tensile crack in a hardening material". J. Mech. Phys. Solids, 16, 1968, p. 13.

20. RICE, J.R. and ROSENGREN, G.F. "Plane strain deformation near a crack tip in a power-law hardening material". J. Mech. Phys. Sol., 1-12.

21. AINSWORTH, R.A. "The assessment of defects in structures of strain hardening material". Eng. Frac. Mech., 19, 1984, p. 633.

22. ANDERSON, T.L., GORDON, J.R. and GARWOOD, S.J. "On the application of R-curves and maximum load toughness to structures." Submitted for presentation at the Nineteenth Symposium on Fracture Mechanics, 30 June - 2 July 1986 San Antonio, Texas, U.S.A..

## APPENDIX

## DERIVATION OF FORMULAE FOR EFFECTIVE NET SECTION STRESS

---

Consider a fully plastic stress distribution acting in a section containing a through-wall flaw (Fig. A1a), a long surface flaw (Fig. A1b), or a long embedded flaw (Fig. A1c). The fully plastic stress distribution is defined in terms of the depth of the tensile zone below mid-thickness, $Z_o$, and a collapse stress, $\sigma_o$, which might be the yield strength, flow stress or otherwise depending on application. The corresponding gross section membrane and bending stresses are determined by consideration of force and moment equilibrium.

For a through-wall flaw (Fig. A1):

$$Wt\ \sigma_{mo} = (W - 2a)\ 2\ Z_o\ \sigma_o \qquad [A1]$$

$$Wt^2\ \sigma_{bo}\ /b = (W - 2a)(t/2 - Z_o)(t/2 + Z_o)\ \sigma_o \qquad [A2]$$

For a long surface flaw (Fig. A1):

$$t\ \sigma_{mo} = 2\ Z_o\ \sigma_o \qquad [A3]$$

$$t^2\ \sigma_{bo}/b = (t/2 - a/2 - Z_o)(t/2 - a/2 + Z_o)\ \sigma_o \qquad [A4]$$

For a long embedded flaw (Fig. A1):

$$t\ \sigma_{mo} = 2\ Z_o\ \sigma_o \qquad [A5]$$

$$\text{if } Z_o < (\tfrac{t}{2} - p - a)$$

$$t^2\ \sigma_{bo}/6 = [(t/2 - a)^2 - Z_o^{\ 2} + 2\ ap]\ \sigma_o \qquad [A6]$$

$$\text{if } Z_o > (\tfrac{t}{2} - p - a)$$

$$t^2 \sigma_{bo}/b \quad = \quad [(t/2)^2 - (a + Z_o)^2] \sigma_o \qquad [A7]$$

For each pair of equations above, $Z_o$ can be eliminated, leading to a quadratic equation in $\sigma_o$, $\sigma_{mo}$ and $\sigma_{bo}$, which can be solved for $\sigma_o$:

$$\sigma_o \quad = \quad f_o (\sigma_{mo}, \sigma_{bo}) \qquad [A8]$$

Consider now any arbitrary set of gross section primary stresses $\sigma_m$ and $\sigma_b$. The effective net section stress is defined as follows:

$$\sigma_n \quad = \quad f_o (\sigma_m, \sigma_b) \qquad [A9]$$

where $f_o$ is the same function as in Eq. [A8]. The above definition ensures that, if $\sigma_m$ and $\sigma_b$ are such as to cause collapse of the net section, then the assumed net section stress distribution will be in equilibrium with the gross section stresses, and the effective net section stress will be equal to the collapse stress.

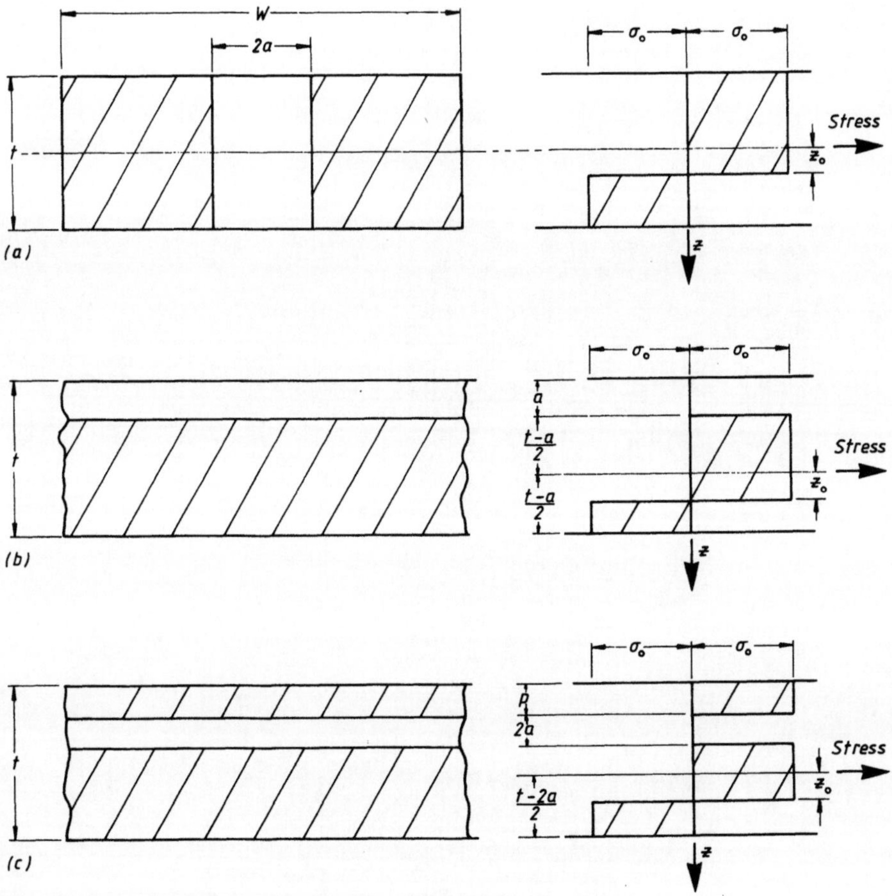


**Fig. A1 Fully plastic stress distributions**

**a) Through-wall flaw**

**b) Long surface flaw**

**c) Long embedded flaw**

# The Prediction of Failure Situations Using the CTOD Concept Based on the Engineering Treatment Model (ETM)

K.-H. SCHWALBE

GKSS-Forschungszentrum Geesthacht GmbH

D-2054 Geesthacht, Federal Republic of Germany

Abstract

Under the assumption of plane stress conditions, the CTOD concept is extended such that a full R-curve methodology arises consisting of an experimental procedure for the determination of the R-curve (details of which are reported in a separate paper) and of the driving force prediction. Predictions of initiation and maximum loads are in reasonable agreement with experimental results.

## Introduction

The application of fracture mechanics concepts to the mechanical behaviour of flawed structures in the regime of non-linear response is complicated by the fact that the driving force - or loading parameter (expressed in terms of J-integral or crack tip opening displacement, CTOD) - is not only a function of the geometry of the part under consideration but also of the material's flow properties (strain hardening curve). Moreover, the variables "geometry" and "material" are not separable which means that for each combination of these variables a full calculation procedure has to be carried out which in general is a finite element calculation since no analytical solutions are known with the exceptions mentioned below.

Since numerical methods are time consuming and expensive it is felt useful to have analytical relationships, i.e. formulae, which show immediately how the relevant parameters influence the driving force and which serve for quick assessments of structural integrity.

The first approach of this kind has been provided by the CTOD Design Curve which gives comprehensive guidance for the application of the CTOD concept to the assessment of structures [ 1, 2]. The CTOD Design Curve relates the CTOD to the applied strain and is supposed to provide conservative assessments.

It can be shown that at least for simple geometries failure predictions can be made by the simple Engineering Treatment Model (ETM) which has recently been proposed for tension loaded panels under plane stress conditions which are symmetrically cracked (centre cracked or double edge cracked) [ 3-5 ]. The ETM provides a full R-curve methodology on the basis of the J-integral or the CTOD. For the latter quantity a special definition related to its experimental determination is used with the symbol $\delta_5$, Fig. 1; for details see Refs.[ 6, 7]. Simple expressions have been derived for J and the CTOD and compared with experimental results.

The present paper describes the extension of the ETM to bending configurations, i.e. single edge notched bend (SENB) and compact (CT) specimens and compares the results with those obtained for the tension configurations. Only the CTOD part of the ETM will be considered. More background information for this subject is given in Ref.[ 8 ].

The experimental data analysed in this paper are from Refs.[ 6, 7].
In essence, the ETM employs known plasticity corrected linear elastic solutions up to the yield load, $F_Y$; beyond this point the condition attained at $F = F_Y$ is extrapolated by a magnification factor derived from the ordinary tensile test.

## Relationship Between J-integral and CTOD

It has been demonstrated in Refs. [ 6,7 ] that the crack tip opening displacement, $\delta_5$, is a valid correlation parameter for stable crack growth which is able to correlate substantially more crack growth than the J-integral. Since J is the more riguourously defined crack field parameter it is worth investigating the relationship between $\delta_5$ and J. In Fig. 2 experimental values of both parameters obtained on 25 specimens are plotted versus each other. Only data points were taken which resulted in "valid" J values according to Refs.[ 6, 7]. Two points are obvious:

- J and $\delta_5$ are linearly related to each other.

- The relation between J and $\delta_5$ is unique in the sense that - within the range of parameters studied - there is no effect of loading mode (tension or bending), specimen size, and crack length.

Thus, $\delta_5$ may at least formally be regarded as a substitute for J. Having demonstrated the adequacy of $\delta_5$ as a parameter describing the behaviour of a crack $\delta_5$ will now be calculated independent of elastic-plastic J solutions.

## Large Scale Yield (Contained Plasticity)

The tools of linear ealstic fracture mechanics can be applied up to the yield load provided that a plasticity correction is introduced. An effective stress intensity factor, $K_{eff}$, can be defined if the effective crack length

$$a_{eff} = a + r_Y \tag{1}$$

with

$$r_Y = \frac{K^2}{(1+n)\ 2\sigma_Y \pi} \tag{2}$$

is inserted into the formulae for the linear elastic stress intensity factor, K. Fig. 3 shows that the thus defined effective stress intensity factor is in very good agreement with

$$K_J = \sqrt{JE} \tag{3}$$

where J is the J-integral experimentally determined on CCT specimens. For the bending configurations, however, a slightly different definition seems to be more appropriate

$$\bar{K} = 0.5(K_{eff} + K) \tag{4}$$

as can be seen from Fig. 4. Numerically, $\bar{K}$ coincides with the proposal of Ref. [ 9 ]. Thus, $K_{eff}$ and $\bar{K}$, respectively, can be used as reasonably accurate loading parameters up to the yield load, $F_Y$.

The non-hardening solutions for the CTOD:

- Dugdale model (for tension configurations only)

$$\delta_D = \frac{8\sigma_Y}{\pi E} a \ln\{ \frac{2W}{\pi a} \arcsin [ \sin \frac{\pi a}{2W} \sec \frac{\pi \sigma}{2\sigma_Y} ]\} \approx \delta_5 \tag{5}$$

or

- small scale yielding solution with plasticity correction

$$\delta = \frac{K_{eff}^2}{E\sigma_Y} \approx \delta_5 \tag{6}$$

for tension and

$$\delta = \frac{\bar{K}^2}{E\sigma_Y} \approx \delta_5 \tag{7}$$

for bending

are suggested to serve as engineering estimates for $\delta_5$: the experimental and predicted values of $\delta_5$ plotted in Fig. 5 support this view. Furthermore, in Fig. 6 Eq(7) shows good agreement with finite element calculations.

Net Section Yield

In the regime of net section yielding, the load-displacement behaviour is supposed to follow a power law the exponent of which is given by the materials strain hardening exponent, n, in the stress-strain law

$$\frac{\sigma}{\sigma_Y} = \left[\frac{\varepsilon}{\varepsilon_Y}\right]^n \quad , \sigma > \sigma_Y \tag{8}$$

In principle, the reference point ($\sigma_Y$, $\varepsilon_Y$) can be any point on the curve, all other points can be determined by linear extrapolation in log log scales. For convenience, the reference point is taken as the yield point with $\sigma_Y = \sigma_{0.2}$. This principle can be applied to the behaviour of a cracked part: if the net section attains the yield condition $F = F_Y$, i.e. if all material points along the net section are at or beyond the yield condition, the behaviour of the cross section as a whole should be given by a function analogous to Eq(8), Fig. 7. A measure of the behaviour of the yielding net section can be given by any displacement along a gage length spanning the yield wedge in Fig. 7, in particular by $\delta_5$. Thus, for the net section yield regime ($F > F_Y$)
we write

$$\frac{F}{F_Y} = \left[\frac{\delta_5}{\delta_Y}\right]^n \tag{9}$$

or

$$\delta_5 = \delta_Y \left[\frac{F}{F_Y}\right]^{1/n} \tag{10}$$

with $\delta_Y$ being the value of $\delta_5$ at $F = F_Y$. This kind of treatment is similar to the "fully plastic solution" employed by Kumar and Shih [ 9 ]. It has been empirically confirmed by experimental results obtained for various materials, like low and medium strength aluminium alloys, ferritic and austenitic steels tested in the form of CCT, DENT, CT, and SENB specimens. Examples for two materials are plotted in Figs. 8 and 9. More examples can be found in Refs. [ 4, 5, 8, 10 ]. The experimental data were taken from Refs. [ 6, 7 ]. As predicted by Eq(10), the following characteristics are evident:

- For a given material, the $\delta_5$ data fall on a single straight line in double logarithmic coordinates, independent of
  - o specimen type,

° specimen size, and

° crack length.

When crack growth occurs, $F_Y$ and $\delta_Y$ have to be determined for the actual crack length. On the other hand, $\delta_Y$ does not change a great deal within the range of crack growth studied so that it was kept constant for convenience.

- The slope of the straight line corresponds to the strain hardening exponent of the material, see material properties in Refs. [ 8 ] and [ 11 ]. It should be noted that the actual values of the material strain hardening exponent were taken from previous Ramberg-Osgood fits [ 11 ] which can differ somewhat from those obtained by a piece-wise power law like Eq(8). For the sake of simplicity, the ETM was based on the piece-wise power law. However, this and the fit of a power law to a real stress-strain curve (which always contains a certain arbitrariness) cause some inconcistency and uncertainty. Therefore, a sensitivity study is planned by which the response of the quantities to be predicted (CTOD and load line displacement) to the fitting procedure of the stress-strain data will be investigated.

Thus, it can be concluded that Eq(10) describes the behaviour of a cracked part in plane stress within a wide range of variables. Whereas in tension this behaviour can be observed also beyond maximum load, bending configurations seem to follow this behaviour only up to the maximum load which is, however, no restriction if we are interested in load controlled situations.

The yield load is an important feature in any elastic-plastic treatment, since it is the "hinge point" between contained yielding and the net section yielding regime with its rapidly increasing deformations. Within the ETM, the yield point marks the condition from which the extrapolation into the net section yield regime can be done if the material's strain hardening exponent is known, see Eq(10). Since up to the yield point ($F \leq F_Y$) the mechanical behaviour is approximately given by the formalisms of linear elastic fracture mechanics (LEFM) with plastic zone correction all elements of the ETM are available if the following items are known:

- The linear elastic solutions for the quantities desired (stress intensity factor, load line displacement, crack tip opening displacement...).

- The yield load, $F_Y$.
- The material's work hardening law as a power law.

For the yield load the following formulae were chosen for the SENB and CT specimens [ 12 ]:

SENB specimen

$$F_Y = A\sigma_Y \frac{BW^2}{S} (1 - \frac{a}{W})^2 \tag{11}$$

$$A = 1.072 \qquad \text{plane stress, } a/W > 0.02 \tag{12}$$

$$A = 1.455 \qquad \text{plane strain, } a/W > 0.296 \tag{13}$$

$$A = 1.455\text{-}3.141\,(0.31\text{-}a/W)^2 \tag{14}$$

plane strain, $a/W < 0.31$

CT specimen

$$F_Y = A\,\sigma_Y\,BW$$

$$A = \sqrt{2.155(1+1.155(a/W)^2)}-(1+1.155(a/W)) \tag{15}$$

plane stress

$$A = 1.155\,[\sqrt{2.7+4,59(a/W)^2} - (1+1.7(a/W))] \tag{16}$$

plane strain

The accuracy of all estimates beyond $F_Y$ depends of course on the accuracy of the yield load solutions. Particular problems arise when the state of stress is somewhere between plane stress and plane strain. Conservative estimates of the load carrying capacity can then be made if plane stress is assumed to prevail. For the analysis of the data of the present paper, plane stress was assumed for all cases. The CCT and DENT specimens were analysed using the simple formula

$$F_Y = \sigma_Y\,2B(W\text{-}a) \tag{17}$$

Formulae of this kind are expected to yield lower bound values since strain hardening should increase the yield load.

According to Eq(10) $\delta_5$ can be calculated if the yield load, $F_Y$, the strain hardening exponent, n, and $\sigma_Y$ are known. Whereas n can be determined by a tensile test, $F_Y$ is given by Eqs(11) through (17); since the tools of linear elastic fracture mechanics can be applied up to the yield load provided that a plasticity correction is introduced, $\delta_Y$ can be related to the effective stress intensity factor as shown above, particularly at $F = F_Y$. Hence, combining Eqs(6, 7, 10) $\delta_5$ can be determined in the net section yielding regime by

$$\delta_5 = \delta_Y \left[\frac{F}{F_Y}\right]^{1/n}$$

$$\approx \left.\frac{K_{eff}^2}{E\,\sigma_Y}\right|_{F_Y} \cdot \left[\frac{F}{F_Y}\right]^{1/n} \quad \text{for tension} \qquad (18)$$

$$\approx \left.\frac{\bar{K}^2}{E\,\sigma_Y}\right|_{F_Y} \cdot \left[\frac{F}{F_Y}\right]^{1/n} \quad \text{for bending} \qquad (19)$$

In the case of tension, the Dugdale type solution (Eq(5)) is also appropriate for determining $\delta_Y$. Fig. 10 shows measured and predicted values for a CT specimen of a pressure vessel steel and for an SENB specimen of an alloy steel.
If Eq(18) is used for bending, $\delta_5$ will be overestimated which in turn will result in an underestimate of the predicted failure load.

## General Yield

The general yield load, $F_{YY}$, may be defined as the load given by Eqs(11)-(17) with $a/W = 0$; it is that load which causes plastic deformation in the gross section of the specimen. The same model for calculating $\delta_5$ can be applied in the general yielding regime since the behaviour of the net section is not affected by the gross section. In Fig. 11 two examples demonstrate that also under general yielding conditions Eq(10) provides a good estimate of $\delta_5$.

## Prediction of Failure

Two failure conditions will now be considered

- initiation of crack growth and
- maximum load.

Initiation of crack growth may be defined in accordance with the European R-curve method under development as a small amount of crack growth, $\Delta a^* = 0.2$ mm. The relevant properties will therefore be provided with the index "0.2". According to Eq(9) the load, $F_{0.2}$, at the initiation of crack growth can be determined by

$$F_{0.2} = F_Y \left[ \frac{\delta_{0.2}}{\delta_Y} \right]^n \tag{20}$$

Table 1 compiles the relevant data of a number of CT, SENB, and CCT specimens of 2024-FC.

In addition, some predictions of maximum load, $F_{max}$, were done using the R-curve methodology, i.e. for a given specimen the driving force curve

$$\delta_5 = \delta_Y \left[ \frac{F}{F_Y} \right]^{1/n} \tag{21}$$

was constructed for that value of F which led to the tangency condition between the driving force curve and the $\delta_5$-R-curve. An example is shown in Fig. 12. The predicted $F_{max}$ was determined as the average of the values for the upper and lower bound of the scatterband of the R-curve.

The compilation in Table 1 shows good agreement between measured and predicted values for $F_{0.2}$ and $F_{max}$, respectively. To visualise the error in predicting $F_{0.2}$ the predicted and experimental values are plotted in Fig. 13. In Fig. 14 experimental and predicted maximum loads for three materials and four specimen types are plotted.

## Conclusions

A previously derived Engineering Treatment Model (ETM) has been applied to tension and bending configurations under plane stress conditions with particular emphasis on bending. The model consists of the following steps:

- The yield load, $F_Y$, has to be determined at which the whole net section undergoes plastic deformation.
- Below $F_Y$ a linear elastic, plasticity corrected treatment is appropriate.
- Beyond $F_Y$ the magnification factor $(F/F_Y)^{1/n}$ (with $F_Y$ accounting for crack growth) serves for determining a characteristic deformation, in particular the CTOD, by the following formalism:

$$\delta_5 = \delta_Y \left[\frac{F}{F_Y}\right]^{1/n}$$

using the actual values of F and $F_Y$. Herein $\delta_Y$ is the $\delta_5$ at $F_Y$ the value of which can be determined using the plasticity corrected small scale yielding solution (Eqs(18, 19)) or the Dugdale type solution (Eq(5)) for tension. ETM has the following desirable characteristics:

- The ETM provides a simple driving force solution for the CTOD independent of the J-integral. A comparison with experimental values determined on four materials in four specimen configurations shows that the ETM has an accuracy sufficient for engineering estimates.
- Predictions of the load at crack initiation and of the maximum load are in good agreement with experimental results.
- The ETM seems to work as well under general yielding conditions as in the net section yielding regime.

If there are any uncertainties with respect to the determination of the strain hardening exponent, a plot of the R-curve test in the spirit of Figs. 8 and 9 may be used to derive an appropriate value for n.

Further work is planned to investigate the mechanical background of the $\delta_5$ type CTOD.

Table 1: Experimental and predicted values of $F_{0.2}$ and $F_{max}$ for 2024-FC

| Specimen SENB + CT specimens | W (mm) | B (mm) | $a_o/W$ | $F_{0.2}$ (N) | $J_{0.2}$ (N/mm2) | $\delta_{0.2}$ (μm) | $F_{max}$ (N) | $F_Y$ (N) | $\delta_Y$ (μm) | prediction $F_{0.2}$ (N) | prediction $F_{max}$ (N) |
|---|---|---|---|---|---|---|---|---|---|---|---|
| BA1 50.5/E3-8 | 50 | 5 | 0.2 | 4600 | 24 | 130 | 6700 | 3136 | 41 | 4576 | 6369 |
| BA1 50.5/D6-8 | 50 | 5 | 0.5 | 2000 | 23 | 135 | 2550 | 1256 | 33 | 1954 | 2478 |
| BA1 100.5/D11-8 | 100 | 5 | 0.5 | 3400 | 23.5 | 160 | 4900 | 2513 | 66 | 3185 | 4625 |
| CA1 100.5/N5-8 | 100 | 5 | 0.702 | 1640 | 27 | 180 | 2020 | 1054 | 52 | 1434 | |
| CA1 100.5/N4-8 | 100 | 5 | 0.506 | 4020 | 23 | 165 | 5380 | 3247 | 103 | 3608 | 5175 |
| CA1 100.5/M3-8 | 100 | 5 | 0.705 | 1390 | 20 | 130 | 1860 | 1031 | 52 | 1402 | 1815 |
| CA1 100.5/M5-8 | 100 | 5 | 0.3 | 7400 | 24 | 165 | 10600 | 7346 | 195 | 6757 | |
| CA1 100.10/O3-13 | 100 | 10 | 0.7 | 3060 | 26 | 145 | 3910 | 2138 | 53 | 2892 | |
| CA1 100.10/M2-13 | 100 | 10 | 0.3 | 15300 | 25 | 155 | 22300 | 14693 | 195 | 13514 | |
| CA1 50.10/L8-13 | 50 | 10 | 0.61 | 3100 | 24 | 155 | 4025 | 1904 | 37 | 2864 | |
| CA1 50.10/L4-13 | 50 | 10 | 0.812 | 910 | 27 | 150 | 1010 | 394 | 16 | 756 | |
| CA1 50.10/L9-13 | 50 | 10 | 0.415 | 6950 | 31 | 180 | 8875 | 4802 | 71 | 5956 | |
| CA1 100.20/M1-23 | 100 | 20 | 0.715 | 5850 | 25 | 155 | 7950 | 3827 | 50 | 5266 | |
| CA1 50.5 /L10-8 | 50 | 5 | 0.506 | 2420 | 26 | 90 | 3000 | 1626 | 41 | 2373 | |
| CCT specimens | | | | | | | | | | | |
| ZA1 100.5/B3-13 | 100 | 5 | 0.2 | 69000 | 20 | 150 | 121500 | 60000 | 86 | 70318 | 123500 |
| ZA1 100.5/A3-13 | 100 | 5 | 0.5 | 46750 | 22 | 140 | 73000 | 37500 | 92 | 43080 | 72200 |
| ZA1 100.5/B5-8 | 100 | 5 | 0.7 | 30000 | 20 | 150 | 46000 | 22500 | 67 | 28392 | 43350 |
| ZA1 50.5/K5-8 | 50 | 5 | 0.508 | 24000 | 16 | 110 | 37600 | 18450 | 45 | 26192 | 37675 |
| ZA1 25.5/D13-8 | 25 | 5 | 0.72 | 9200 | n.d. | 145 | 12800 | 5250 | 32 | 8244 | 11125 |

Average $\delta_{0.2}$ = 147 μm

LIST OF SYMBOLS

| | |
|---|---|
| $a$ | crack length |
| $B$ | specimen thickness |
| $E$ | Young's modulus |
| $F$ | applied force |
| $J$ | J-integral |
| $K$ | linear elastic stress intensity factor |
| $K_{eff}$ | K with $a \rightarrow a_{eff}$ (Eq(1)) |
| $\bar{K}$ | average of K and $K_{eff}$ |
| $n$ | strain hardening exponent |
| $W$ | width of CT or SENB specimen,<br>half-width of CCT specimen |
| $\delta$ | crack tip opening displacement |
| $\delta_5$ | $\delta$ measured over a gage length of 5 mm |
| $\varepsilon$ | strain |
| $\varepsilon_Y$ | yield strain |
| $\sigma$ | stress |
| $\sigma_Y$ | yield stress |

Index

| | |
|---|---|
| Y | value at incipient net section yield |
| YY | value at incipient gross section (general) yield |

REFERENCES

[ 1] Harrison, J.D., M.G. Dawes, G.L. Archer, and M.S. Kamath: The COD approach and its application to welded structures, Research Report 55/1978/E; The Welding Institute, 1978.

[ 2] BSI PD 6493: 1980, Guidance on some methods for the derivation of acceptance levels for defects in fusion welded joints, British Standards Institution, 1980.

[ 3] Schwalbe, K.-H.: A simple estimation model for a centre cracked panel in tension, Proceedings of ICF International Symposium on Fracture Mechanics (Beijing), Science Press, Beijing, China, 1983.

[ 4] Schwalbe, K.-H.: A simple engineering treatment of centre cracked tension panels in the regime of non-linear fracture mechanics under plane stress conditions, GKSS 84/E/38, GKSS Research Centre, Geesthacht, 1984.

[ 5] Schwalbe, K.-H.: Some aspects of the crack tip opening displacement concept, to appear in a Special Issue on Fracture Mechanics of the Journal of the Aeronautical Society of India.

[ 6] Hellmann, D., and K.-H. Schwalbe: Geometry and size effects on $J_R$ and $\delta_R$-curves under plane stress conditions, 15th National Symposium on Fracture Mechanics, College Park, MD, 1982, in: ASTM STP 833, 1984, pp. 577-605.

[ 7] Schwalbe, K.-H., and D. Hellmann: Correlation of stable crack growth with the J-integral and the crack tip opening displacement, effects of geometry, size, and material, progress report on an investigation of the elastic-plastic R-curve methodology, GKSS 84/E/37, GKSS Research Centre, Geesthacht, 1984.

[ 8] Schwalbe, K.-H.: Extension of the Engineering Treatment Model (ETM) to bending configurations under plane stress, to appear as a GKSS report.

[ 9] Kumar, V., and C.F. Shih: Fully plastic crack solutions, estimation scheme, and stability analysis for the compact specimen, in: ASTM STP 700, pp. 406-438, 1980.

[10] Aeberli, K.E.: Report in preparation.

[11] Hellmann, D., and K.-H. Schwalbe: On the experimental determination of CTOD based R-curves, this workshop.

[12] Royer, J., J.M. Tissot, A. Pelissier-Tanon, P. le Poac, and D. Miannay:
J-integral determination and analysis for small test specimens and their usefulness for estimating fracture toughness, in: ASTM STP 668, pp. 334-357, 1979.

[13] Amstutz, H., and T. Seeger: Bruchmechanische Analyse eines austenitischen Stahles (X6CrNi1811) an zwei Standardrißscheiben mit ebenem Spannungszustand, F1-2/1984, Technical University of Darmstadt.

[14] Heerens, J.: Unpublished results.

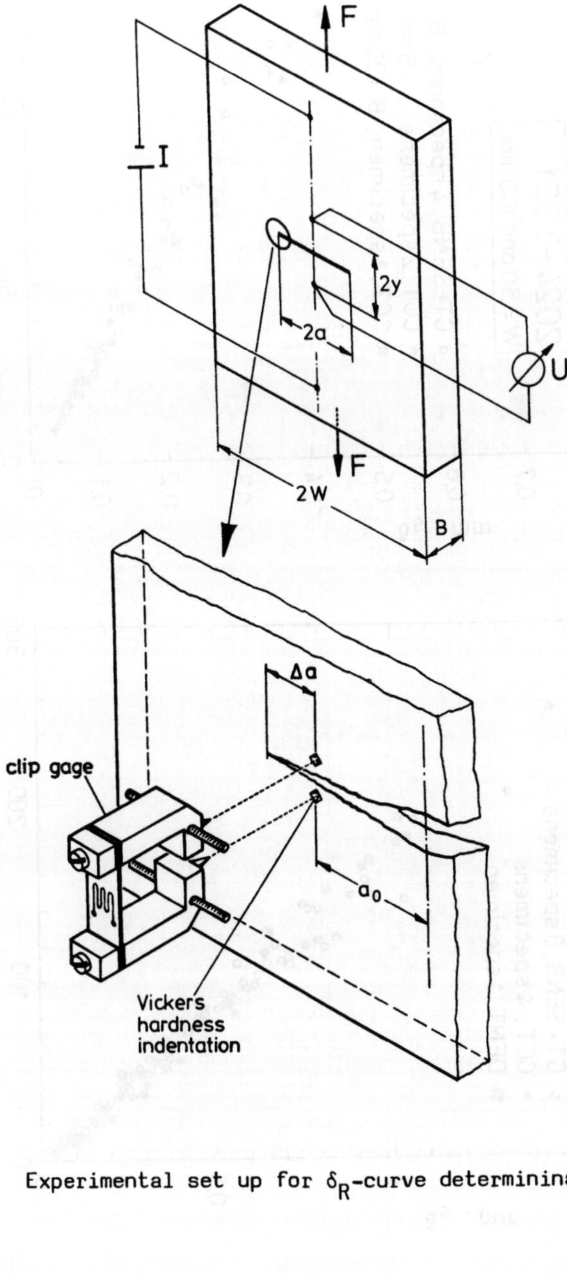


Fig. 1: Experimental set up for $\delta_R$-curve determinination.

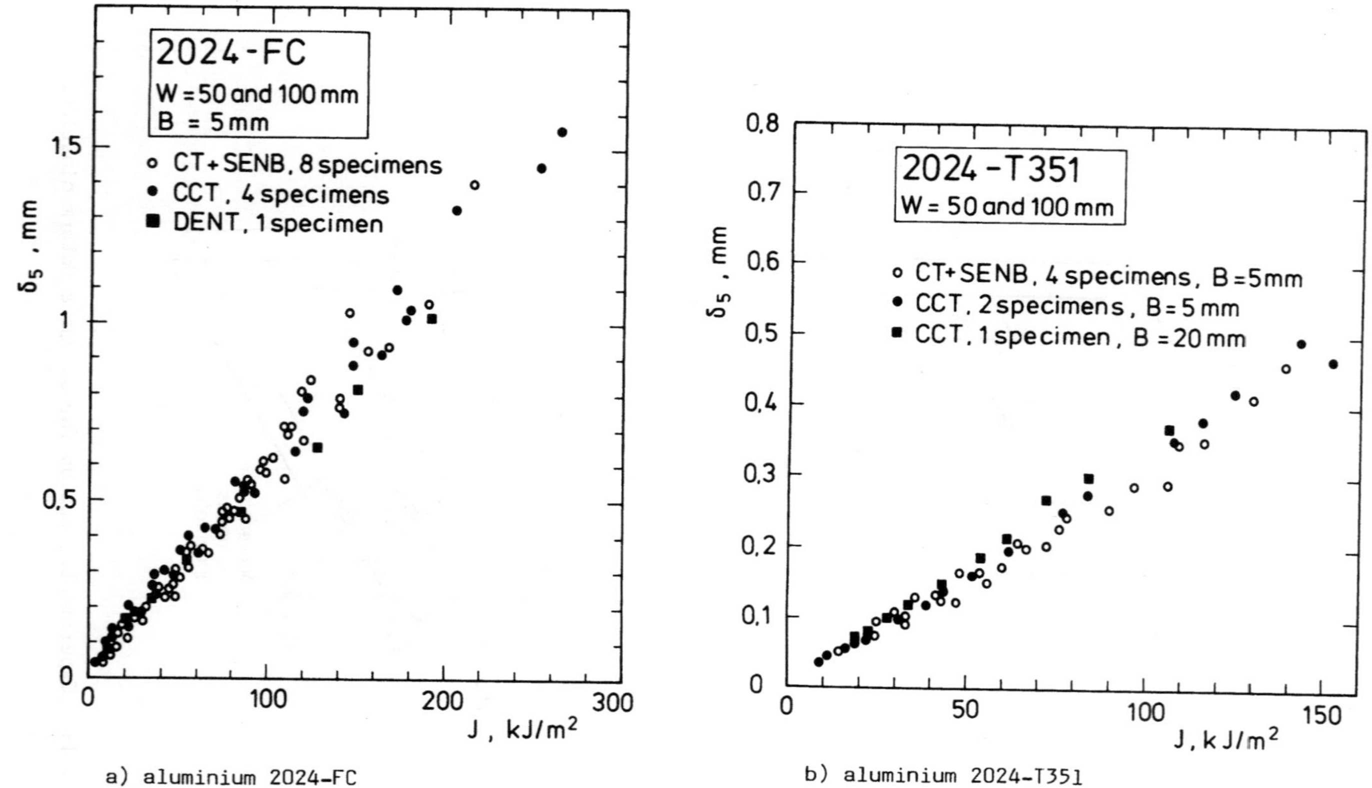


a) aluminium 2024-FC

b) aluminium 2024-T351

Fig. 2: Experimentally determined relationship between $\delta_5$ and the J-integral for bending and tension configurations demonstrating the independence of geometry.

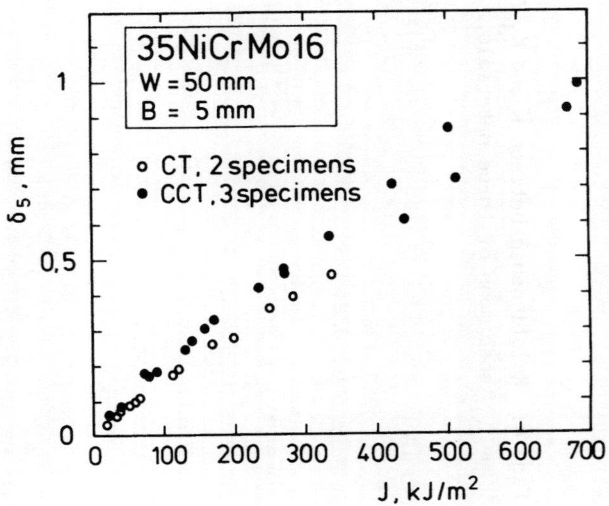


Fig. 2 c) steel 35NiCrMo16

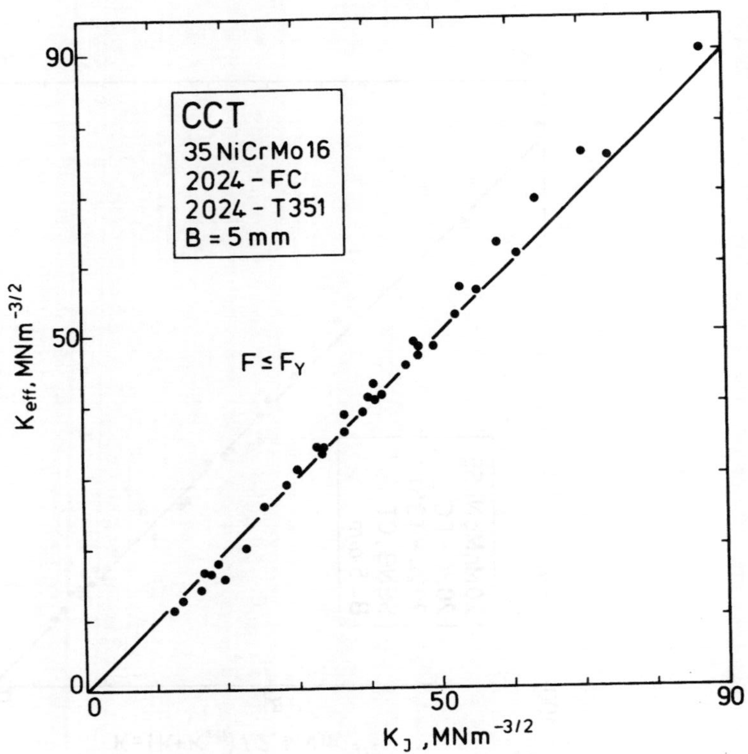


Fig. 3: Relationship between plasticity corrected stress intensity factor, $K_{eff}$, and $K_J$ determined via the J-integral; CCT specimens of three materials.

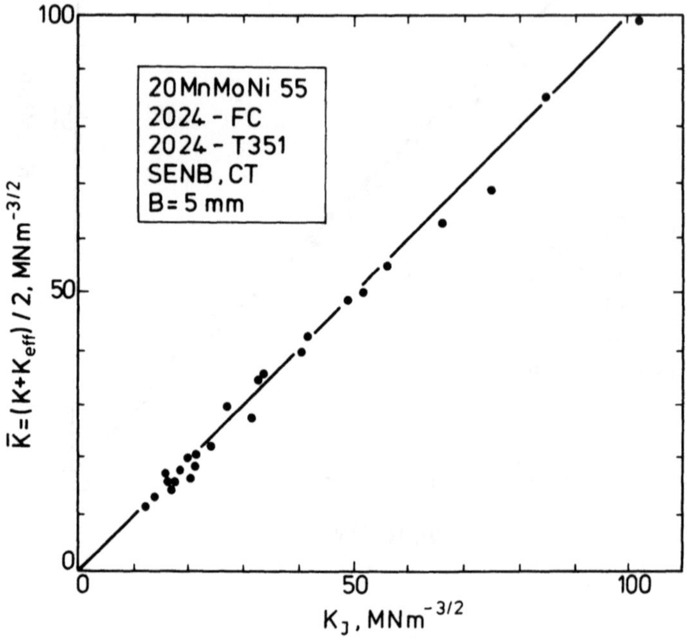


Fig. 4: Relationship between $\bar{K}$ and $K_J$ for SENB and CT specimens of three materials.

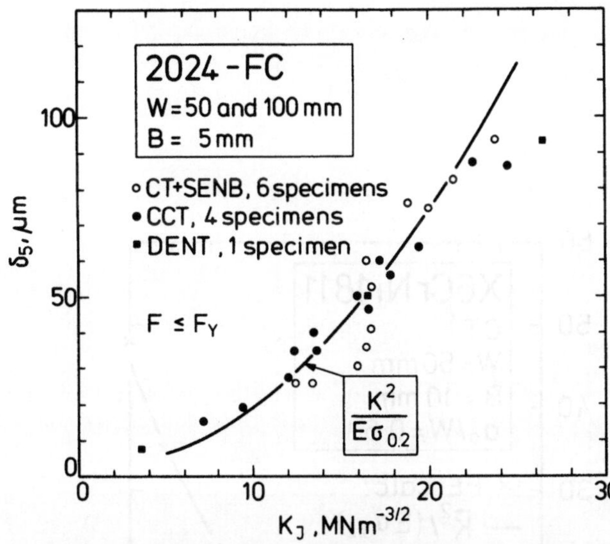


a) aluminium 2024-FC

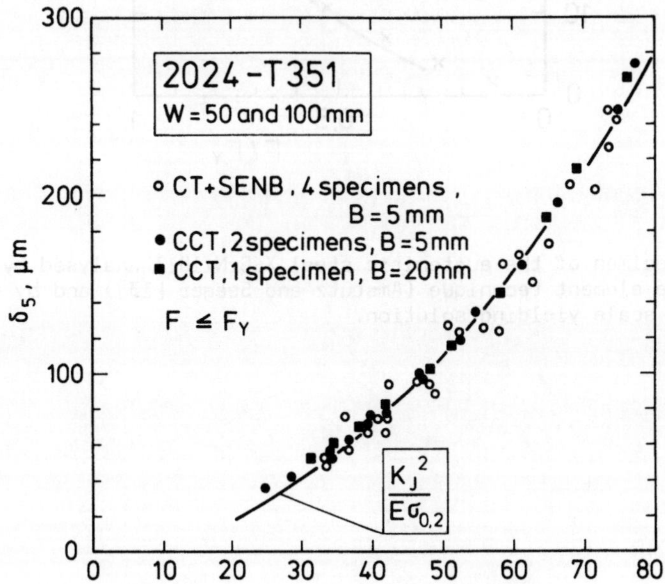


b) aluminium 2024-T351

Fig. 5: Experimental $\delta_5$ as a function of $K_J$ along with the theoretical curve

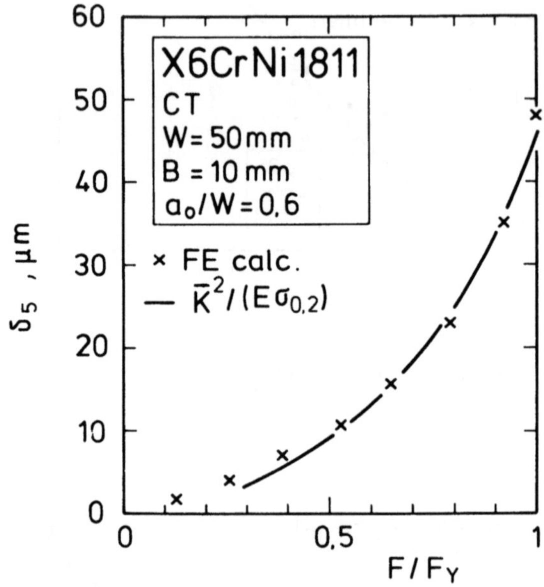


Fig. 6: CT specimen of the austenitic steel X6CrNi1811 analysed by the finite element technique (Amstutz and Seeger [13]) and by the small scale yielding solution.

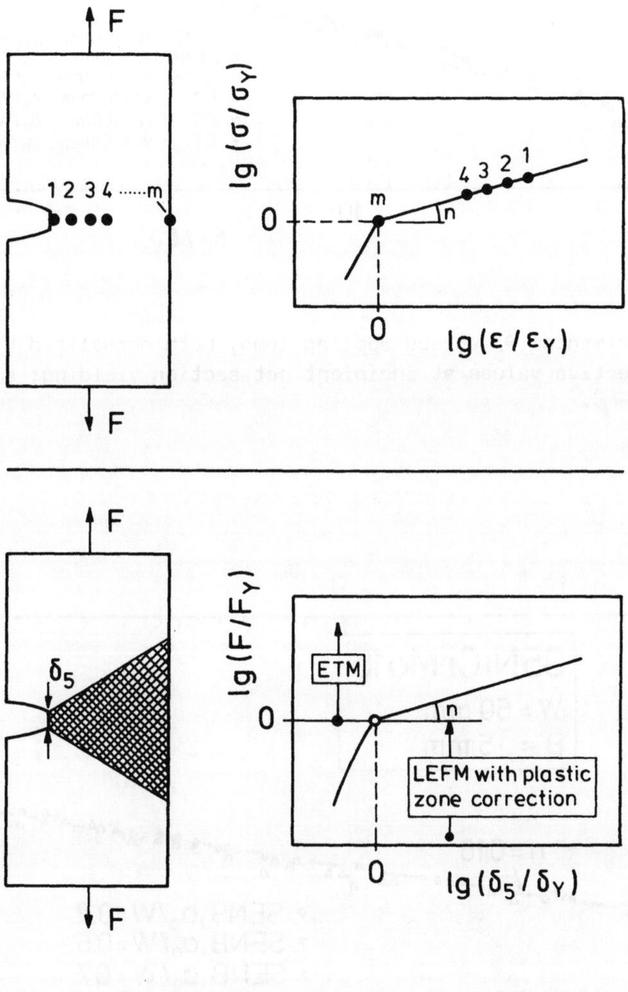


Fig. 7: Transfer of the strain-hardening law to the yielding net section.

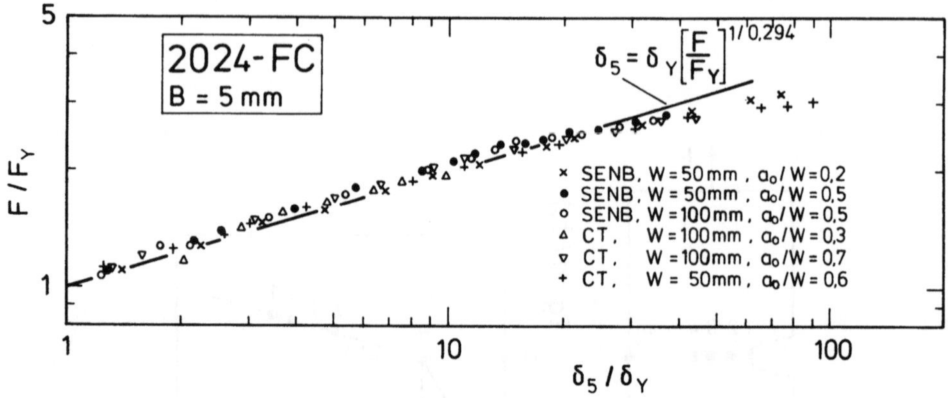


Fig. 8: Experimental $\delta_5$ versus applied load, both normalized by their respective values at incipient net section yielding; aluminium 2024-FC.

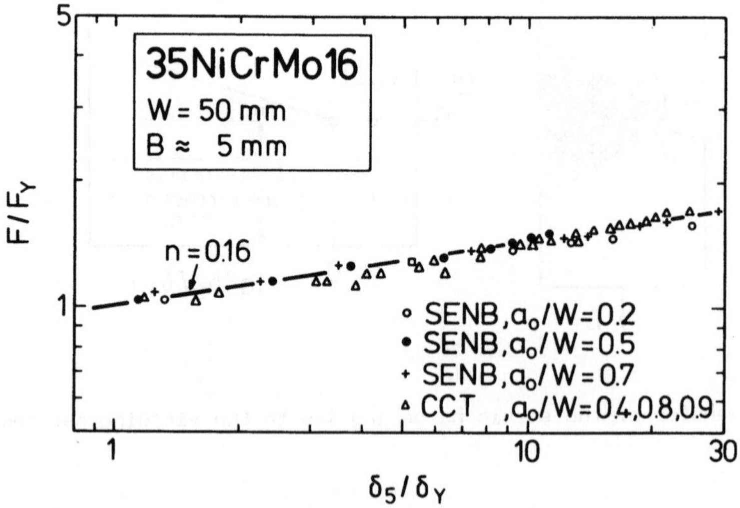


Fig. 9: Same as Fig. 8; steel 35NiCrMo16.

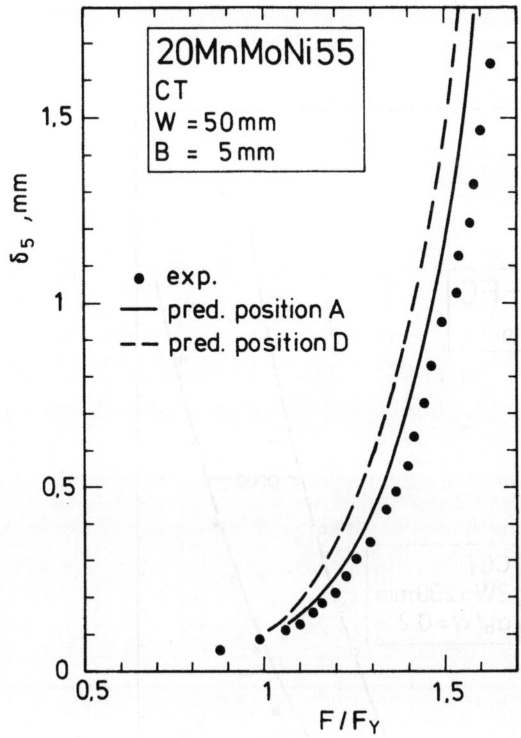


a) for a CT specimen of the pressure vessel steel 20MnMoNi55 [14]; calculations were done using stress-strain data from two different depth positions in the block embracing the position of the CT-specimen; see Ref. [8];

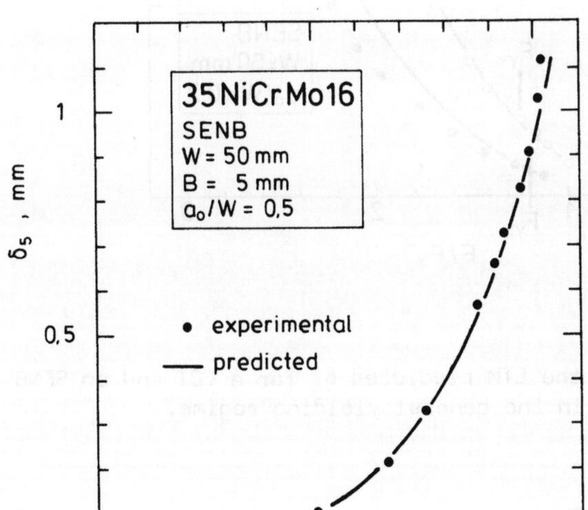


b) for an SENB specimen of the alloy steel 35NiCrMo16.

Fig. 10: Experimental and by the ETM predicted $\delta_5$

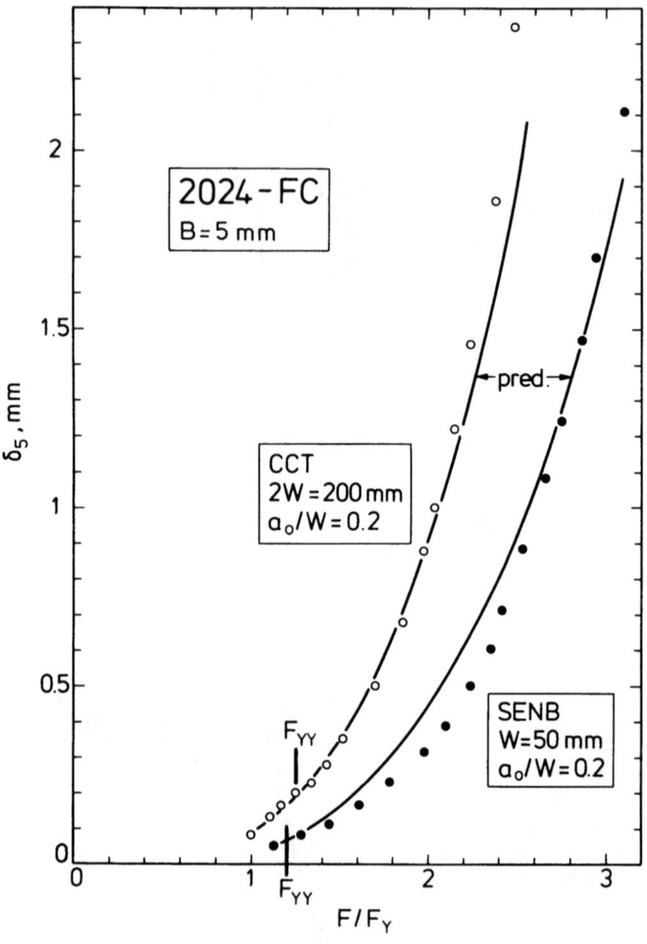


Fig. 11: Experimental and by the ETM predicted $\delta_5$ for a CCT and an SENB specimen of 2024-FC in the general yielding regime.

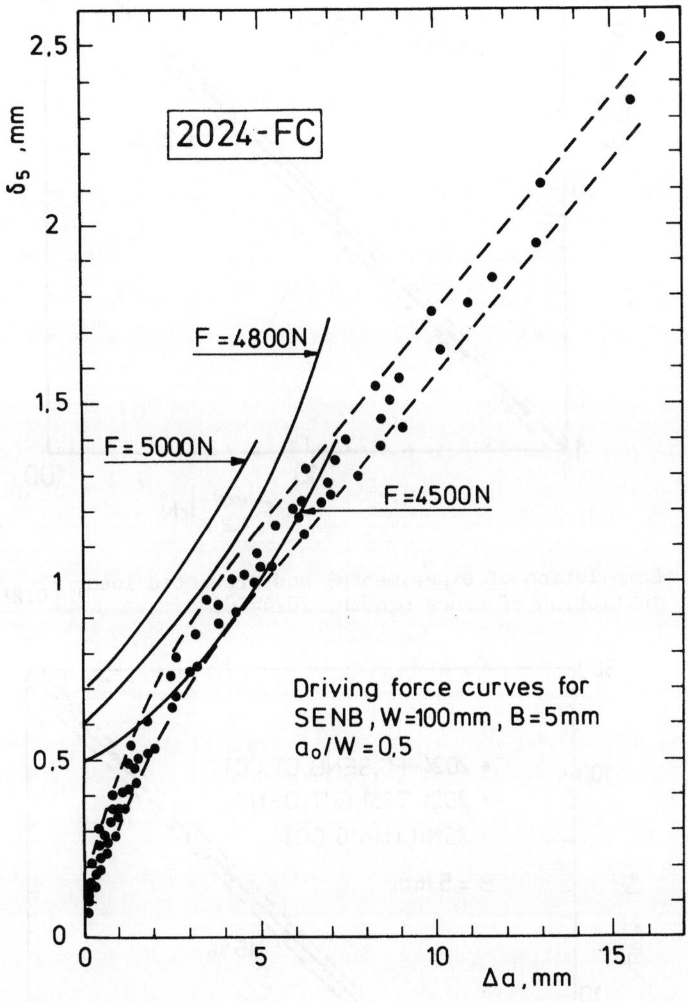


Fig. 12: Instability prediction for a bend specimen of 2024-FC using the $\delta_5$-R-curve and the driving force according to the ETM.

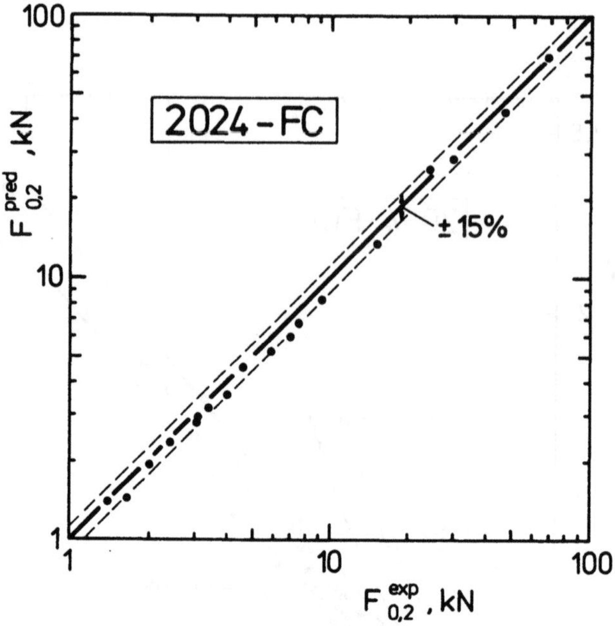


Fig. 13: Compilation of experimental and predicted load, $F_{0.2}$, at initiation of crack growth; 2024-FC.

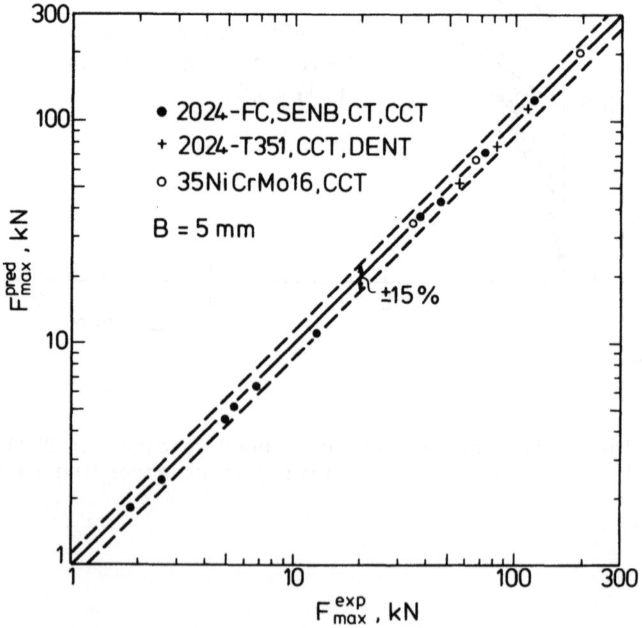


Fig. 14: Compilation of experimental and predicted maximum loads for a variety of materials and specimen geometries.

# Fracture Mechanics Tests on Axially Cracked Pipe Sections Subjected to Internal Pressure

K.E. AEBERLI*, D. SCHULZE**, P. MORAWIETZ**, H. FUHLROTT**, J. HEERENS*, K.-H. SCHWALBE*

* GKSS-Forschungszentrum Geesthacht GmbH, 2054 Geesthacht

** Rheinisch-Westfälischer TÜV, 4300 Essen

Summary

As a continuation of investigations into the validity of the crack tip opening displacement, $\delta_5$, as a crack field parameter, axially cracked pipe sections were internally pressured and $\delta_5$-R-curves were recorded. A comparison ot these with $\delta_5$-R-curves of similar compact tension specimens showed close coincidence largely independent of both initial crack length and of plastic bulging in the cracked region of the pipe sections.

## 1. Introduction

The recent development of fracture mechanics has shown that, in many cases, its procedure can be successfully applied to the modelling of failure conditions in structural sections /1/. Generally use is made of initiation concepts whereby the relevant material properties ($K_o \approx K_{Ic}$, $\delta_o$, $J_o \approx J_{Ic}$) are, within certain limits, independent of size and geometry /2/. However the study of stable crack growth reveals that, due to the "R-curve effect" (i.e. the increase of crack growth resistance with crack growth), safety margins beyond crack initiation exist, which can be quantified by means of the R-curve method /3/.

If P designates a crack field parameter (K, $\delta$, J) and R the crack resistance (expressed as K, $\delta$ or J) then the condition for the instability of a specimen or a section is given when:

$$\frac{\partial P}{\partial a} \geq \frac{\partial R}{\partial a} \tag{1}$$

Results from a current comprehensive research program being conducted at GKSS into the effects of size and geometry on the R-curve have led to the conclusion that, under plane stress conditions, the crack tip opening displacement, $\delta_5$, measured at the original fatigue crack tip is a more suitable correlation parameter for crack growth than the J-integral (determined by the usual method). (For the definition and measurement of $\delta_5$ see Refs. /4, 5/). It has been shown that, for a given thickness, B, the $\delta_5$-R-curve can be considered to be a "valid" R-curve as long as a certain fraction, $\alpha$ , of the ligament (determined by the material and type of loading (tension or bending)) is not exceeded /4/, /5/, i.e.

$$\Delta a \leqq \alpha(W-a_o) \qquad (2)$$

If a section is to be assessed according to the R-curve concept a necessary prerequisite is that the material-dependent R-curve can be reproduced for the section under consideration. This provided the instigation for a research program undertaken jointly by GKSS and the Rheinisch-Westfälischer TÜV. An internally pressurized pipe section having an axial crack was chosen to represent a structure which on one hand is relatively simple to evaluate but on the other hand is important for fracture mechanics considerations.

From this investigation it was expected to provide answers to the following questions:

- Do R-curves determined from internally pressurized cylinders coincide with those obtained from standard laboratory specimens?
- Can the maximum pressure be predicted by simple means using the R-curve method?

The present report describes the preparatory work, gives a survey of the intended program and presents the first results.

## 2. Material, Specimen Preparation and Test Program

The material chosen was the 2024 aluminium-copper-magnesium alloy used in the major part of the research described in /4/ and /5/. The material was investigated in its as-recieved condition T351, and an oven-cooled, overaged condition, FC. The respective tensile properties are given in Table 1.

5 and 10 mm thick, 80 mm wide compact tension (CT) specimens and 350 mm long pipe sections having 5 and 10 mm wall thickness and an external diameter of 70 mm were machined out of the same 100 mm thick plate. By this means it was expected that the CT specimens and the pipe sections would be in the same condition and have the same material properties.

Axial cracks of the same orientation as those in the CT specimens (i.e. T-L) were spark-eroded into the pipe sections. The relevant specimens were subsequently heat treated using the procedure described in /4/.

A test series was planned to determine $\delta_5$-R-curves for the two specimen types. The procedure described in detail in /4/ and /5/ was used to provide $\delta_5$-R-curves for the CT-specimens. These $\delta_5$-R-curves were to be compared with those of the corresponding pipe sections to discover whether a prediction of the behaviour of a pipe section was possible based on the respective $\delta_5$-R-curve from a CT-specimen.

## 3. Experimental Procedure

The pipe sections were loaded internally by oil pressure generated either by means of a servohydraulic piston or else by using a hand pump. In order to maintain the oil pressure it was necessary to make the pipe oil-tight.

A servohydraulic piston was used to provide a cyclic pressure to pre-fatigue the pipe sections. The sections were then instrumented and the R-curve tests carried out. The final crack extensions were marked by ink and post-fatiguing after which the pipe sections were pulled apart. The crack extensions were measured using a light microscope.

The experimental details are given in the following sections.

### 3.1 Oil-tight sealing of the pipe sections

The pipe sections were sealed at each end by means of circular end-plates and O-rings. Originally the pipe sections were machined with flanges for connecting the end plates. The pipe section W3 was thus capped using bolts and split tension rings. Calculations showed that a flange thickness of 30 mm was sufficient to support the maximum shear stresses likely to occur during these tests.

The tests on the pipe section W3 showed that the flanges were unnecessary, it being possible to pull the end plates on to the pipe ends by means of three threaded rods placed symmetrically about the pipe section. This also considerably reduced the radial constraint at the pipe ends but at the same time introduced an axial constraint. The remaining pipe sections were tested using this configuration.

The notch and the subsequent crack were also required to be oil-tight. This was achieved by a special procedure developed by TÜV-Essen whereby the pipe section was fitted with a rubber inner tube in the region of the expected crack. Under internal pressure this made the crack oil-tight without affecting the fracture behaviour of the pipe section.

### 3.2.1 Prefatiguing the pipe sections

The pipe sections were pre-fatigued at a stress ratio of R = 0.1 with maximum stress intensity factors of $K_{max} = 13\ MNmm^{-3/2}$ and $6\ MNmm^{-3/2}$ for the T351 and FC conditions respectively. This corresponded to half the stress intensity required to initiate crack growth in a monotonic test on a CT-sample of the same material. The stress intensities were calculated using the formula for an infinitely wide centre-cracked plate corrected by the Folias factor $M_F$ /6/. This factor takes into account the elastic bulging of the pipe wall in the region of the crack. Thus for a crack length 2a:

$$K = \sigma\sqrt{\pi a} \cdot M_F \tag{3}$$

where $\sigma = p'R_i/t$ for a thin walled tube of inner radius, $R_i$, wall thickness, t, and under an internal pressure, p. The internal pressure was reduced with increasing crack length to keep within the stress intensity limits. The final fatigue crack length usually lay between 5 and 10 mm.

### 3.2.2 Post-fatiguing the pipe sections

After completion of the test the final crack extensions were marked with ink. The section of the pipe wall arround the crack was then cut out and holes for bolts drilled above and below the centre of the crack. It was then either cooled in liquid $N_2$ or fatigued to mark the final crack extension, and subsequently pulled apart.

Observation of the crack surfaces showed that post-fatiguing had caused a certain amount of fretting. This tended to mask the position of the final crack tip which was however revealed by the ink.

## 3.3 Instrumentation for the R-curve test

The following quantities were measured as a function of the internal pressure.

a) The crack length using the potential method.

b) The crack tip opening displacement ( $\delta_5$) at both fatigue crack tips.

c) The crack opening displacement at the centre of the spark-machined notch.

A fully-instrumented pipe section is shown in Fig. 1.

### 3.3.1 Crack length measurement using the D.C. potential method

The layout is drawn schematically in Fig. 2. The current I is supplied by a constant current source through leads attached by screws to two diametrically opposite points on the pipe circumference set at 90° to the crack plane. The potential is taken from solder pins tapped into two holes situated symmetrically about the spark-machined notch at points 8 mm from its longitudinal axes. Both pairs of connections lie half-way along the pipe section at the middle of the notch.

The measured potential increases with the crack length. The usual Johnson formula /7/ is not suitable in this case and thus a calibration curve had to be determined. A pipe section was wired for this purpose and the potential measured as a function of crack length. Starting with a 20 mm notch the crack was extended in mm steps up to 60 mm using a hacksaw. The crack length, a, was plotted as a function of the normalized potential $U/U_o$ where $U_o$ was the potential at a = 20 mm. This curve was approximated by a polynominal using the least squares method (Fig. 3) in order to facilitate the subsequent data processing.

In the case of an R-curve test the calibration curve is used to determine the potential ratio $(U/U_o)^*$ from the initial crack length $a_o$. Since the calibration curve is normalized by the potential $U_o$ at a = 20 mm a fictitious reference potential, $U_o'$, must be found so that

$$\frac{U_o^*}{U_o'} = \left(\frac{U}{U_o}\right)^* \qquad (3)$$

where $U_o^*$ is the potential measured at $a_o$. The calibration curve based on $U_o'$ can now be used to determine any crack length from the corresponding potential.

### 3.3.2 Measurement of the crack tip opening displacement at the fatigue crack tips

The crack tip opening displacements were measured using two $\delta_5$ clip-gages located by hardness indents on either side of the fatigue crack tips. A full description of the use of these clip-gauges is given in /8/.

### 3.3.3 Measurement of the crack opening displacement at the middle of the spark-machined crack

A standard clip-gauge was located between two knife-edges set into the pipe wall at the middle of the spark-eroded notch. The calibration of the clip-gauge is similar to that of the $\delta_5$-clip-gauges /8/.

## 3.4 Data Processing

The analogue measurement signals were amplified if necessary by a pre-amplifier before being read by digital voltmeters. The voltmeters provided the analogue-digital conversion required for the computer input. A generalized data recording program was used to read in the test data and store it on disquettes. The analysis of the data was carried out using a fracture mechanics data evaluation program.

## 4. Results

Two fractured pipe sections are shown in Fig. 4. The annealed pipes fractured in plane stress whereas the T351 pipes fractured in plane strain. A typical plane strain fracture surface is shown in Fig. 5. It can be seen that bulging has taken place in the crack region and the fronts of both the fatigue ($a_o$) and final ($a_e$) cracks are curved.

The $\delta_5$-R-curves of the 5 mm thick T351 specimens are shown in Fig. 6 (for the CT-specimens), Fig. 7 (pipe sections) and Fig. 8 (CT and pipe section curves superimposed) to demonstrate the dependence of the $\delta_5$-R-curves on the position of the specimens within the original plate. Superimposed $\delta_5$-R-curves for 10 mm thick T351 specimens, and 5 mm, and 10 mm thick annealed specimens are drawn in Figs. 9, 10 and 11 respectively.

In the case of the pipe sections all $\delta_5$ values are taken as the average of the $\delta_5$ measurements at the two fatigue crack tips and the crack extensions are taken as half the respective total crack extensions (e.g. the final crack extension in Fig. 5 is $\Delta a_e = (2a_e - 2a_o)/2$).

## 5. Discussion

### 5.1 The effect of the fracture surface on the potential drop method

As previously mentioned the crack fronts of the fatigue and the final cracks are curved. Since the potential calibration curve is based on a straight crack front the error in the calculated crack length will increase with increasing crack tunnelling. A linear correction factor based on the difference between the microscopically measured final crack extension and that calculated using the potential method, was used as a first approximation.

This is likely to be inaccurate since, firstly, the variation of the crack profile is not a linear function of the crack extension and, secondly, the measured potential feigns a crack length which is shorter than that resulting from a linear average taken across the width of the fracture surface.

It is not certain whether the truerelation between the potential drop and the crack extension can be deduced from a single test. However at present no data is available for an exact correlation of the crack profile with the potential measurements.

## 5.2 The effect of the hoop stress gradient on the crack front

It can be seen in Fig. 5 that the crack front ist not symmetrical whereby the greatest tunnelling occurs nearer to the inside of the pipe wall. This is probably caused by the hoop stress gradient across the wall, the stress at the inner surface being greater than that at the outer surface. A combination of the stress gradient and the variation of the triaxial stress state along the crack front determine the position af maximum tunnelling.

## 5.3 The effect of crack extension asymmetry on the $\delta_5$ measurements

Although the crack does not always propagate symmetrically the average value of the two $\delta_5$ measuring clips must be used in the $\delta_5$-R-curves. This is because the potential measurement cannot distinguish between the amounts of crack propagation at the two crack ends. Consequently a plot of the individual $\delta_5$-clip outputs against an averaged crack length is meaningless. There ist however some justification for plotting the average values as one is trying to determine the fracture characteristics of the pipe section as a whole. Further support for using the average values of both $\delta_5$ and $\Delta a$ is given in /9/.

In this respect it should be pointed out that there are some restrictions to the validity of this assumption which are dictated by the curvature of the $\delta_5$-R-curves and the asymmetry of the crack extension. If it is fair to assume that the individual $\delta_5$-R-curves for the two crack tips are the same and thus independent of the initial crack length, then the two crack ends result in points at different positions along the common $\delta_5$-R-curve. The averaging method used above would always provide an exact $\delta_5$-R-curve if the curve were linear. This is not the case in practise, which means that the average $\delta_5$-R-curve will always lie somewhat below the true $\delta_5$-R-curve. The smaller the radius of curvature and the greater the asymmetry, the greater will be the underestimation of the $\delta_5$-R-curve at any point. In practical terms the error will be small near initiation and for large crack extensions since the asymmetry is small in the first case and the radius of curvature is large in the second case. Between these limits the error may well reach a maximum depending on the shape of the $\delta_5$-R-curve and the rate of change of the crack propagation asymmetry.

### 5.4 A comparison of the $\delta_5$-R-curves of the CT-specimens and the pipe sections

It can be seen from Figs. 10 and 11 that for the 5 mm and 10 mm annealed specimens respectively the $\delta_5$-R-curves of the CT specimens and the pipe sections correspond well in spite of the pronounced bulging which ocurred in the region of the crack. The coincidence between the $\delta_5$-R-curves of the CT specimens and the pipes is remarkable. Obviously the effect of the plastic bulging of the pipes has little effect on the $\delta_5$ measurement.

At first glance this does not appear to be the case for the T351 specimens (Figs. 8 and 9). However, a closer examination of the $\delta_5$-R-curve for the 5 mm CT specimens (Fig. 6) provides the explanation.

It was found that the $\delta_5$-R-curves of the CT-specimens showed a through-thickness dependence based on their original positions within the plate. Thus specimens taken from positions near the plate surfaces exhibited greater crack growth resistance than those from positions near to the centre of the plate. This corresponded well with the microhardness values which were higher in the centre of the plate than at the surfaces. The positions of the CT-specimens W53 and W49 corresponded to those of the pipe walls (i.e. the crack location). It can be seen in Figs. 8 and 9 that the coincidence of the $\delta_5$-R-curves of the aforementioned CT-specimens with those of the corresponding pipe section is very good.

Figs. 7 and, to a lesser extent, 11, show that varying the initial crack length had little effect on the $\delta_5$-R-curves of the pipe sections (W8 and W3). In fact in one case, due to pre-fatiguing problems the initial crack was highly asymmetric (a ratio of 2 to 1).

## 6. Conclusions

1. The $\delta_5$-R-curves of the axially cracked pipe sections tested coincide closely with those of the corresponding CT-specimens.
2. Changes of initial crack length and crack asymmetry have little effect on the $\delta_5$-R-curves of the pipe sections.
3. Plastic bulging in the region of the crack has no marked effect on the $\delta_5$-R-curves of the pipe sections.

## References


1. Rolfe, S.T. and Barson, J.M.: Fracture and Fatigue Control in Structures, Application of Fracture Mechanics; Prentice-Hall, Englewood Cliffs, N.J., 1977
2. Schwalbe, K.-H.; Überlegungen zur Anwendung der Bruchmechanik - Möglichkeiten und derzeitige Grenzen, Berichtsband der 15. Sitzung des Arbeitskreises "Bruchvorgänge", Deutscher Verband für Materialprüfung, Berlin, 1983.
3. Schwalbe, K.-H., Setz, W., Schwarmann, L., Geier, W. Wheeler, C., Rooke, D., de Koning, A.V., and Eastbrook, J.; Anwendung der Bruchmechanik auf Querschnitte geringer Dicke, Fortschritts-Bericht VDI-Z., Reihe 18, Nr. 9, 1980.
4. Hellmann, D., and Schwalbe, K.-H.; Geometry and Size Effects on $J_R$ and $\delta_R$ Curves Under Plane Stress Conditions, in: ASTM STP 833, 1984, pp. 577-605.
5. Hellmann, D., and Schwalbe, K.-H.; Stabiles Rißwachstum bei global nichtlinearem Werkstoffverhalten und ebenem Spannungszustand; GKSS-Annual Report 1982, pp. 20-33.
6. Folias, E.S.; "The Prediction of Catastrophic Failures in Pressurized Vessels", Prospects of Fracture Mechanics, Sih, Van Elst, Broek (eds), Noordhoff, 1974.
7. Johnson, H.H.; Calibrating the Electric Potential Method for Studying Slow Crack Growth, Materials Research and Standards, Sept. 1965, 442-445.

8. Hellmann, D.; Geometrieeinfluß auf das stabile Rißwachstum bei nichtlinearem Werkstoffverhalten, GKSS-Report 83/E/38, 1983.

9. Hellmann, D., and K.-H. Schwalbe; R-Curve Behaviour of Double-Edge Notched Tension Specimens in Plane Stress, in preparation.

## Nomenclature

| | |
|---|---|
| $a$ | crack length |
| $a_o$ | fatigue pre-crack length |
| $\Delta a$ | amount of crack growth |
| $M_F$ | Folias factor |
| $p$ | internal pressure |
| $P$ | crack field parameter |
| $R$ | crack growth resistance |
| $R_i$ | inner radius of pipe |
| $t$ | wall thickness of pipe |
| $U$ | potential drop |
| $U_o$ | $U$ at $a = a_o$ |
| $W$ | specimen width, half length of pipe |
| $\delta_5$ | crack tip opening displacement measured at the original fatigue crack tip over a gage length of 5 mm |

Table 1: Material Al-Cu-Mg 2024 Alloy

1 T351 as-received condition

2 FC oven cooled, overaged condition

Properties

| Condition | $R_{p0.2}$ MPA | $R_M$ MPA | Work Hardening Exponent N |
|---|---|---|---|
| T351 | 317 | 440 | 0.09 |
| FC | 75 | 217 | 0.29 |

| Condition | Sample Type | Designation | B, mm | W, mm | $A_o/W$ |
|---|---|---|---|---|---|
| T351 | CT | W 50, W 53, W 56 | 5 | 80 | 0.50 |
| T351 | pipe | W 3 | 5 | 175 | 0.27 |
| | | W 15 | 5 | 175 | 0.10 |
| T351 | CT | W 48, W 49 | 10 | 80 | 0.50 |
| T351 | pipe | W 9 | 10 | 175 | 0.11 |
| FC | CT | W 54, W 55 | 5 | 80 | 0.50 |
| FC | pipe | W 18 | 5 | 175 | 0.12 |
| FC | CT | W 47 | 10 | 80 | 0.50 |
| FC | pipe | W 8 | 10 | 175 | 0.17 |
| | | W 16 | 10 | 175 | 0.11 |

Fig. 1: A fully instrumented pipe section

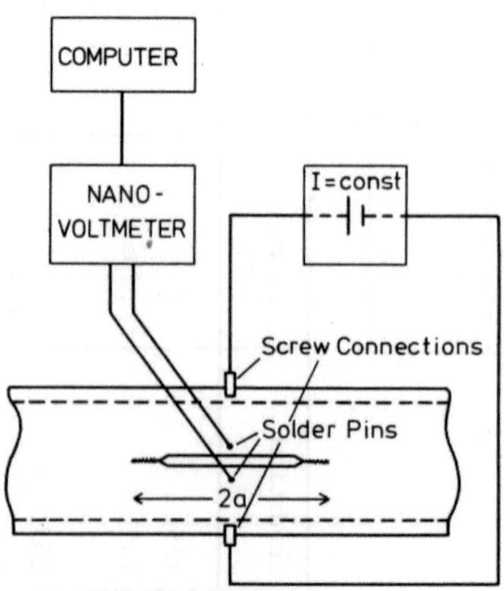


Fig. 2: Diagram of the potential measurement circuit

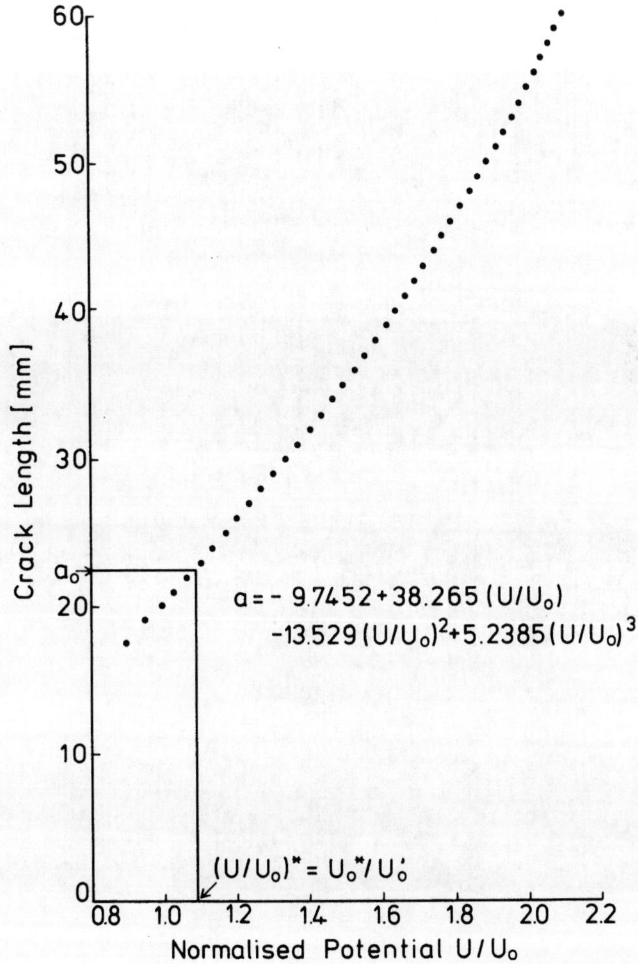


Fig. 3: The potential calibration curve $a = f(U/U_o)$

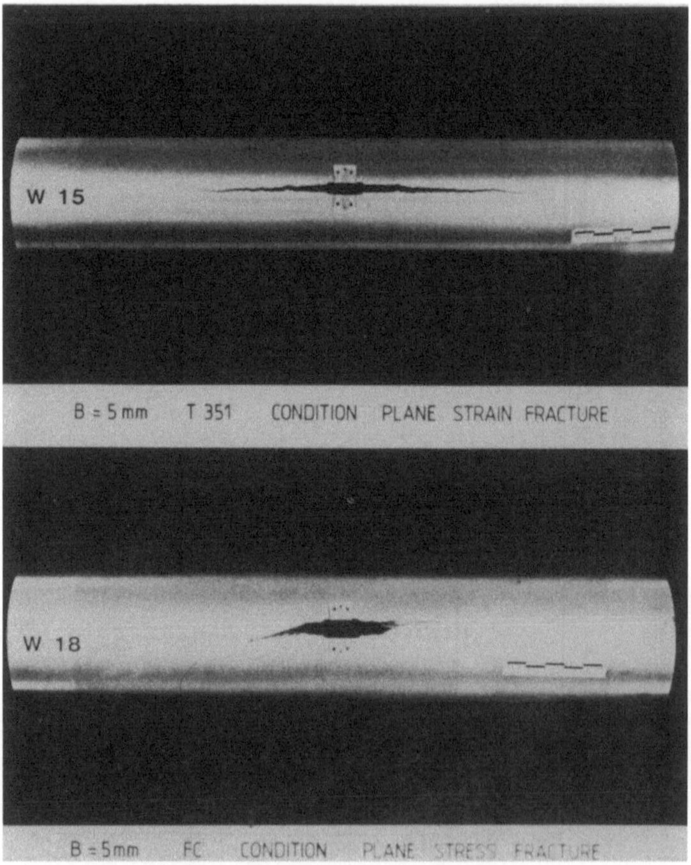


Fig. 4: Pipe sections after a test

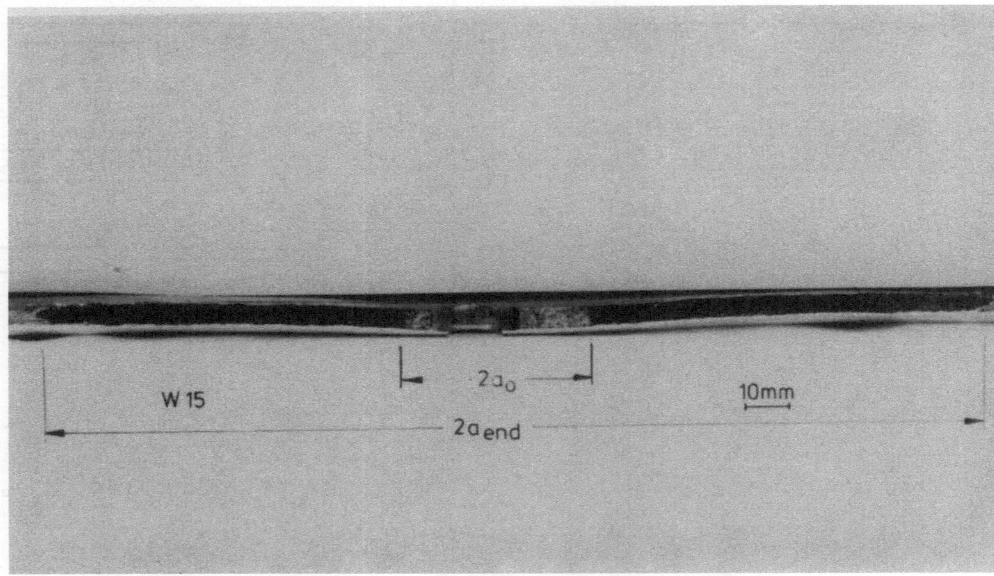


Fig. 5: A typical fracture surface of a pipe section

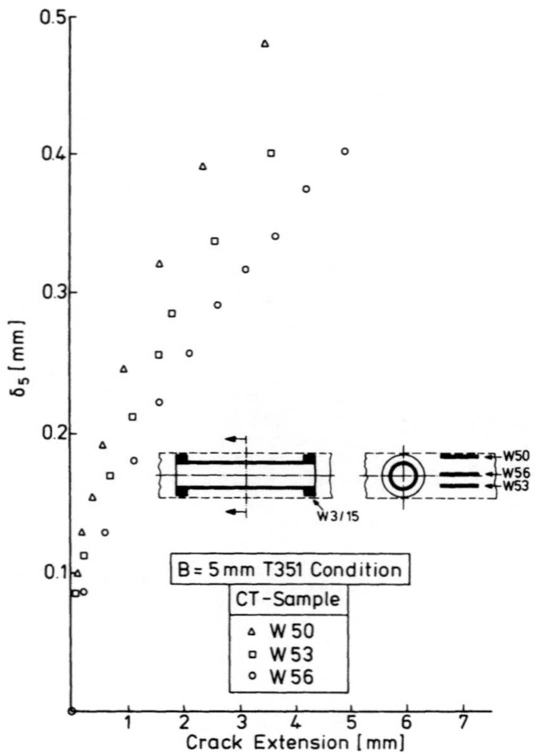


Fig. 6: Positional dependence of $\delta_5$-R-curves of T351 CT-specimens (B = 5 mm).

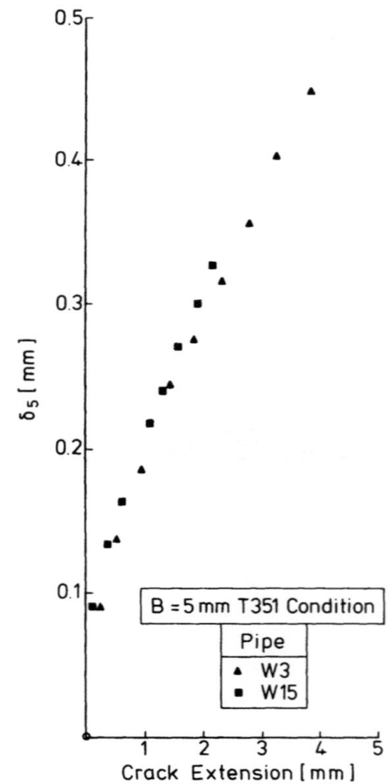


Fig. 7: Initial crack length independence of $\delta_5$-R-curves of a pipe sections (T351, B = 5 mm).

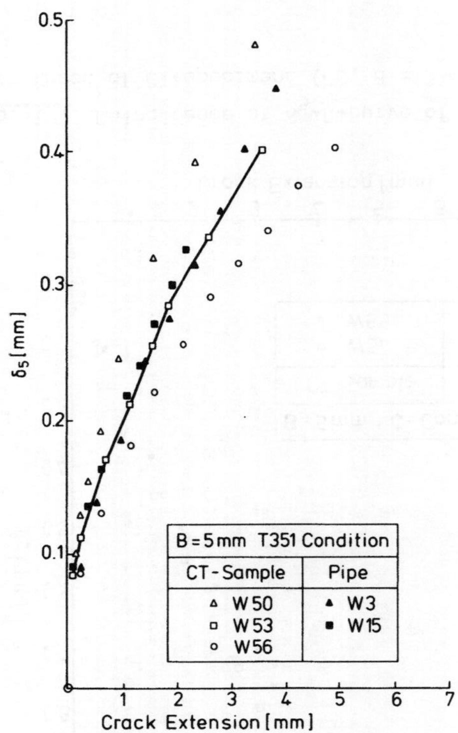


Fig. 8: Coincidence of $\delta_5$-R-curves of pipe sections with that of correctly chosen CT-specimen (T351, B = 5 mm).

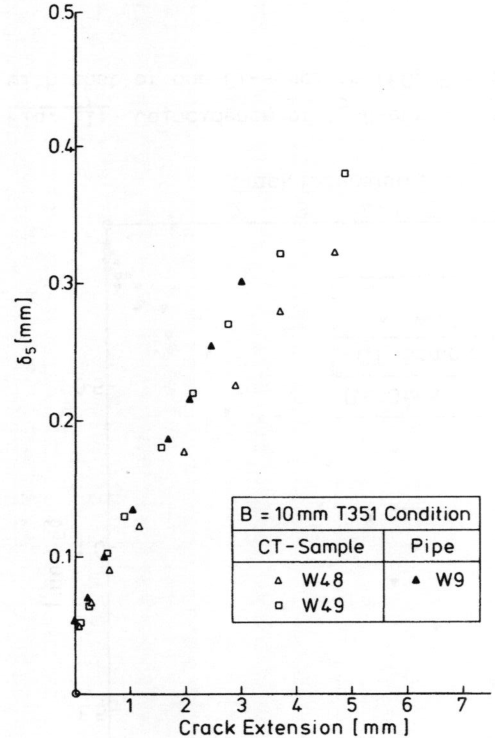


Fig. 9: Coincidence of $\delta_5$-R-curve of pipe section with that of correctly chosen CT-specimen (T351, B = 10 mm).

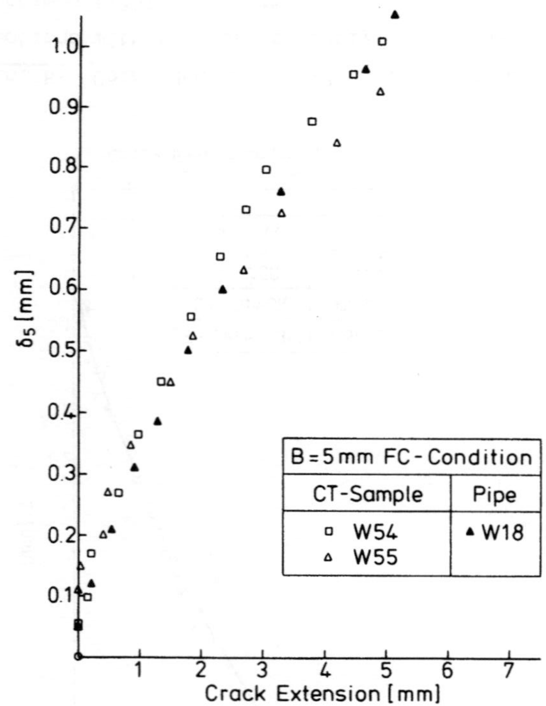


**Fig. 10:** Coincidence of $\delta_5$-R-curve of pipe section with those of CT-specimens (FC, B = 5 mm).

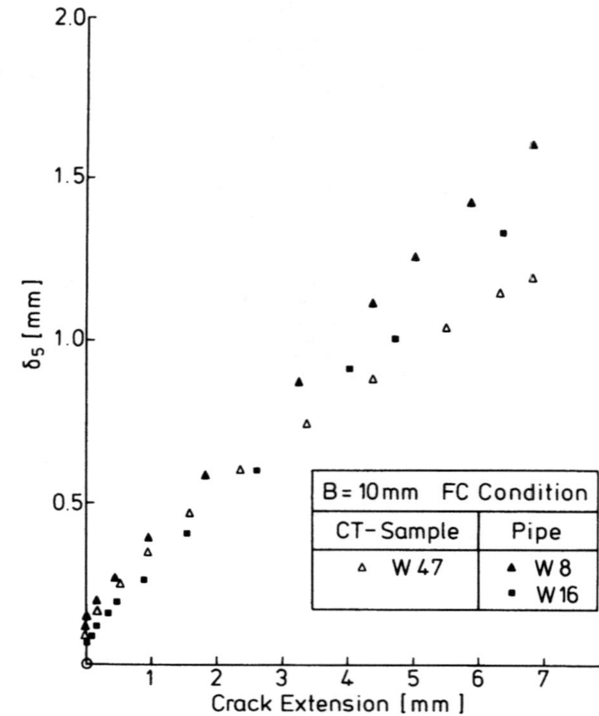


**Fig. 11:** Coincidence of $\delta_5$-R-curves of pipe sections with that of one CT-specimen (FC, B = 10 mm).

MIX
Papier aus verantwortungsvollen Quellen
Paper from responsible sources
FSC® C105338

If you have any concerns about our products,
you can contact us on
**ProductSafety@springernature.com**

In case Publisher is established outside the EU,
the EU authorized representative is:
**Springer Nature Customer Service Center GmbH**
**Europaplatz 3, 69115 Heidelberg, Germany**

Printed by Libri Plureos GmbH
in Hamburg, Germany